# Herbs and Spices - New Perspectives in Human Health and Food Industry

*Edited by Eva Ivanišová*

Published in London, United Kingdom

Herbs and Spices - New Perspectives in Human Health and Food Industry
http://dx.doi.org/10.5772/intechopen.1001733
Edited by Eva Ivanišová

Contributors
Amrit Acharya, Anamaria Cozma-Petruţ, Aurel Vasiu, Biba Vikas, Brijesh Kumar Singh, Carmen Dana Şandru, Diana Olah, Emeje Martins, Emöke Páll, Eva Ivanisova, Gabriel Gati, Gheorghiţă Duca, Gita Ghimire, Isao Okunishi, Jose Prakash Dharmian, Köbölkuti Lorand, Maadh Abdulwahab Alfahad, Madhulika Tripathi, Manish Kumar Verma, Marina Spînu, Melaku Tafese Awulachew, Mihai Băieş, Mujianto Mujianto, Nusrat Iqbal, Ogu Eneojo, Palanisamy Dhanapriya, Prabin Shakya, Prakash Ramakrishnan, Sabina Gayibova, Senthilnathan Palaniyandi, Sevgi Kolayli, Sukumaran Anil, Thirunavukkarasu Selvamuthukumaran, Ulugbek Gayibov, Vasile Cozma, Vishnupriya K. Sweety



First published in London, United Kingdom, 2024 by IntechOpen
IntechOpen is the global imprint of INTECHOPEN LIMITED, registered in England and Wales, registration number: 11086078, 167-169 Great Portland Street, London, W1W 5PF, United Kingdom

British Library Cataloguing-in-Publication Data
A catalogue record for this book is available from the British Library

Additional hard and PDF copies can be obtained from orders@intechopen.com

Herbs and Spices - New Perspectives in Human Health and Food Industry
Edited by Eva Ivanišová
p. cm.
Print ISBN 978-0-85466-779-6
Online ISBN 978-0-85466-778-9
eBook (PDF) ISBN 978-0-85466-780-2

For EU product safety concerns:
IN TECH d.o.o., Prolaz Marije Krucifikse Kozulić 3, 51000 Rijeka, Croatia,
info@intechopen.com or visit our website at intechopen.com.

# Meet the editor

Ing. Eva Ivanišová, Ph.D., is an assistant professor at the Institute of Food Sciences, Faculty of Biotechnology and Food Sciences, Slovak University of Agriculture in Nitra. She is also a scientific garant of Food Incubator in Research Centrum AgroBioTech, Slovak University of Agriculture in Nitra. Her field of study is food technology, especially plant raw materials and foodstuffs, foods with added value (enriched with medicinal plants), lesser-known types of fruits and vegetables, as well as alternative food sources such as edible insects and flowers. Dr. Ivanišová has published more than 490 scientific works in international and domestic journals.

# Contents

# Preface

In recent years, there has been a growing interest in the multifaceted benefits of herbs and spices. These natural ingredients, which have been integral to culinary traditions across the globe for centuries, are now being rediscovered for their significant contributions to human health and their transformative impact on the food industry. This book, *Herbs and Spices – New Perspectives in Human Health and Food Industry*, explores the diverse roles that these botanical treasures play in our lives.

The journey of herbs and spices began in ancient civilizations, where they were revered not only for their flavor-enhancing properties but also for their medicinal qualities. From the healing properties of turmeric in Ayurveda to the aromatic allure of cinnamon in traditional Chinese medicine, these natural substances have a rich history of use in promoting health and well-being. Today, modern science is uncovering the molecular mechanisms behind these age-old remedies, providing new insights into how herbs and spices can contribute to disease prevention and overall health.

In the food industry, the application of herbs and spices extends beyond flavoring. They are essential in the development of functional foods, natural preservatives, and innovative culinary creations. The quest for clean labels and natural ingredients has propelled the demand for these plant-derived substances, making them a focal point in food technology and product development. This book delves into how herbs and spices are reshaping the food industry, driving trends towards sustainability, health consciousness, and culinary innovation.

This volume brings together the expertise of researchers to provide a comprehensive overview of the latest advancements and applications of herbs and spices. It covers a wide range of topics, including the pharmacological properties of various spices, their role in traditional and modern medicine, innovative uses in food preservation and flavor enhancement, and their economic and cultural significance.

As you embark on this exploration of herbs and spices, we hope to inspire a deeper appreciation for these extraordinary plants. Whether you are a student, researcher, or industry professional, we believe this book will provide valuable insights and stimulate further inquiry into the remarkable world of herbs and spices. We extend our gratitude to all the contributors who have shared their knowledge and passion in this endeavor. It is our sincere hope that this book will serve as a valuable resource and a source of inspiration for all who seek to harness the power of herbs and spices in improving human health and advancing the food industry.

**Eva Ivanišová**
Institute of Food Sciences,
Slovak University of Agriculture in Nitra,
Nitra, Slovak Republic

Section 1

# Food Industry

Chapter 1

# A Miracle Food Supplement Obtained from Beehives: Propolis

*Sevgi Kolayli*


## Abstract

In recent years, propolis has garnered substantial global attention as a dietary supplement, owing to its multifaceted nature and diverse biological properties. Derived from beehives, this natural product, characterized by its sticky, waxy consistency, and aromatic scent, is a complex amalgamation of plant and animal origins. Produced by honeybees through the enzymatic processing of resinous substances from herbal extracts, propolis serves an array of structural and functional roles, from insulating beehives to fortifying their defense against diseases. Raw propolis, collected from hives, is harnessed for various formulations as a food supplement following extraction using different solvents. Ethanol (70%) stands as the preferred solvent for propolis extraction, with oil-based and water-based extracts also viable options. Non-toxic propolis extracts are commonly consumed, often encapsulated within natural polymers like chitosan, pectin, alginate, and dextran. Renowned for its diverse array of biological activities, propolis showcases a broad spectrum of functionalities. These encompass antioxidant, antimicrobial, antiviral, anti-inflammatory, anti-tumoral, anti-diabetic, immuno-modulating, hepatoprotective, and neuroprotective properties. This book chapter aims to delve into propolis' composition and its widespread popularity as a dietary supplement. Furthermore, it will explore the multifaceted impact of propolis on human health.


**Keywords:** food supplement, propolis, antioxidant, phenolic, health

## 1. Introduction

Over the past four to five decades, there has been a significant surge in people's inclination toward herbal and bee-derived products for both preventive and therapeutic purposes. Bee-derived substances like honey, pollen, propolis, bee pollen, bee bread, and royal jelly, classified as hybrid plant-animal products, boast diverse applications as complementary health products [1, 2]. Propolis, with its pleasant aroma, finds extensive application in various industries, including food, beverages, and consumer goods [3]. The utilization of herbal extracts, medicines, and natural products, alongside conventional treatments, has notably expanded, gaining recognition as traditional and complementary medicine as well as dietary supplements. The natural substance resembling a paste, is sourced by scraping it from beehives. It is a dark and adhesive material formed through honeybees processing different resinous elements from nature along with their own secretions. This amalgam is created by blending diverse natural secondary metabolites, beeswax, and various enzymes. Serving as a vital insulating material for the hive, propolis exhibits robust antibacterial, antiviral, and anti-inflammatory properties that contribute to the colony's health [4].

The diversity of propolis sources corresponds to geographical, climatic, regional, and botanical variations in production areas [5–9]. Regions within temperate climate zones, such as Europe, North America, New Zealand, and Western Asia, are conducive to propolis production, generating what is referred to as poplar temperate propolis. This resinous material, procured and processed by western honeybees, exhibits variable colors including green, yellow, red, and brown. It is sourced from leaves, flowers, buds, stems, fruits, and bark crevices of various tree species, such as poplar, pine, chestnut, alder, birch, willow, eucalyptus, fir, and acacia [9–11]. Propolis obtained from beehives is not used as ham propolis but as a food supplement after extraction. Its biological benefits have been discovered, and interest in propolis is increasing day by day.

### 1.1 History

Throughout history, propolis has held a significant place in treating wounds, diseases, and diverse ailments due to its potent antimicrobial, anti-inflammatory, and wound-healing properties. Cultures across various ancient civilizations incorporated propolis into their medical practices, rooted in observed benefits and traditional knowledge. The therapeutic use of propolis in apitherapy can be traced back thousands of years, with evidence of its distribution in ancient Egyptian practices, even being employed in the mummification process and transactional activities. It was also valued for wound care and its potential health benefits [11, 12].

During ancient Greek times, renowned philosophers like Hippocrates recognized propolis' healing capabilities, utilizing it as an antiseptic agent for treating wounds. In ancient Rome, propolis was highly regarded for its therapeutic potential, particularly in topical applications for wound healing and managing various health issues. Traditional Chinese medicine also emphasizes propolis, employing it for treating a spectrum of ailments. Recorded formulas integrated propolis with herbal mixtures for topical use, addressing skin problems and sports-related issues [1].

In contemporary times, modern apitherapy acknowledges propolis as a vital natural elixir, continuing its widespread usage. Today, propolis is recognized as a crucial fortifying agent incorporated into diverse formulations due to its antioxidant, antimicrobial, and anti-inflammatory properties.

### 1.2 Composition of propolis

Propolis, a complex substance harvested by bees from plant resins, encompasses diverse compounds including tree components, resins, beeswax, and essential oils. Bees gather resins from various botanical sources like buds and bark, forming the foundational element of propolis essential for its health benefits. This resinous part, termed the balsamic component, comprises a secondary metabolite shell rich in polyphenols, with variations in polyphenol types and quantities depending on the plant sources. Ethanol at concentrations of 70–80% stands as the most effective solvent for extracting propolis polyphenols, yielding a mixture of hydrophilic and hydrophobic compounds [1, 13, 14].

Phenolic acids such as caffeic acid, ferulic acid, coumaric acid, gallic acid, syringic acid, benzoic acid, and protocatechuic acid are extracted using high-polarity solvents, while flavonoids like rutin, quercetin, pinocembrin, and hesperetin are obtained using low-polarity solvents. However, the prevalent solvent for both polarities' compounds remains at 65–70% ethanol.

The first step when preparing propolis extracts is to pulverize the raw propolis extracted from beehives and then extract it by maceration using a 70% ethanol solution. After being filtered, it is used either directly as drops or after being converted into encapsulated products with different encapsulation agents (**Figure 1**). Yet, a contemporary challenge revolves around the use of ethanolic propolis extracts due to concerns regarding ethanol's side effects and religious sensitivities. As alternatives, propolis extracts and capsules can be developed using environmentally friendly solvents like water, glycerol, various polyalcohols (PPG and PEG), and natural agents like honey and olive oil. Although extracts in these green solvents may not match ethanolic extracts in quantity, they still serve as significant sources of polyphenols. Additionally, encapsulated derivatives of ethanolic propolis extracts, prepared using various agents such as alginate, chitosan, dextran, cellulose, pectin, inulin, gelatin, whey proteins, and arabinogalactans, are utilized as food supplements [14, 15].

The potent and distinct scent of propolis is attributed to its volatile compounds. This fragrance can vary, sometimes appearing sweeter, more bitter, herbal, or slightly pungent, potentially due to propolis originating from different sources or a blend of various plant extracts. The types and quantities of volatiles in propolis, typically emanating a woody resin aroma, differ based on the flora it's collected from. However, the primary volatiles in propolis include essential oils like terpenes, fatty acids, esters, aldehydes, and ketones [8, 15].

Volatile compounds, although a small portion of propolis's overall mass, play a crucial role in maintaining the hygiene, antibacterial, and antiviral properties within the beehive. They contribute to fostering a healthy hive environment. While there's limited research on this aspect, some studies suggest that these compounds exhibit potent antibacterial and antiviral activities. However, the complete spectrum of functions carried out by these volatile compounds in propolis is yet to be fully understood due to the limited available studies in this area [8, 16].

The volatile components found in propolis, comprising a diverse array of compounds like terpenes, terpenoids, sesquiterpenes, short-chain fatty acids, alkanes, and alkenes, exhibit qualitative and quantitative variations based on the local flora where it's collected. Within most temperate regions, sesquiterpenes are the dominant volatile oils, followed by aromatic compounds like benzyl acetate, benzyl alcohol, and benzyl benzoate. For instance, European propolis typically contains β-eudesmol as a major constituent, which is identified as a significant sesquiterpene in essential oils derived from leaf buds of *Pinus spp*. Conversely, in propolis from Southeastern Anatolia, major essential oils like α-terpinene and α-terpineol were detected. These compounds, shaping the aromatic characteristics of propolis, showcase a wide spectrum of biological activities. They possess antimicrobial, antiviral, and antioxidant properties, alongside serving as insect repellents [8, 14, 16].

The wax fraction, which constitutes the third part of propolis and can be extracted with diethyl ether or chloroform, comprises long-chain nonpolar compounds, predominantly beeswax. This portion primarily consists of wax produced by bees, contributing significantly to hive insulation and the mummification of deceased bees and other insects. Propolis typically contains between 4% and 20% of this wax fraction [17, 18].

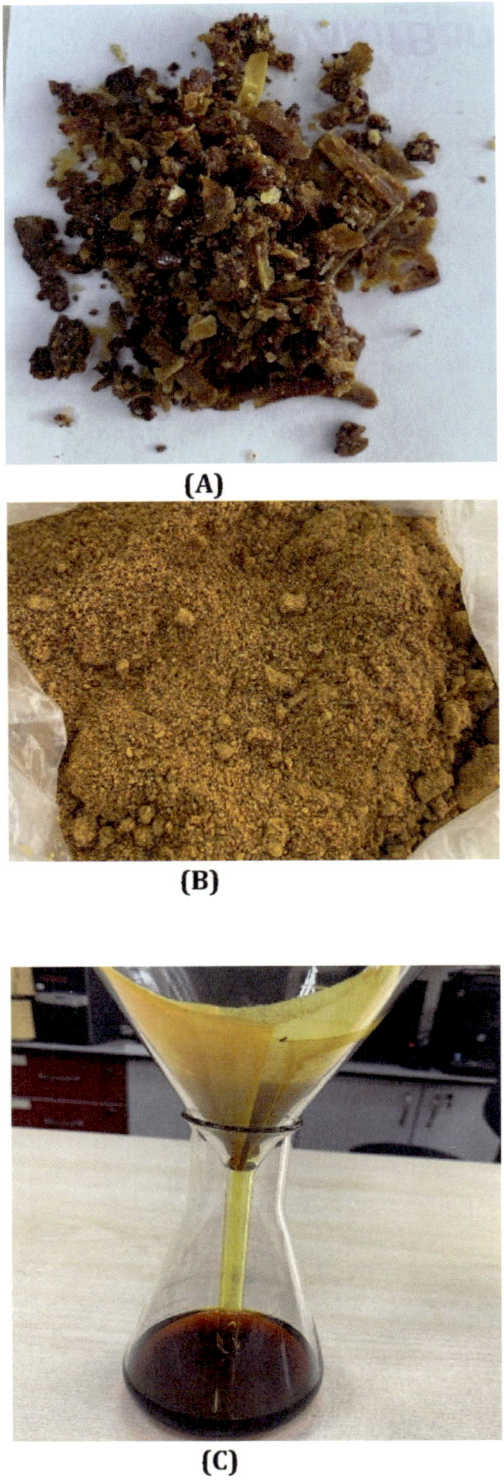


**Figure 1.**
*(A) raw propolis, (B) powder propolis, (C) ethanolic extract.*

The wax component of propolis holds substantial value and finds diverse applications across industries. It serves as an encapsulation agent in cosmetics and pharmaceuticals, as well as in the production of candles, fragrances, and nutritional supplements. Moreover, due to its latent heat storage capacity, wax is utilized in solar energy systems [19].

### 1.3 Food supplement properties

Interest in propolis extracts has surged significantly in recent years, leading to the development of diverse products formulated with various solvents and packaging methods, serving as potent reinforcing agents. The quality and standards of these propolis products, whether in extract form or encapsulated with natural polymers, are regulated by multiple authorities, including International Honey Commission (IHC), International Organization for Standardization (ISO-24381:2023), among other countries [12, 20, 21]. The primary attributes that distinguish propolis are its robust antioxidant and antimicrobial properties. Numerous scientific studies endorse propolis as the natural product boasting the highest antioxidant activity [21, 22].

While propolis boasts diverse bioactive compounds like polyphenols, flavonoids, and phenolic acids, it is not considered a significant nutritional source. Despite trace amounts of minerals, vitamins, or proteins present in its structure, their contribution to nutritional value is negligible. Although propolis might contain some amino acids attributed to the inclusion of pollen, used by bees in the propolis-making process alongside plant resins, its protein content is notably lower compared to other sources. The focus on propolis does not typically emphasize its protein content due to its relatively meager quantities [23, 24].

### 1.4 Natural antioxidant source

Oxidation is an inevitable reaction chain for all living and inanimate objects exposed to oxygen. It is caused by the continued presence of reactive oxidant species formed as a result of oxidative stress, and with the progression of inflammation, treatable disease, diabetes, cancer, or neurodegenerative disease develops [25].

The most important triggering agent of aging, cardiovascular diseases, neurodegenerative diseases, and cancer is oxidative stress. However, living conditions, smoking and alcohol use, exposure to radioactive light, and poor nutritional habits are known to increase oxidative stress. However, living conditions, smoking and alcohol use, exposure to radioactive light, and poor nutritional habits are known to increase oxidative stress. Any effect that reduces, slows down, or prevents oxidative stress is called antioxidant. Many free oxygen radicals and their derivatives (SOR) formed by oxidative stress cause the oxidation of macromolecules such as lipids, proteins, and DNA in the environment. This causes many damages ranging from cardiovascular diseases to cancer. Endogenous and exogenous antioxidants are biomolecules that reduce oxidative stress, scavenge free oxygen radicals, or prevent the formation of radicals. Epidemiological investigations have consistently demonstrated that populations adhering to dietary patterns rich in antioxidants exhibit a reduced risk of developing neurodegenerative diseases, including cardiovascular diseases and Alzheimer's [26–28]. Plant foods rich in ascorbic acid, α-tocopherol, β-carotene, and various polyphenols are also rich sources of antioxidants. Propolis is an important antioxidant source rich in polyphenols. Studies have shown that propolis has a higher antioxidant value than herbal sources. It shows that it has a higher polyphenol content than fruits such as blueberries, aronia, pomegranates, oranges, apples, rose hips, and strawberries [28–32]. The antioxidant capacity of propolis is closely related to the amount of polyphenol contents. **Table 1** presents a comparative analysis of the total phenolic content (TPC) and total flavonoid content (TFC) levels among select antioxidant-rich natural products.

In **Table 1**, propolis is one of the natural products with the highest total phenolic substance amounts. The closest bee products to propolis are bee bread and bee pollen, but propolis contains approximately 3–4 times higher polyphenols than these two products. The fruit closest to propolis is aronia, and green and black tea extracts also contain high polyphenols. Depending on the total amount of phenolic substances, the total antioxidant capacities of these products were found to be significantly higher in propolis.

As a matter of fact, studies have shown that antioxidant-rich propolis plays an important role in preventing diseases caused by oxidative stress [25]. In various experimental animal studies on oxidative stress modeling, it has been reported that

| | TPC | TFC | Reference |
|---|---|---|---|
| Raw propolis Anatolian | 110–190 mg GAE/g | 16–59 mg QUE/g | [11] |
| Brown Brazilian propolis | 55.74 mg GAE/g | 30.89 mg QUE/g | [32] |
| Red Brazilian propolis | 91.32 mg GAE/g | 31.48 mg QUE/g | [32] |
| Green Brazilian propolis | 90.55 mg GAE/g | 59.45 mg QUE/g | [32] |
| Anatolian bee pollen | 7–31 mg GAE/g | 1.24–4.60 mg QUE/g | [33] |
| Bee bread | 11.90–32.40 | — | [34] |
| Rhododendron honey | 16.10 mg GAE/g | 1.00 QUE/g | [35] |
| Chestnut honey (*Castania sativa* L) | 0.85 mg GAE/g | 0.02 mg QUE/g | [36] |
| Figus (*Ficus carica* L.) | 0.35–0.48 mg GAE/g | 0.01–0.03 | [37] |
| Aronina (*Aronia melanocarpa*) | 14.94–50.02 mg GAE/ g dry matter | 5.70–15.27 mg QUE/g dry matter | [31] |
| Blueberry (*Vaccinium spp.*) | 1.71–4.35 mg GAE/ g dry matter | | [30] |
| Grape's pulp (*Vitis labrusca)* | 3.57 mg GAE/ g dry matter | 0 < 35 mg QUE/g dry matter | [38] |
| Grape's seed (*Vitis labrusca)* | 0.20 mg GAE/ g | 0.030 mg QUE/g dry matter | [38] |
| Green tea extract | 59.8 mg Cat/ g dry matter | — | [28] |
| Black tea extract | 59.8 Cat/ g dry matter | — | [28] |
| Ginkgo extract | 10–40 mg GAE/g | — | [39] |
| Olive fruits | 19.95–26.57 mg GAE/g | | [40] |
| Apricot pollen | — | 10.29 ± 1.16 | [41] |

**Table 1.**
*Comparison of some important natural products in terms of total phenolic substances.*

propolis reduces stress induced by agents such as hydrogen peroxide, paraquat, $CCl_4$, the ischemia-reperfusion model, heavy metals, pesticides, and other xenobiotics [42, 43]. In the partial diabetes-oxidative stress model induced by streptozotocin (STZ), propolis administration notably mitigated protein and lipid oxidation, reflected in the reduced levels of malondialdehyde (MDA) and protein carbonyl (PC). Furthermore, propolis treatment led to an elevation in tissue glutathione (GSH) levels. Concurrently, interleukin and C-reactive protein (CRP) levels exhibited a decrease following propolis intervention [44]. In a clinical study, administered propolis (1000 mg/day) and melatonin (20 mg/day) to primary pneumo-sepsis patients for 10 days, observing reduced inflammation, improved organ function, and enhanced gastrointestinal tolerance. While survival rates increased, they were not statistically significant [45].

### 1.5 Mouth and dental health

Propolis is an important antibacterial and anti-inflammatory agent, especially in oral hygiene and some oral and gum diseases. One of the oral and dental diseases for which propolis is most used is gingivitis. Gingivitis is a common gum disease characterized by inflammation of the gums, often caused by the accumulation of plaque on the teeth. It manifests through redness, swelling, and potential bleeding of the gums during brushing or flossing. If left untreated, gingivitis can progress to more severe periodontal diseases, leading to potential tooth loss. Gingivitis has the potential to progress into periodontitis, where both gingival and bone tissues suffer destruction. Adequate mechanical plaque removal may not be practiced by most of the population. It has been reported in many clinical studies that patients diagnosed with periodontitis received a positive response to treatment with propolis supplementation [46–49]. Numerous comparative studies with propolis and chlorhexidine have shown that ethanolic propolis extracts are as effective as chlorhexidine. It has been shown to prevent the formation of dental plaque. It has been reported in many studies that propolis toothpastes and propolis mouthwashes are valuable for oral and dental health [50, 51].

### 1.6 Natural antibacterial and antiviral effects

It has been proven by numerous studies that propolis extracts show antimicrobial activity against a wide range of pathogenic microorganisms. It has been shown to be effective against infections of the gastrointestinal tract such as the upper respiratory tract and stomach-intestinal tract in humans, and against microorganisms resistant to various antibiotics. Although it has not been determined exactly which compound the antimicrobial activity of propolis comes from, it has been shown that this activity is due to the phenolic acids, flavonoids, and volatile compounds it contains. It is stated in the literature that these natural antimicrobial compounds show different inhibition mechanisms against bacterial species. Antimicrobial activity varies depending on the type of propolis. However, bacteriostatic, and bactericidal action mechanisms vary depending on the botanical properties of propolis. The capacity of propolis to prevent antibacterial sensing and antibiofilm formation indicates that it has a strong potential, especially against antibiotic-resistant bacteria. Additionally, propolis can stimulate the body's immune system, activating natural defense mechanisms to fight infection [52, 53]. There are many *in vitro* and clinical studies showing that propolis is also effective on *Helicobacter pylori*, which causes ulcers in the stomach. *Helicobacter pylori,* one of the few bacteria that can survive in an acidic stomach environment, is treated with double or triple antibiotics. Studies have shown that propolis extracts stop the proliferation of bacteria. It has also been shown that it inhibits the urease enzyme that bacteria secrete into the extracellular environment in order to adapt to the acidic stomach environment [54, 55]. It is also reported that propolis has an antibacterial effect against other bacteria that can live in the gastrointestinal tract, such as *Campylobact*er *spp, Escherichia coli*, and *Salmonella* spp. [55]. Additionally, propolis extracts are also used as antifungal agents against various *Candida spp*, *Aspergillus spp,* and *Cryptococcus neoformans*. It has been reported in many studies that it stops the growth of species [56, 57].

Propolis is a highly preferred natural supplement due to its potential anti-viral effects against various types of viruses. Many studies have shown that propolis may be effective against several viruses, such as influenza viruses, herpes simplex virus (HSV), HIV, and hepatitis B and C viruses [58]. Molecular docking studies against the COVID-19 virus have shown that flavonoids prevent the COVID-19 virus from binding to ACE receptors. It has also been confirmed that propolis has a significant adjuvant effect in COVID-19 patients. However, the effect of propolis on a particular type of virus may vary depending on factors such as its concentration and application methods. Therefore, more research is needed to better understand the antiviral effects of propolis [22, 59]. In fact, with the COVID-19 pandemic, propolis production and sales in the world have peaked.

### 1.7 Propolis and cancer

There exists a plethora of cell cultures and experimental studies demonstrating the potential of propolis as a natural agent in cancer treatment. Extracts from propolis have exhibited the ability to impede the growth of diverse cancer cells and induce apoptosis in cell culture settings. These studies indicate that propolis may hinder or arrest cancer development by modulating cell cycle pathways, activating caspase enzymes, and inhibiting angiogenesis. Moreover, its antioxidant and anti-inflammatory properties suggest potential in thwarting cancer initiation or curbing the metastatic spread of cancer cells. However, the absence of human clinical studies hinders definitive conclusions on the efficacy of propolis in cancer treatment. Therefore, further extensive research is imperative to establish conclusive evidence regarding the role of propolis in cancer therapy. Nevertheless, numerous clinical studies have demonstrated propolis' adjunctive role in cancer treatments [60, 61]. The findings obtained so far show that propolis is promising in cancer treatment as a complementary medicine tool.

### 1.8 Propolis and cardiovascular

Propolis serves as a significant dietary supplement utilized in both preventing and managing cardiovascular diseases, although its precise mechanisms in this context remain somewhat elusive. Its multifaceted properties involve various mechanisms contributing to its positive impact. Acting as a potent antioxidant, propolis plays a crucial role in inhibiting LDL oxidation, thereby reducing oxidative damage. This antioxidative quality is pivotal in mitigating oxidative stress within vascular walls, diminishing the harmful effects of free radicals, and potentially preventing cardiovascular conditions like atherosclerosis. Additionally, propolis displays anti-inflammatory characteristics, effectively reducing vascular inflammation and contributing to vascular health by regulating inflammatory responses in the vascular system. Studies suggest that propolis influences lipoprotein mechanisms, resulting in decreased LDL levels, elevated HDL levels, and fostering cardiovascular protection. Propolis demonstrates its preventive effect against cardiovascular diseases by inhibiting the lipase enzyme while activating the paraxonase enzyme [61, 62]. Additionally, propolis has been associated with blood-thinning effects and has been reported to play a role in lowering blood pressure, further contributing to its potential to support cardiovascular health [63–65].

### 1.9 Anti-inflammatory properties

Studies have extensively demonstrated propolis's significant anti-inflammatory properties, which operate through diverse mechanisms. Propolis exhibits a capacity to inhibit cyclooxygenase (COX) enzymes, consequently reducing prostaglandin production. This inhibition plays a pivotal role in alleviating symptoms associated with inflammation and pain. Moreover, propolis, endowed with potent antioxidant attributes, effectively neutralizes free radicals, thereby mitigating inflammation. Its ability to curtail nitric oxide synthesis (iNOS) contributes to controlling the inflammatory process. By diminishing the concentration of key inflammatory cytokines like IL-1b, IL-6, and TNF-a, propolis acts to attenuate the overall severity of inflammation. Additionally, propolis showcases regulatory effects on the immune system, further bolstering its capability to modulate inflammatory responses [44, 63, 66].

## 2. Conclusion

Propolis stands as an extraordinary natural adjunct, showcasing a rich tapestry of therapeutic potential and diverse biological attributes. Its historical roots in ancient medicinal practices have transcended time, finding relevance in contemporary dietary supplementation. The multifaceted properties of propolis span a broad spectrum, notably encompassing robust antioxidant, antimicrobial, and anti-inflammatory characteristics. These attributes underscore its potential contributions to various facets of human health, spanning oral health, cardiovascular disease prevention, and ongoing explorations into potential cancer treatments. Propolis, revered for its potent antioxidant prowess against oxidative damage, is anticipated to gain further prominence in averting diseases spawned by oxidative stress. Its anti-aging efficacy positions it as a prime contender in the realm of apitherapy and the cosmetic industry. Looking forward, propolis emerges as a field ripe for continued investigation. Future trends in propolis research aspire to unearth deeper insights into its mechanisms of action and multifarious applications across diverse industries. The prospective utilization of propolis beyond the realms of health, extending into cosmetics, food, and material sciences, hints at its remarkable versatility. However, materializing these potentials mandates rigorous clinical studies to authenticate and harness the full spectrum of propolis' benefits. Further research initiatives serve as the linchpin in unlocking its potential across varied industries and maximizing its impact on human health and beyond. As an herbal derivative boasting medicinal properties akin to herbal plants and spices, propolis joins the ranks of natural products rich in secondary metabolites derived entirely from nature. This aligns is not only with the evolution of natural medicine comprehension but also as a harbinger of innovative formulations in functional foods, dietary supplements, and medical applications.

## Author details

Sevgi Kolayli
Faculty of Science, Department of Chemistry, Karadeniz Technical University, Trabzon, Türkiye

*Address all correspondence to: skolayli@ktu.edu.tr; skolayli61@yahoo.com

## References

[1] Kuropatnicki AK, Szliszka E, Krol W. Historical aspects of propolis research in modern times. Evidence-based Complementary and Alternative Medicine. 2013;**2013**. DOI: 10.1155/2013/964149

[2] Burdock GA. Review of the biological properties and toxicity of bee propolis (propolis). Food and Chemical Toxicology. 1998;**36**(4):347-363. DOI: 10.1016/S0278-6915(97)00145-2

[3] Tomaszewski M, Dein M, Novy A, Hartman TG, Steinhaus M, Luckett CR, et al. Quantitation and seasonal variation of key odorants in propolis. Journal of Agricultural and Food Chemistry. 2019;**67**(5):1495-1503. DOI: 10.1021/acs.jafc.8b05965

[4] Ikeda NY, Ambrosio CM, Miano AC, Rosalen PL, Gloria EM, Alencar SM. Essential oils extracted from organic propolis residues: An exploratory analysis of their antibacterial and antioxidant properties and volatile profile. Molecules. 2021;**26**(15):4694. DOI: 10.3390/molecules26154694

[5] Seidel V, Peyfoon E, Watson DG, Fearnley J. Comparative study of the antibacterial activity of propolis from different geographical and climatic zones. Phytotherapy Research. 2008;**22**(9):1256-1263. DOI: 10.1002/ptr.2480

[6] Falcão SI, Vilas-Boas M, Estevinho LM, Barros C, Domingues MR, Cardoso SM. Phenolic characterization of northeast Portuguese propolis: Usual and unusual compounds. Analytical and Bioanalytical Chemistry. 2010;**396**:887-897. DOI: 10.1007/s00216-009-3232-8

[7] Alvear M, Santos E, Cabezas F, Pérez-SanMartín A, Lespinasse M, Veloz J. Geographic area of collection determines the chemical composition and antimicrobial potential of three extracts of chilean propolis. Plants. 2021;**10**(8):1543. DOI: 10.3390/plants10081543

[8] Bankova V, Popova M, Trusheva B. Propolis volatile compounds: Chemical diversity and biological activity: A review. Chemistry Central Journal. 2014;**8**(1):1-8. DOI: 10.1186/1752-153X-8-28

[9] Salatino A, Fernandes-Silva CC, Righi AA, Salatino MLF. Propolis research and the chemistry of plant products. Natural Product Reports. 2011;**28**(5):925-936. DOI: 10.1039/C0NP00072H

[10] Pobiega K, Kraśniewska K, Gniewosz M. Application of propolis in antimicrobial and antioxidative protection of food quality–a review. Trends in Food Science and Technology. 2019;**83**:53-62. DOI: 10.1016/j.tifs.2018.11.007

[11] Kolaylı S, Birinci C, Kara Y, Ozkok A, Samancı AET, Sahin H, et al. A melissopalynological and chemical characterization of Anatolian propolis and an assessment of its antioxidant potential. European Food Research and Technology. 2023a;**249**(5):1213-1233. DOI: 10.1007/s00217-023-04208-x

[12] Braakhuis A. Evidence on the health benefits of supplemental propolis. Nutrients. 2019;**11**(11):2705. DOI: 10.3390/nu11112705

[13] Bankova V. Recent trends and important developments in propolis research. Evidence-based Complementary and Alternative

Medicine. 2005;**2**:29-32. DOI: 10.1093/ecam/neh059

[14] Kolaylı S, Ureyen EU, Kara Y. The antimicrobial, anti-quorum sensing, and anti-biofilm activities of ethanolic propolis extracts used as food supplements. Biology Bulletin. 2022;**49**(Suppl 3):S21-S30

[15] Kasote D, Bankova V, Viljoen AM. Propolis: Chemical diversity and challenges in quality control. Phytochemistry Reviews. 2022;**21**(6):1887-1911. DOI: 10.1007/s11101-022-09816-1

[16] de Oliveira MS, Cruz JN, Ferreira OO, Pereira DS, Pereir NS, Oliveira MEC, et al. Chemical composition of volatile compounds in Apis mellifera propolis from the northeast region of Pará state, Brazil. Molecules. 2021;**26**(11):3462. DOI: 10.3390/molecules26113462

[17] Negri G, Marcucci C, Salatino A, Salatino MLF. Comb and propolis waxes from Brazil (states of São Paulo and Paraná). Journal of the Brazilian Chemical Society. 2000;**11**:453-457. DOI: 10.1590/S0103-50532000000500004

[18] Vilas-Boas M, Lopes M, Nunes LF, Pereyra A, Kunert C, Beck G, et al. Assessing the performance of analytical methods for propolis–a collaborative trial by the international honey commission. Journal of Apicultural Research. 2023;**62**(3):542-555. DOI: 10.1080/00218839.2022.2053320

[19] Cuce PM, Kolayli S, Cuce E. Enhanced performance figures of solar cookers through latent heat storage and low-cost booster reflectors. International: Journal of Low-Carbon Technologies. 2020;**15**(3):427-433. DOI: 10.1093/ijlct/ctz079

[20] Bogdanov S, Martin P, Lullmann C. Harmonised methods of the international honey commission. Swiss Bee Research Centre, FAM, Liebefeld. 2002;**5**:1-62

[21] Vilas-Boas M, Lopes M, Nunes LF, Pereyra A, Kunert C, Beckh G, et al. Assessing the performance of analytical methods for propolis–A collaborative trial by the international honey commission. Journal of Apicultural Research. 2020;**62**(3):542-555. DOI: 10.1155/2020/7538232

[22] Kurek-Górecka A, Keskin Ş, Bobis O, Felitti R, Górecki M, Otręba R-SA. Comparison of the antioxidant activity of propolis samples from different geographical regions. Plants. 2022;**11**(9):1203. DOI: 10.3390/plants11091203

[23] Kara Y, Can Z, Kolaylı S. What should be the ideal solvent percentage and solvent-Propolis ratio in the preparation of ethanolic Propolis extract? Food Analytical Methods. 2022;**15**(6):1707-1719. DOI: 10.1007/s12161-022-02244-z

[24] Al Naggar Y, Sun J, Robertson A, Giesy JP, Wiseman S. Chemical characterization and antioxidant properties of Canadian propolis. Journal of Apicultural Research. 2016;**55**(4):305-314. DOI: 10.1080/00218839.2022.2053320

[25] Lugrin J, Rosenblatt-Velin N, Parapanov LL. The role of oxidative stress during inflammatory processes. Biological Chemistry. 2014;**395**(2):203-230. DOI: /10.1515/hsz-2013-0241

[26] Grosso G, Buscemi S, Galvano F, Mistretta A, Marventano S, Vela VL, et al. Mediterranean diet and cancer: Epidemiological evidence and mechanism of selected aspects. BMC Surgery. 2013;**13**:1-9. DOI: 10.1186/1471-2482-13-S2-S14

[27] Mahmod AI, Haif SK, Kamal A, Al-Ataby IA, Talib WH. Chemoprevention effect of the Mediterranean diet on colorectal cancer: Current studies and future prospects. Frontiers in Nutrition. 2022;**9**:924192. DOI: 10.3389/fnut.2022.924192

[28] Rababah TM, Hettiarachchy NS, Horax R. Total phenolics and antioxidant activities of fenugreek, green tea, black tea, grape seed, ginger, rosemary, gotu kola, and ginkgo extracts, vitamin E, and tert-butylhydroquinone. Journal of Agricultural and Food Chemistry. 2004;**52**(16):5183-5186. DOI: 10.1021/jf049645z

[29] Engin SP, Mert C. The effects of harvesting time on the physicochemical components of aronia berry. Turkish Journal of Agriculture and Forestry. 2020;**44**(4):361-370. DOI: 10.3906/tar-1903-130

[30] Kim JG, Kim H, Kim SJ, Park KS. Fruit quality, anthocyanin and total phenolic contents, and antioxidant activities of 45 blueberry cultivars grown in Suwon, Korea. Journal of Zhejiang University Science B. 2013;**14**:793-799. DOI: 10.1631/jzus.B1300012

[31] Tolić MT, Landeka Jurčević I, Panjkota Krbavčić I, Markovi K, Vahčić N. Phenolic content, antioxidant capacity and quality of chokeberry (*Aronia melanocarpa*) products. Food Technology and Biotechnology. 2015;**53**(2):171-179. DOI: 10.17113/ftb.53.02.15.3833

[32] Andrade JKS, Denadai M, de Oliveira CS, Nunes ML, Narain N. Evaluation of bioactive compounds potential and antioxidant activity of brown, green and red propolis from Brazilian northeast region. Food Research International. 2017:101. DOI: 10.1016/j.foodres.2017.08.066

[33] Kolayli S, Birinci C, Kanbur ED, Ucurum O, Kara Y, Takma C. Comparison of biochemical and nutritional properties of bee pollen samples according to botanical differences. European Food Research and Technology. 2023b:1-12. DOI: 10.1007/s00217-023-04428-1

[34] Kalaycıoğlu Z, Kanbur ED, Kolaylı S, Erim FB. Antioxidant activities, aliphatic organic acid and sugar contents of Anatolian bee bread: Characterization by principal component analysis. European Food Research and Technology. 2023;**249**:5. DOI: 10.1007/s00217-023-04218-9

[35] Yildiz O, Gurkan H, Sahingil D, Degirmenci A, Er Kemal M, Kolayli S, et al. Floral authentication of some monofloral honeys based on volatile composition and physicochemical parameters. European Food Research and Technology. 2022;**248**(8):2145-2155. DOI: 10.1007/s00217-022-04037-4

[36] Kolayli S, Can Z, Yildiz O, Sahin H, Karaoglu SA. A comparative study of the antihyaluronidase, antiurease, antioxidant, antimicrobial and physicochemical properties of different unifloral degrees of chestnut (*Castanea sativa* mill.) honeys. Journal of Enzyme Inhibition and Medicinal Chemistry. 2016;**31**(sup3):96-104. DOI: 10.1080/14756366.2016.1209494

[37] Bayrak Ç, Birinci C, Kemal M, Kolayli S. The phenolic composition and antioxidant properties of figs (*Ficus carica* L.) grown in the Black Sea region. Plant Foods for Human Nutrition. 2023;**78**(3):539-545. DOI: 10.1007/s11130-023-01089-z

[38] Kavgacı M, Yukunc GO, Keskin M, Can Z, Kolaylı S. Comparison of phenolic profile and antioxidant properties of pulp and seeds of two different

grape types (*Vitis vinifera* L. and *Vitis labrusca* L.) grown in Anatolia: The amount of resveratrol of grape samples. Chemistry Africa. 2023:1-7. DOI: 10.1007/s42250-023-00660-0

[39] Sati P, Pandey A, Rawat S, Rani A. Phytochemicals and antioxidants in leaf extracts of Ginkgo biloba with reference to location, seasonal variation and solvent system. Journal of Pharmacy Research. 2013;**7**(9):804-809

[40] Özcan MM, Fındık S, AlJuhaimi F, Ghafoor K, Babiker EE, Adiamo OQ. The effect of harvest time and varieties on total phenolics, antioxidant activity and phenolic compounds of olive fruit and leaves. Journal of Food Science and Technology. 2019;**56**:2373-2385. DOI: 10.1007/s13197-019-03650-8

[41] Huseynova A, Alakbarli AA. Spectrophotometric analysis of flavonoid quantity in pollen of *Amygdalus communis* L. and determination of biomarkers. Journal of Apitherapy and Nature. 2023;**6**(1):22-29. DOI: 10.35206/jan.1318471

[42] Abass AO, Kamel NN, Khalifa WH, Goud GF, El-Manylawi MAF, Mehaisen GM, et al. Propolis supplementation attenuates the negative effects of oxidative stress induced by paraquat injection on productive performance and immune function in Turkey poults. Poultry Science. 2017;**96**(12):4419-4429. DOI: 10.3382/ps/pex248

[43] Laaroussi H, Bakour M, Ousaaid D, Ferreira-Santos P, Genisheva Z, El Ghouizi A, et al. Protective effect of honey and propolis against gentamicin-induced oxidative stress and hepatorenal damages. Oxidative Medicine and Cellular Longevity. 2021;**2021**. DOI: 10.1155/2021/9719906

[44] Zulhendri F, Lesmana R, Tandean S, Christoper A, Chandrasekaran K, Irsyam I, et al. Recent update on the anti-inflammatory activities of propolis. Molecules. 2022;**27**(23):8473. DOI: 10.3390/molecules27238473

[45] Pahlavani N, Malekahmadi M, Sedaghat A, Rostami A, Alkadir OKA, Taifi A, et al. Effects of melatonin and propolis supplementation on inflammation, oxidative stress, and clinical outcomes in patients with primary pneumosepsis: A randomized controlled clinical trial. Complementary Medicine Research. 2022;**29**(4):275-285. DOI: 10.1159/000523766

[46] Park JY, Ko KA, Lee JY, Oh JW, Lim HC, Lee DW, et al. Clinical and immunological efficacy of mangosteen and propolis extracted complex in patients with gingivitis: A multi-centered randomized controlled clinical trial. Nutrients. 2021;**13**(8):2604. DOI: 10.3390/nu13082604

[47] Pereira EMR, da Silva JLDC, Silva FF, De Luca MP, Lorentz TCM, Santos VR. Clinical evidence of the efficacy of a mouthwash containing propolis for the control of plaque and gingivitis: A phase II study. Evidence-Based Complementary and Alternative Medicine. 2011;**2021**. DOI: 10.1155/2011/750249

[48] Bretz WA, Paulino N, Nör JE, Moreira A. The effectiveness of propolis on gingivitis: A randomized controlled trial. The Journal of Alternative and Complementary Medicine. 2014;**20**(12):943-948. DOI: 10.1089/acm.2013.0431

[49] Kurek-Górecka A, Kłósek M, Pietsz G, Czuba ZP, Kolayli S, Can Z, et al. The phenolic profile and anti-inflammatory effect of ethanolic extract of polish propolis on activated human gingival fibroblasts-1 cell line. Molecules.

2023;**28**(22):7477. DOI: 10.3390/molecules28227477

[50] Kalay TS, Kara Y, Karaoglu SA, Kolayli S. Evaluation of stabilized chlorine dioxide in terms of antimicrobial activity and dentin bond strength. Combinatorial Chemistry & High Throughput Screening. 2022;**25**(9):1427-1436. DOI: 10.2174/1386207324666210816121255

[51] Vanni R, Waldner-Tomic NM, Belibasakis GN, Attin T, Schmidlin PR, Thurnheer T. Antibacterial efficacy of a propolis toothpaste and mouthrinse against a supragingival multispecies biofilm. Oral Health & Preventive. Dentistry. 2015;**13**(6). DOI: 10.3290/j.ohpd.a34372

[52] Akca AE, Akca G, Topçu FT, Macit E, Pikdöken L, Özgen IŞ. The comparative evaluation of the antimicrobial effect of propolis with chlorhexidine against oral pathogens: An in vitro study. BioMed Research International. 2016;**2016**

[53] Afata TN, Dekebo A. Chemical composition and antimicrobial effect of western Ethiopian propolis. Chemistry & Biodiversity. 2023;**20**(2):e202200922. DOI: 10.1002/cbdv.202200922

[54] Governa P, Romagnoli G, Albanese P, Rossi F, ManettiF BM. Effect of in vitro simulated digestion on the anti-*helicobacter pylori* activity of different Propolis extracts. Journal of Enzyme Inhibition and Medicinal Chemistry. 2023;**38**(1):2183810. DOI: 10.1080/14756366.2023.2183810

[55] Baltas N, Karaoglu SA, Tarakci C, Kolayli S. Effect of propolis in gastric disorders: Inhibition studies on the growth of *Helicobacter pylori* and production of its urease. Journal of Enzyme Inhibition and Medicinal Chemistry. 2016;**31**(sup2):46-50

[56] Freires IA, Queiroz VCPP, Furletti VF, Ikegaki M, de Alencar SM, Duarte MCT, et al. Chemical composition and antifungal potential of Brazilian propolis against Candida spp. Journal de Mycologie Medicale. 2016;**26**(2):122-132. DOI: 10.1016/j.mycmed.2016.01.003

[57] Diba K, Mousavi B, Mahmoudi M, Hashemi J. In-vitro anti-fungal activity of Propolis alcoholic extract on *Candida spp*. and *aspergillus spp*. Tehran University Medical Journal. 2010;**68**:2

[58] Zulhendri F, Chandrasekaran K, Kowacz M, Ravalia M, Kripal K, Fearnley J, et al. Antiviral, antibacterial, antifungal, and antiparasitic properties of propolis: A review. Food. 2021;**10**(6):1360. DOI: 10.3390/foods10061360

[59] Kurek-Gorecka A, Walczyńska-Dragon K, Felitti R, Nitecka-Buchta A, Baron S, Olczyk P. The influence of propolis on dental plaque reduction and the correlation between dental plaque and severity of COVID-19 complications—A literature review. Molecules. 2021;**26**(18):5516. DOI: 10.3390/molecules26185516

[60] Hermansyah D, Zulhendri F, Perera CO, Firsty NN, Chandrasekaran K, Abdulah R, et al. The potential use of propolis as an adjunctive therapy in breast cancers. Integrative Cancer Therapies. 2022;**21**:15347354221096868. DOI: 0.1177/15347354221096868

[61] Liu CC, Hsu JM, Kuo LK, Chuu CP. Caffeic acid phenethyl ester as an adjuvant therapy for advanced prostate cancer. Medical Hypotheses. 2013;**80**(5):617-619. DOI: 10.1016/j.mehy.2013.02.003

[62] Vekić J, Ivanišević J, Zeljković A, Spasojević-Kalimanovska V,

Bogavac-Stanojević N, Mihajlović M, et al. Effect of propolis and N-acetylcysteine supplementation on lipoprotein subclasses distribution and paraoxonase 1 activity in subjects with acute respiratory infection. Journal of Medical Biochemistry. 2020;**39**(4):467. DOI: 10.5937/jomb0-24695

[63] Sugita C, Yamashita A, Tsutsumi S, Kai H, Sonoda T, Yoshida H, et al. Brazilian propolis (AF-08) inhibits collagen-induced platelet aggregation without affecting blood coagulation. Journal of. Natural Medicines. 2021;**75**(4):975-984. DOI: DOI.10.1007/s11418-021-01518-w

[64] Silva H, Francisco R, Saraiva A, Francisco S, Carrascosa C, Raposo A. The cardiovascular therapeutic potential of propolis—A comprehensive review. Biology. 2021;**10**:27. DOI: 10.3390/biology10010027

[65] Kolaylı S, Sahin H, Malkoc M, Cakir H, Kara Y, Bektas Ö, et al. Does ethanolic propolis extract affect blood clotting parameters? Progress in Nutrition. 2022;**24**:1. DOI: 10.23751/pn.v24i1.12017

[66] Paulino M, Alvareda E, Iribarne F, Miranda P, Espinosa V, Aguilera S, et al. Toward the understanding of the molecular basis for the inhibition of COX-1 and COX-2 by phenolic compounds present in Uruguayan propolis and grape pomace. Journal of Biomolecular Structure and Dynamics. 2016;**34**(12):2643-2657. DOI: 10.1080/07391102.2015.1124808

Chapter 2

# Indonesia Spices Used in Global Cooking for Human Health

*Mujianto Mujianto*

## Abstract

Cooking spices produced on an industrial scale and derived from these spices have authentic Indonesian flavors; besides being nutritious, cooking spices from these spices is also healthy. One source of raw materials for this original Indonesian cooking spice is overfermented tempeh. The research results show that overfermented tempeh flour has great potential as a source of raw material for the cooking seasoning industry. Analysis of the main components of overfermented tempeh flour with 10 cooking spices circulating in Indonesia resulted in a very close similarity, namely three main components. The first main component is eight cooking spices.

**Keywords:** cooking seasoning industry, overfermented tempeh, principal component analysis, similarity, authentic Indonesia flavors

## 1. Introduction

Indonesia has a very diverse range of spice and herbal plants. Indonesia is known as "Indonesia the Mother of Spices." There are seven types of superior Indonesian spices for export purposes, namely pepper, nutmeg, vanilla, cloves, ginger, cinnamon, and turmeric. Herbal or leafy spices are mostly used to flavor food and drink, depending on whether the leaves are fresh or dried. Herbal spices are utilized not just as flavorings but also for their nutritional, antioxidant, antibacterial, and therapeutic qualities. Certain herbs and soybeans, including soy sauce, fish sauce, and others, are also utilized as condiments in a lot of cooking dishes. The second-oldest industry in the world, spices, has been around since before the time of ancient Egypt. The Arab countries ruled it for centuries, and Marco Polo, Vasco de Gama, and Christopher Columbus' voyages were prompted by their quest for spices [1]. The essential oils that are derived from the delicate stems, leaves, and flowering tops are used to flavor liquors, soft drinks, beverages, and pharmaceutical preparations, in addition to being employed in cosmetics, perfumeries, and toiletries [2].

Soybeans have wide genetic diversity due to their ability to adapt to agroclimates [3]. The processed Indonesian soybean product that is known throughout the world is tempeh; because tempeh is made from soybeans, its nutritional value and potential as a functional food is very large. Tempeh is an ethnic Indonesian fermented food [4]. The 40-hour fermentation process increased the quantities of free amino acids, proteins releasable during in vitro digestion, and total protein. Quinoa tempeh's profile of


IntechOpen

essential amino acids matched the reference pattern. It is possible to suggest the extended tempeh-style fermentation of quinoa as a means of producing food with additional value [5]. *R. microsporus* (RM), *R. oryzae* (RO), or *R. stolonifer* (RS) were used to temper the mixture. When combined, the various forms of tempeh seem to offer unique advantages for colon health, and tempeh RS may have particularly positive impacts on aspects crucial to colon health [6]. According to one perspective, any food can be considered functional as long as it contains nutrients and has a physiological impact. A food that positively impacts one or more specific bodily processes in a way that goes beyond sufficient nutritional effects and is relevant to either an enhanced state of health and well-being or a decreased risk of disease is referred to as functional food in Europe by the Functional Food Science Association [7].

## 2. Methodology

### 2.1 Material

Overfermented tempeh was purchased from a local market in Sidoarjo Regency. Overfermented tempeh is then dried by sun drying. Overfermented dried tempeh with a water content of less than 10% is then made into rejected tempeh flour. Ten kinds of cooking spices were purchased from several shops in the cities of Surabaya and Sidoarjo. These cooking spices were then coded to maintain the confidentiality of the producers and, at the same time, enforce research ethics.

### 2.2 Method

Overfermented tempeh flour and ten kinds of cooking spices are then packaged with a certain code to observe the absorbance pattern using infrared spectroscopy (Fourier transform infrared) at the Chemistry Laboratory, Faculty of Mathematics and Natural Sciences, Institut Sepuluh Nopember Surabaya. Observing the size of flour particles using the Zise Particle Analyzer in the Physics Laboratory, while observing the size of the particles using scanning electron microscopy (SEM) in the Atsiri Laboratory, Faculty of Mathematics and Natural Sciences, Brawijaya University. Observation of fatty acid patterns using GC-FID at the Pharmacy Laboratory, Faculty of Pharmacy, Airlangga University. Proximate observations were carried out at the Agricultural Products Technology Laboratory, Faculty of Agricultural Technology, Brawijaya University. Color observations using the color analysis version 4 application from www.leizer-soft.com.

## 3. Results and discussion

Several properties of the chemicals that make up food flavors can be swiftly and precisely detected using Fourier transform infrared (FTIR) absorbance patterns. The material under examination may exist in a solid, liquid, or emulsion state. In the GC-FID profile study utilizing methanol as a solvent, gas chromatography (GC-FID) is outfitted with an ionized smoke detector (flame ionization) for volatile flavor components. Thirteen components were identified by GC/MS studies of the essential oil. The principal components of the oil were carvacrol (14%), γ-terpinene (18.34%), p-cymene (18.32%), and thymol (29.45%) [8]. **Table 1** below displays the findings of the investigation into the flavor attributes of overfermented tempeh flour.

Characterization of essential oils (EO) from *R. officinalis* and *L. dentata* using GC-FID [9]. Food adulteration can be quickly and accurately detected and predicted with mid-infrared spectroscopy (MIR), which is frequently used in conjunction with chemometric approaches (**Figure 1**) [10].

With an absorbance value of 0–30% or n (A) of 542 points (29.01%), the distribution of absorbance values for overfermented tempeh flour lies between a wavelength of 400 and 2250 $cm^{-1}$, whereas the distribution of absorbance values for 4500 and 2250 $cm^{-1}$ has an absorbance value of 30–55% or n (B) of 532 points (28.48%). With absorbance values ranging from 0–30% or n (C) up to 376 points (20.13%), the distribution of absorbance values for overfermented tempeh flour lies between wavelengths of 2250 to 4500 $cm^{-1}$, whereas the distribution of absorbance values lies at length. Waves having an absorbance value of 0–30% or n (D) between 400 and 2250 $cm^{-1}$. **Table 2** below displays the findings of the examination of the Fourier transform infrared (FTIR) absorbance values distribution for overfermented tempeh flour (**Figure 2**).

**Table 2** displays the Fourier transform infrared (FTIR) analysis results, which indicate that the overfermented tempeh flour has 14 wave peaks totaling 552,541 unit area. At the apex of the wave, the biggest area is 1058.96 $cm^{-1}$, with an area of 100,228 area units, and the smallest area is 406.99 $cm^{-1}$, with 4693 area units. In

| NO | Parameter | Unit | Methods | Results analysis |
|---|---|---|---|---|
| | Peaks FTIR dan GC-FID | | | |
| 1 | Peaks FTIR | pc | *Gaussian* Model | 14 |
| 2 | The area under the Peaks FTIR | SL | *Gaussian* Model | 552,541 |
| 3 | Peaks GC-FID | pc | GC-FID Agilent | 37 |
| 4 | The Area Under the Peaks GC-FID | SL | GC-FID Agilent | 87,014,24 |

**Table 1.**
*Overfermented tempeh flour's flavor character analysis results.*

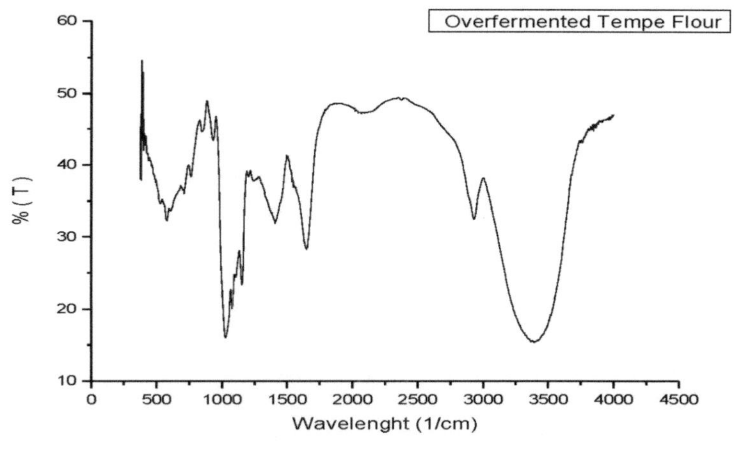


**Figure 1.**
*Fourier transform infrared (FTIR) absorbance pattern of overfermented tempeh flour.*

| Wavelength ($cm^{-1}$) | Absorbance Point 0–30% | Absorbance Point 30–55% |
|---|---|---|
| 400–2.250 $cm^{-1}$ | n (A) = 542 (29,01%) | n (B) = 532 (28,48%) |
| 2.250–4.500 $cm^{-1}$ | n (D) = 418 (22,38%) | n (C) = 376 (20,13%) |

**Table 2.**
*Distribution of Fourier transform infrared (FTIR) absorbance values for overfermented tempeh flour.*

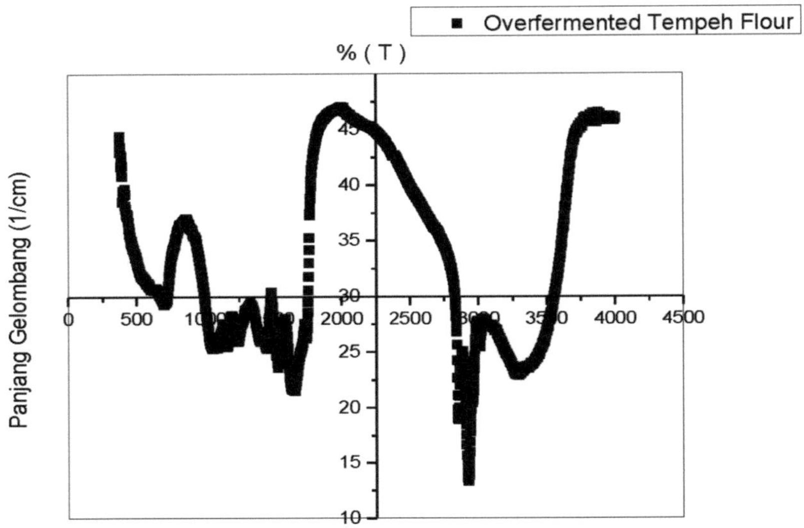


**Figure 2.**
*Fourier transform infrared (FTIR) absorbance pattern of overfermented tempeh flour in four quadrants.*

overfermented tempeh flour, the ether functional group has the largest area proportion (31.39%), followed by the furan or benzofuran functional group (4.13%), and the alkyl functional group (18.8%), alkene (18.79%), saturated alkyl (13.76%), ester (10.57%), and terminal alkyne (triple bond) (4.18%) (**Table 3**).

Overfermented tempeh flour has seven functional groups, namely alkene, ether, ester, saturated alkyl, furan, or benzofuran, alkyl, and terminal alkyne functional groups. Planting settings, ontogenic variations, and species of walnuts all affect the fatty acid profiles of the kernels. To identify the primary markers that chemometric analysis can utilize to differentiate between various walnuts [11].

The results of Fourier transform infrared peaks analysis using the Gaussian and Lorentz Fit methods are as shown in **Figures 3** and **4** show no different results. "Yo" for the Gaussian method is 42,47,594 while "Yo" for the Lorenzt Fit method is 43,11,747.

## 3.1 Functional properties of overfermented tempeh flour

The results of the analysis of the proximate content of overfermented tempeh flour are shown in **Table 4**. Overfermented tempeh flour has antioxidant power to reduce free radicals, the antioxidant power of overfermented tempeh flour is 160.18 mg/ml so that overfermented tempeh flour has functional properties. Overfermented tempeh flour which has functional properties as a source of natural and healthy nutrient is one

| No | Peaks | Area | Percent | Chemical Formula | Functional Group |
|---|---|---|---|---|---|
| 1 | 406.99 | 4.693 | 0.85 | RCH=CHR | Alkene |
| 2 | 702.11 | 10.207 | 1.85 | RCH=CHR (cis) | Alkene |
| 3 | 1058.96 | 100.228 | 18.14 | R = C-O-C | Ether |
| 4 | 1163.11 | 35.328 | 6.39 | R-COOCH3 | Ester |
| 5 | 1238.34 | 73.22 | 13.25 | R = C-O-C | Ether |
| 6 | 1404.22 | 40.728 | 7.37 | R-CH2CO | Saturated alkyl |
| 7 | 1456.3 | 20.991 | 3.80 | R-OCH3 | Saturated alkyl |
| 8 | 1523.82 | 22.847 | 4.13 | C=C | Furan atau benzofuran |
| 9 | 1637.62 | 31.932 | 5.78 | RCH=CH2 | Alkene |
| 10 | 1739.85 | 20.962 | 3.79 | R-C-CO = R | Ester |
| 11 | 2854.74 | 99.206 | 17.95 | R-CH2 | Alkyl |
| 12 | 2926.11 | 48.168 | 8.72 | R-CH2 | Alkyl |
| 13 | 3010.98 | 20.946 | 3.79 | R = CH-R | Alkene |
| 14 | 3269.45 | 23.085 | 4.18 | C = C (rangkap 3) | Terminal alkyne |
| Total area | | 552.541 | 100.00 | | |

**Table 3.**
*Fourier transform infrared analysis of functional groups of overfermented tempeh flour.*

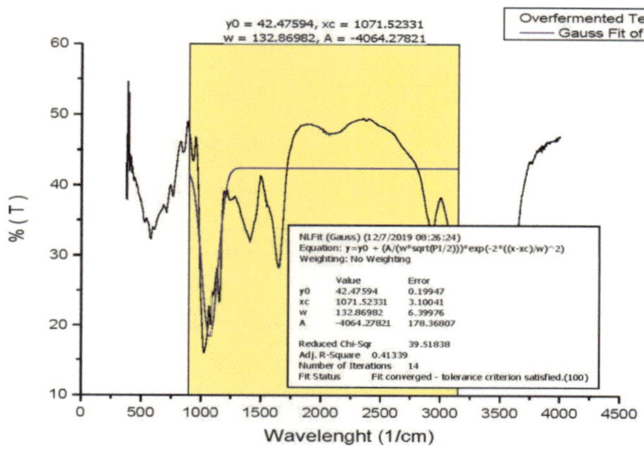


**Figure 3.**
*Results of peaks Fourier transform infrared analysis of Overfermented tempeh flour using the Gaussian method.*

that has a low protein molecular weight and has high reducing sugars so that it has the potential to be a source of Millard products to produce savory properties (umami).

## 3.2 Color of overfermented tempeh flour

Results of color analysis of overfermented tempeh flour and overfermented tempeh groats flour before the enzymatic hydrolysis process using the color analysis version 4 application, developer Roy Leizer. www.leizer-soft.com, based on the

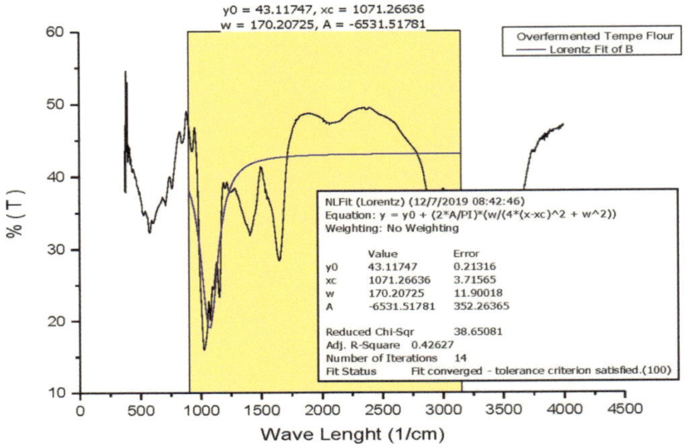


**Figure 4.**
*Results of peaks Fourier transform infrared analysis of Overfermented tempeh flour using the Lorentz fit method.*

| No | Parameter | Unit | Methods | Analysis results |
|---|---|---|---|---|
| A | Kadar Proksimat | | | |
| 1 | Ash Content | % | AOAC 2012 | 2.33 |
| 2 | Crude Protein | % | AOAC 2012 | 48.61 |
| 3 | Crude Fat | % | AOAC 2012 | 24.49 |
| 4 | Crude Fiber | % | SNI-01-2891-1992 | 5.66 |
| 5 | Antioksidan | mg/ml | IC 50 | 160.18 |
| 6 | Total Sugar | % | Anthrone-Sulfate | 1.06 |
| 7 | Reducing Sugar | % | *Somogyi–Nelson* | 0.11 |
| 8 | Starch | % | Hidrolisis Asam | 8.64 |
| 9 | Color | % | Color Analysis | 65.12 |
| B | Profile GC-MSD dan GC-FID | | | |
| 1 | Asam Palmitat | % | GC-MSD Agilent 6980 N | 9.71 |
| 2 | Asam Linoleat | % | GC-MSD Agilent 6980 N | 66.67 |
| 3 | Asam Oleat | % | GC-MSD Agilent 6980 N | 19.16 |
| 4 | Asam Stearat | % | GC-MSD Agilent 6980 N | 3.61 |
| 5 | Asam Arakat | % | GC-MSD Agilent 6980 N | 0.31 |
| 6 | Asam Behenat | % | GC-MSD Agilent 6980 N | 0.36 |
| 7 | Asam Lignoserat | % | GC-MSD Agilent 6980 N | 0.18 |

**Table 4.**
*Character analysis results of functional properties of Overfermented tempeh flour.*

results of the analysis, overfermented tempeh flour has five basic colors, namely raw umber at 61.82%, dark tan at 23.56%, bole at 5.11%, army green at 5.01%, and dark olive green at 4.40%. Overfermented tempeh groats flour has 7 (seven) basic colors, namely raw umber at 73.75%, dark tan at 14.15%, bole at 5.90%, brass at 2.05%, army green at 1.63%, seal brown at 1.49%, and dark olive green at 4.40% (**Figure 5**).

Overfermented Tempeh Flour

Color Analysis Results before Hydrolysis

| Color Name | Color Group | Red | Green | Blue | Hex | Percent | Group Total |
|---|---|---|---|---|---|---|---|
| | | | | | | | |
| Raw Umber | Orange And Brown | 123 | 93 | 41 | 7B5D29 | 0.00 | 61.8205 |
| Raw Umber | Orange And Brown | 123 | 93 | 41 | 7B5D29 | 33.54 | |
| Dark Brown | Orange And Brown | 99 | 69 | 33 | 634521 | 12.72 | |
| Chamoisee | Orange And Brown | 165 | 125 | 82 | A57D52 | 9.34 | |
| Bistre | Orange And Brown | 66 | 48 | 33 | 423021 | 1.92 | |
| Dark Lava | Orange And Brown | 66 | 65 | 41 | 424129 | 1.76 | |
| Desert | Orange And Brown | 189 | 158 | 99 | BD9E63 | 1.60 | |
| Pastel Brown | Orange And Brown | 140 | 113 | 74 | 8C714A | 0.92 | |
| Copper | Orange And Brown | 165 | 125 | 57 | A57D39 | 0.00 | |
| Golden Brown | Orange And Brown | 140 | 97 | 41 | 8C6129 | 0.00 | |
| Shadow | Orange And Brown | 140 | 117 | 82 | 8C7552 | 0.00 | |
| Dark Tan | Orange-Yellow | 140 | 121 | 74 | 8C794A | 0.00 | 23.5582 |
| Dark Tan | Orange-Yellow | 140 | 121 | 74 | 8C794A | 23.56 | |
| Bole | Red-Orange | 107 | 85 | 57 | 6B5539 | 0.00 | 5.1124 |
| Bole | Red-Orange | 107 | 85 | 57 | 6B5539 | 5.11 | |
| Army Green | Yellow-Green | 90 | 81 | 41 | 5A5129 | 0.00 | 5.0059 |
| Army Green | Yellow-Green | 90 | 81 | 41 | 5A5129 | 5.01 | |
| Dark Olive Green | Green | 99 | 89 | 57 | 635939 | 0.00 | 4.4043 |
| Dark Olive Green | Green | 99 | 89 | 57 | 635939 | 4.36 | |

Overfermented Tempeh Groats Flour

Color Analysis Results before Hydrolysis

| Color Name | Color Group | Red | Green | Blue | Hex | Percent | Group Total |
|---|---|---|---|---|---|---|---|
| | | | | | | | |
| Raw Umber | Orange And Brown | 123 | 93 | 41 | 7B5D29 | 0.00 | 73.7476 |
| Raw Umber | Orange And Brown | 123 | 93 | 41 | 7B5D29 | 23.79 | |
| Dark Brown | Orange And Brown | 82 | 56 | 24 | 523818 | 17.61 | |
| Desert | Orange And Brown | 189 | 158 | 107 | BD9E6B | 9.24 | |
| Chamoisee | Orange And Brown | 156 | 109 | 74 | 9C6D4A | 8.70 | |
| Bistre | Orange And Brown | 66 | 44 | 24 | 422C18 | 6.71 | |
| Dark Lava | Orange And Brown | 74 | 60 | 41 | 4A3C29 | 2.94 | |
| Pastel Brown | Orange And Brown | 148 | 105 | 74 | 94694A | 1.24 | |
| Copper | Orange And Brown | 165 | 134 | 57 | A58639 | 1.15 | |
| Burlywood | Orange And Brown | 214 | 178 | 132 | D6B284 | 1.13 | |
| Sepia | Orange And Brown | 99 | 77 | 16 | 634D10 | 0.42 | |
| Shadow | Orange And Brown | 140 | 130 | 90 | 8C825A | 0.28 | |
| Golden Brown | Orange And Brown | 140 | 101 | 33 | 8C6521 | 0.27 | |
| Fawn | Orange And Brown | 214 | 178 | 107 | D6B26B | 0.16 | |
| Earth Yellow | Orange And Brown | 214 | 174 | 99 | D6AE63 | 0.03 | |
| Desert Sand | Orange And Brown | 222 | 207 | 156 | DECF9C | 0.02 | |
| Sunset | Orange And Brown | 239 | 211 | 156 | EFD39C | 0.02 | |
| Bronze | Orange And Brown | 198 | 142 | 41 | C68E29 | 0.01 | |
| Khaki | Orange And Brown | 206 | 186 | 140 | CEBA8C | 0.01 | |
| Indian Yellow | Orange And Brown | 206 | 162 | 74 | CEA24A | 0.00 | |
| Ochre | Orange And Brown | 189 | 130 | 33 | BD8221 | 0.00 | |
| Dark Tan | Orange-Yellow | 156 | 142 | 74 | 9C8E4A | 0.00 | 14.1460 |
| Dark Tan | Orange-Yellow | 156 | 142 | 74 | 9C8E4A | 12.13 | |
| Sand | Orange-Yellow | 206 | 182 | 115 | CEB673 | 1.55 | |
| Sand Dune | Orange-Yellow | 156 | 117 | 49 | 9C7531 | 0.42 | |
| Bole | Red-Orange | 107 | 81 | 49 | 6B5131 | 0.00 | 5.8974 |
| Bole | Red-Orange | 107 | 81 | 49 | 6B5131 | 5.78 | |
| Auburn | Red-Orange | 90 | 56 | 16 | 5A3810 | 0.05 | |
| Brass | Yellow | 189 | 154 | 74 | BD9A4A | 0.00 | 2.0513 |
| Brass | Yellow | 189 | 154 | 74 | BD9A4A | 1.58 | |
| Dark Khaki | Yellow | 198 | 178 | 107 | C6B26B | 0.47 | |
| Army Green | Yellow-Green | 82 | 65 | 24 | 524118 | 0.00 | 1.6292 |
| Army Green | Yellow-Green | 82 | 65 | 24 | 524118 | 1.63 | |
| Seal Brown | Red | 41 | 24 | 8 | 291808 | 0.00 | 1.4852 |
| Seal Brown | Red | 41 | 24 | 8 | 291808 | 1.18 | |
| Dark Sienna | Red | 57 | 28 | 0 | 391C00 | 0.30 | |
| Dark Olive Green | Green | 99 | 89 | 57 | 635939 | 0.00 | 0.9862 |
| Dark Olive Green | Green | 99 | 89 | 57 | 635939 | 0.95 | |

**Figure 5.**
*The color analysis results before hydrolysis.*

### 3.3 Results of GC-MS and GC-fid peaks observations

The results of observations of overfermented tempeh flour using gas chromatography-mass spectrometry (GC-MS) produced seven peaks with a total area of 2,638,026,318 unit area; the second peak had the largest area proportion of 66.67%, are the peaks for linoleic acid compounds. The third peaks are 19.16% oleic acid compounds, the first peaks are 9.71% palmitic acid compounds, the fourth peaks are 3.61% stearic acid compounds, the sixth peak is 0.36% behenic acid compound, and the seventh peak is 0.18% lignoceric acid compound. Systematic techniques were utilized for the isolation and detection of taste-active umami proteolytics, while gas chromatography-olfactometry-mass spectrometer (GC-O-MS) was utilized to identify the compounds with fragrance (**Table 5**) [12].

Analysis of overfermented tempeh flour using GC-MS is specifically only for the fatty acid profile using hexane solvent (pure analyzer) with an operating temperature between 1700C and 2700C. The observation results for profiling all compounds in overfermented tempeh flour using methanol solvent (pure analysis) 37 peaks.

The total area under the curve using gas chromatography flame instrument detector (GC-FID) analysis is 87,014.23978 area units; the highest area proportion is 1.8864% or 1641.46252 area units while the smallest area proportion is 0.0054% or 4.66513 units wide. Based on the GC-FID analysis above, overfermented tempeh flour contains 36 compounds that are soluble in methanol. The area proportion of 95.16% or 82,805.6 is the proportion for the methanol compound (pure analysis) as a solvent (**Table 6**).

### 3.4 Characteristics of overfermented tempeh flour particles

Overfermented tempeh flour has a specific gravity of 2050 gr/ml or greater than the specific gravity of water of 1000 gr/ml while a polydispersity index of 0.700 indicates that the particle size population of tempeh flour is not uniform. Zeta potential is used to measure the behavior of colloidal stability of particles. The zeta potential value of rejected tempeh flour particles is −28,000 mv ($\pm$ 10 to $\pm$30) including the instability category for producing colloids. The fungus bonds the soybeans more tightly when more yeast is applied to the soybean particles, improving their durability [13]. The findings indicated that lupin tempeh had a greater moisture content—between 60 and 66%—than soy tempeh (58.90%). In the meantime, the lupin tempeh's firmness ranged from 21.38 to 30.89 N. The majority of tempes made from $\leq$50% of particles sized $\geq$5 mm (L3, L5, and L6) showed poor firmness and fragrance quality [14]. In comparison to soy tempeh ($218\times10^3$ CFU/g), the mean total count of colony in lupin tempeh was $3.0\times10^4$ CFU/g, suggesting a slower rate of mycelia growth on lupin beans. On the other hand, lupin tempeh, which was made from entire beans (L4 and L7), was just as widely accepted as soy tempeh overall (**Table 7**) [14].

### 3.5 Investigating overfermented tempeh flour particle properties with scanning electron microscopy

The findings of examining the properties of overfermented tempeh flour particles (TTA01) at a 500 tempeh fermentation magnification using scanning electron microscopy (SEM). Using a particle magnification of 5000, the white hue of the overfermented tempeh flour particles is further emphasized (**Figures 6** and 7).

| Header | Peak | R.T. | First | Max | Last | PK TY | Height | Area | Pct Max | Pct Total |
|---|---|---|---|---|---|---|---|---|---|---|
| 1 | 1 | 5.947 | 737 | 768 | 792 | BB | 4,547,629 | 256,027,244 | 14.56 | 9.705 |
| 2 | 2 | 10.158 | 1182 | 1319 | 1325 | BV 6 | 9,124,989 | 1,758,757,164 | 100 | 66.669 |
| 3 | 3 | 10.333 | 1325 | 1342 | 1371 | VB 5 | 6,821,032 | 505,447,609 | 28.74 | 19.16 |
| 4 | 4 | 10.899 | 1377 | 1416 | 1433 | BB 4 | 1,210,552 | 95,208,682 | 5.41 | 3.609 |
| 5 | 5 | 16.126 | 2088 | 2100 | 2120 | BB 6 | 243,436 | 8,231,317 | 0.47 | 0.312 |
| 6 | 6 | 20.199 | 2620 | 2633 | 2651 | BB 3 | 310,982 | 9,604,974 | 0.55 | 0.364 |
| 7 | 7 | 23.546 | 3061 | 3071 | 3094 | BB 5 | 131,181 | 4,749,328 | 0.27 | 0.18 |

**Table 5.**
*GC-MS peaks observation results for overfermented tempeh flour.*

| No | Peaks | Retention time | Type | Width (min) | Area | Percent area |
|---|---|---|---|---|---|---|
| 1 | 1 | 2.853 | PB.S | 0.0800 | 82805.6 | 95.1633 |
| 2 | 2 | 3.318 | BV.X | 0.0657 | 598.64459 | 0.6880 |
| 3 | 3 | 3.684 | VV.X | 0.0640 | 24.6538 | 0.0283 |
| 4 | 4 | 5.327 | VV.X | 0.1305 | 5.88565 | 0.0068 |
| 5 | 5 | 6.225 | PB | 0.0752 | 15.55542 | 0.0179 |
| 6 | 6 | 8.921 | BB | 0.0971 | 4.66513 | 0.0054 |
| 7 | 7 | 10.362 | PB | 0.0618 | 7.53214 | 0.0087 |
| 8 | 8 | 10.818 | BB | 0.1745 | 11.33715 | 0.0130 |
| 9 | 9 | 14.409 | VV | 0.0593 | 36.49489 | 0.0419 |
| 10 | 10 | 14.856 | VV | 0.0829 | 7.62057 | 0.0088 |
| 11 | 11 | 15.641 | VV | 0.1591 | 31.83162 | 0.0366 |
| 12 | 12 | 17.398 | VV | 0.0809 | 343.87631 | 0.3952 |
| 13 | 13 | 17.892 | VV | 0.0588 | 25.2536 | 0.0290 |
| 14 | 14 | 18.141 | VB | 0.0660 | 16.55139 | 0.0190 |
| 15 | 15 | 18.519 | BV | 0.0917 | 12.08102 | 0.0139 |
| 16 | 16 | 18.985 | VV | 0.0722 | 25.11616 | 0.0289 |
| 17 | 17 | 19.970 | VV | 0.0866 | 6.40986 | 0.0074 |
| 18 | 18 | 20.494 | VV | 0.1109 | 25.03601 | 0.0288 |
| 19 | 19 | 20.726 | VV | 0.0617 | 213.65863 | 0.2455 |
| 20 | 20 | 21.012 | VV | 0.0993 | 102.80613 | 0.1181 |
| 21 | 21 | 21.484 | VV | 0.0863 | 10.35454 | 0.0119 |
| 22 | 22 | 21.701 | VV | 0.0982 | 181.90768 | 0.2091 |
| 23 | 23 | 21.980 | VV | 0.0893 | 16.73873 | 0.0192 |
| 24 | 24 | 22.134 | VV | 0.0919 | 21.06925 | 0.0242 |
| 25 | 25 | 22.323 | VV | 0.1026 | 8.77752 | 0.0101 |
| 26 | 26 | 22.733 | VV | 0.1793 | 66.29128 | 0.0762 |

| No | Peaks | Retention time | Type | Width (min) | Area | Percent area |
|---|---|---|---|---|---|---|
| 27 | 27 | 23.211 | VV | 0.0782 | 89.39465 | 0.1027 |
| 28 | 28 | 23.511 | VV | 0.0996 | 1641.4625 | 1.8864 |
| 29 | 29 | 23.887 | VV | 0.0990 | 172.37405 | 0.1981 |
| 30 | 30 | 24.322 | VV | 0.1576 | 23.64201 | 0.0272 |
| 31 | 31 | 24.649 | VV | 0.1293 | 23.04631 | 0.0265 |
| 32 | 32 | 25.219 | VV | 0.2204 | 11.32734 | 0.0130 |
| 33 | 33 | 25.560 | VV | 0.1249 | 11.27103 | 0.0130 |
| 34 | 34 | 25.841 | VV | 0.1106 | 7.97517 | 0.0092 |
| 35 | 35 | 26.263 | VV | 0.0855 | 76.24365 | 0.0876 |
| 36 | 36 | 26.446 | VV | 0.1048 | 156.85835 | 0.1803 |
| 37 | 37 | 26.808 | VB | 0.1391 | 174.89563 | 0.2010 |
| | | | | | 87014.2398 | |

**Table 6.**
*GC-FID peaks observation results for overfermented tempeh flour.*

| No | Particle parameters | Analysis results |
|---|---|---|
| 1 | Z-Average (r.nm) | 344.280 |
| 2 | Polydispersity Index | 0.700 |
| 3 | Mean Count Rate (kcps) | 207.680 |
| 4 | Derived Count Rate (kcps) | 16,463.240 |
| 5 | Zeta-Potential (mV) | −28.000 |
| 6 | Mobilitas (μmcm/Vs) | −2.195 |
| 7 | Conductivity (mS/cm) | 0.046 |
| 8 | Mass weight Protein (kDA) | 4,050,000.000 |
| 9 | % Polydispersity | 88.000 |
| 10 | Berat jenis (g/ml) | 2.050 |
| 11 | 10% Diameter (μm) | 20.995 |
| 12 | 50% Diameter (μm) | 68.450 |
| 13 | 90% Diameter (μm) | 340.740 |
| 14 | Specific surface ($cm^2/g$) | 881.090 |

**Table 7.**
*Characteristics of overfermented tempeh flour particles.*

### 3.6 Principal component analysis (PCA) of overfermented tempeh flour with ten Indonesian cooking spices

Overfermented tempeh flour (TTA) and ten cooking spices that are popular in Indonesia were the subjects of a principal component analysis that included 20,691

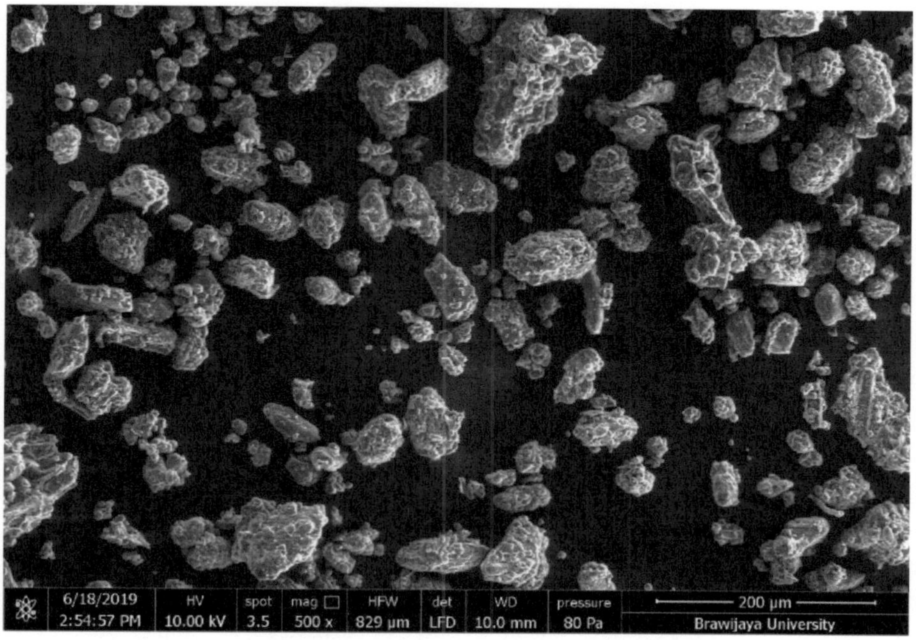


**Figure 6.**
*Features of overfermented tempeh flour particles as determined by 500 scanning electron microscopy (SEM).*

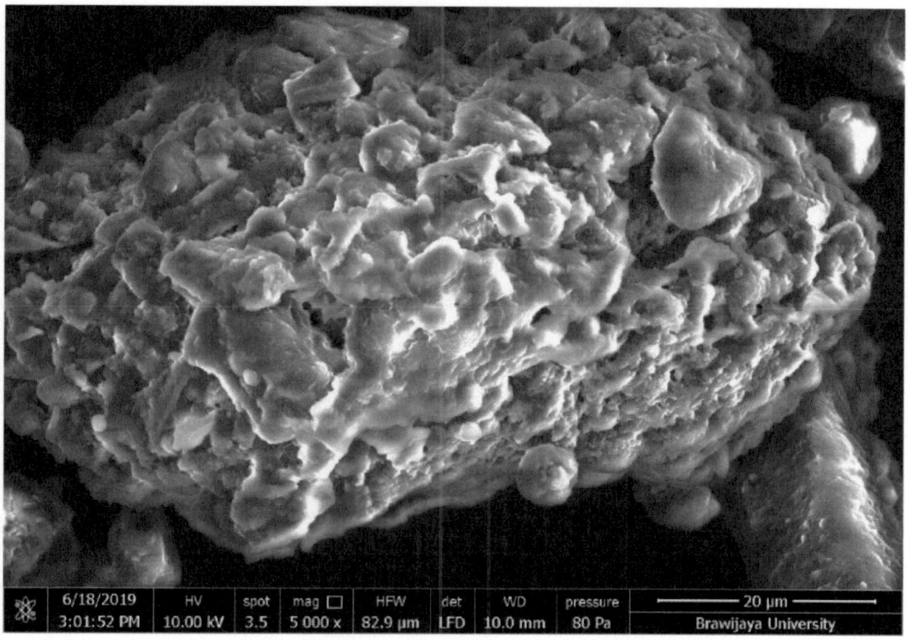


**Figure 7.**
*Features of overfermented tempeh flour particles as determined by 5000 scanning electron microscopy (SEM).*

absorbance point observation data (%) at wavelengths ranging from 400 to 4500 $cm^{-1}$. Multiway principal component analysis (MPCA) was developed for the detection of fermentation fault-batch (**Table 8**) [15].

| No | Types of cooking spices | Code |
|---|---|---|
| 1 | Overfermented Tempeh Flour | TTA01 |
| 2 | Balado Flavored Cooking Spices | BM01 |
| 3 | Curry Flavored Cooking Spices | BM02 |
| 4 | Indofood Magic Cooking Spices | BM03 |
| 5 | Straw Mushroom Cooking Seasoning | BM04 |
| 6 | Mama's Cooking Spices | BM05 |
| 7 | Chinese Vetsin Cooking Spices | VC |
| 8 | Beef Broth Cooking Spices | BM07 |
| 9 | Chicken Stock Cooking Spices | BM08 |
| 10 | Totole Mushroom Powder Cooking Spices | BM09 |
| 11 | Mono Sodium Glutamate | MSG |

**Table 8.**
*Codes for types of cooking spices in principal component analysis.*

Overfermented tempeh flour has an average absorbance (%) using Fourier transform infrared (FTIR) spectroscopy of 33.9847% ± 8.41071, and this average absorbance was obtained from 1881 observation points at a wavelength between 400 and 4500 $cm^{-1}$. The results of observations of the average absorbance (mean), standard deviation (SD) of absorbance, and observation points (N) for 10 (ten) cooking spices circulating in Indonesia can be seen in **Table 9**.

Based on the data in **Table 9**, the average absorption of nine cooking spices, namely Balado Flavored Cooking Spices (BM01), Curry Flavored Cooking Spices (BM02), Indofood Magic Cooking Spices (BM03), Straw Mushroom Cooking Spices (BM04), Mama Cooking Spices (BM05), Chinese Vetsin Cooking Spices (VC), Beef Broth Cooking Spices (BM07), Chicken Stock Cooking Spices (BM08), and Totole Mushroom Powder Cooking Spices (BM09) have an average absorbance (%) from 31.0120 to 35.4486% while the average absorbance for monosodium glutamate (MSG) is 26.7945%. Spices can be added singly or in combination to improve food products' flavor and color [1]. The quality of these diets may also be significantly influenced by the herbs and spices that are used in them [16]. Using gas chromatography and gas chromatography–mass spectrometry, the chemical contents of a few chosen agarwood oil samples were examined. Principal component analysis (PCA), a chemometric technique, was used to analyze their pattern recognition characteristics [17].

From 20,691 observation data, the Keiser–Meyer–Oikin (KMO) test result was 0.811, which is a number greater than 0.500, indicating that the validity of the observation data can be explained or is legitimate data (**Table 10**).

According to the test results, the variable can be further anticipated and investigated if the KMO value is greater than 0.5, namely 0.811 > 0.500. Barlett's test <0.05, or 0.000 < 0.050, indicates a significant degree of correlation between the variables. Factor 1 has eight variables, while factor 2 has six variables when the main components are rotated. The Kaiser–Meyer–Olkin (KMO) test was employed in principal component analysis to assess the appropriateness of sampling. Between 0 and 0.5 is the minimum, between 0.5 and 0.7 is the medium, between 0.7 and 0.8 is good, between 0.8 and 0.9 is very good, and 0.9 and above is regarded as perfect. The KMO coefficient is undesirable in these ranges (**Figure 8**) [18].

| No | Code | Average | Standard deviation | Analysis n | Missing n |
|---|---|---|---|---|---|
| 1 | TTA01 | 33.98470 | 8.41071 | 1881 | 0 |
| 2 | BM01 | 33.66460 | 4.21051 | 1881 | 0 |
| 3 | BM02 | 35.44860 | 3.04244 | 1881 | 0 |
| 4 | BM03 | 31.86430 | 7.12663 | 1881 | 0 |
| 5 | BM04 | 32.31660 | 8.22289 | 1881 | 0 |
| 6 | BM05 | 34.04980 | 3.38931 | 1881 | 0 |
| 7 | VC | 33.18230 | 5.39659 | 1881 | 0 |
| 8 | BM07 | 32.11040 | 5.00854 | 1881 | 0 |
| 9 | BM08 | 31.56530 | 7.81160 | 1881 | 0 |
| 10 | BM09 | 31.01200 | 6.73196 | 1881 | 0 |
| 11 | MSG | 26.79450 | 10.66356 | 1881 | 0 |

**Table 9.**
*Data description of overfermented tempeh flour (TTA01) and 10 (ten) cooking spices in Indonesia.*

| Kaiser–Meyer–Olkin Measure of Sampling Adequacy | | 0.811 |
|---|---|---|
| Bartlett's Test of Sphericity | Approx. chi-square | 63,586.279 |
| | df | 55 |
| | Sig. | 0.000 |

**Table 10.**
*KMO test results and Barlett's FTIR absorbance test for Overfermented tempeh flour and ten cooking spices.*

The figure above illustrates the link between the eigenvalues and the main components. It demonstrates that the eigenvalues are nearly zero up to the second main component (F2), and the cumulative percentage of variance explained up to this point is equal to 94.771%.

**Table 11** shows that the overall variance is 8.974 (81.584%) for the first (one) main component, 1.296 (11.784%) for the second, 0.482 (4.386%) for the third, and 0.105 (0.9588%) for the fourth, up to 10.857 (99.389%) for the fifth main component (F1, F2, F3, and F4). The first main component, 8.974 > 1, and the second main component, 1.296 > 1, are two of the four main components that were generated and have eigenvalues greater than 1 (**Table 12**).

After extracting 11 variables, the factoring process yields 4 primary components. The Varimax method is used to perform variable rotation in order to pinpoint a variable's precise location within the main component. **Table 13** displays the effects of varying rotation.

The results of the rotation of the matrix components produce three main components. The variables that form the main component with a loading value of more than 0.5 are the first main component (F1) formed by eight variables, namely BM04 (0.946), BM08 (0.938), BM07 (0.865), BM05 (0.850), BM03 (0.807), BM01 (0.753), BM09 (0.646), and BM02 (0.499). The second principal component (F2) is formed by 6 (six) variables, namely MSG (0.944), VC (0.900), BM09 (0.727), TTA01 (0.567), BM01 (0.554), and BM03 (0.494). The third main component (F3) is formed by 2

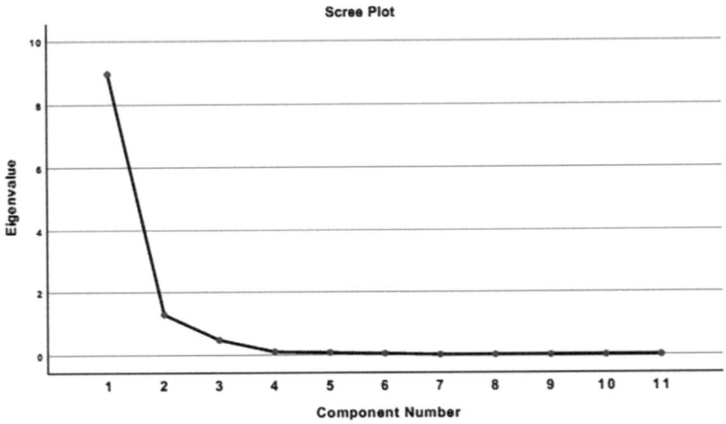


**Figure 8.**
*Relationship between eigenvalue and main components.*

| No | Component | Total | Initial eigenvalues % of variance | Cumulative % | Extraction sums of squared | |
|---|---|---|---|---|---|---|
| | | | | | Total | % Of variance |
| 1 | 1 | 8.974 | 81.584 | 81.854 | 8.974 | 81.584 |
| 2 | 2 | 1.296 | 11.784 | 93.368 | 1.296 | 11.784 |
| 3 | 3 | 0.482 | 4.386 | 97.754 | 0.482 | 4.386 |
| 4 | 4 | 0.105 | 0.958 | 98.711 | 0.105 | 0.958 |
| 5 | 5 | 0.075 | 0.678 | 99.389 | | |
| 6 | 6 | 0.042 | 0.383 | 99.772 | | |
| 7 | 7 | 0.010 | 0.086 | 99.860 | | |
| 8 | 8 | 0.007 | 0.062 | 99.922 | | |
| 9 | 9 | 0.005 | 0.042 | 99.964 | | |
| 10 | 10 | 0.003 | 0.024 | 99.989 | | |
| 11 | 11 | 0.001 | 0.011 | 100.000 | | |

**Table 11.**
*Relationship between Main components and Total variance values and cumulative Total variance values.*

(two) variables, namely BM02 (0.731) and TTA01 (0.692). The three main components when written in mathematical equations are as follows:

$$F1 = 0{,}946\ BM04 + 0{,}938\ BM08 + 0{,}865\ BM07 + 0{,}850\ BM05 + 0{,}807\ BM03 + 0{,}753\ BM01 + 0{,}646\ BM09 + 0{,}499\ BM02 \quad (1)$$

$$F2 = 0{,}944\ MSG + 0{,}900\ VC + 0{,}727\ BM09 + 0{,}567\ TTA01 + 0{,}554\ BM01 + 0{,}494\ BM03 \quad (2)$$

$$F3 = 0{,}731\ BM02 + 0{,}692\ TTA01 \quad (3)$$

| No | Cooking spice code | Component | | | |
|---|---|---|---|---|---|
| | | 1 | 2 | 3 | 4 |
| 1 | BM03 | 0.976 | −0.102 | −0.141 | |
| 2 | BM09 | 0.958 | 0.171 | −0.219 | |
| 3 | BM01 | 0.941 | | −0.254 | −0.170 |
| 4 | BM07 | 0.935 | −0.319 | 0.117 | |
| 5 | BM05 | 0.929 | −0.297 | 0.127 | |
| 6 | BM08 | 0.916 | −0.369 | | 0.118 |
| 7 | BM04 | 0.907 | −0.399 | | |
| 8 | BM02 | 0.901 | 0.102 | 0.384 | −0.150 |
| 9 | TTA01 | 0.871 | 0.257 | 0.373 | |
| 10 | VC | 0.818 | 0.555 | | |
| 11 | MSG | 0.760 | 0.621 | −0.104 | 0.146 |

**Table 12.**
*Matrix components before rotation of overfermented tempeh flour (TTA01) and ten cooking spices.*

| No | Cooking spice code | Component | | | |
|---|---|---|---|---|---|
| | | 1 | 2 | 3 | 4 |
| 1 | BM04 | 0.946 | 0.229 | 0.216 | |
| 2 | BM08 | 0.938 | 0.261 | 0.220 | |
| 3 | BM07 | 0.865 | 0.236 | 0.431 | |
| 4 | BM05 | 0.850 | 0.253 | 0.431 | |
| 5 | BM03 | 0.807 | 0.494 | 0.254 | 0.173 |
| 6 | BM01 | 0.753 | 0.554 | 0.161 | 0.282 |
| 7 | MSG | 0.203 | 0.944 | 0.246 | |
| 8 | VC | 0.268 | 0.900 | 0.305 | 0.101 |
| 9 | BM09 | 0.646 | 0.727 | 0.193 | 0.109 |
| 10 | BM02 | 0.499 | 0.438 | 0.731 | 0.128 |
| 11 | TTA01 | 0.400 | 0.567 | 0.692 | |

**Table 13.**
*Matrix components after rotation of overfermented tempeh flour (TTA01) and ten cooking spices.*

Information.

F1 = first main component, which is similar to straw mushroom cooking seasoning (BM04).

F2 = second main component, which is similar to monosodium glutamate (MSG).

F3 = third main component, which is similar to curry-flavored cooking spices (BM02).

## 4. Conclusion

Based on the results of the analysis of the main components of ten cooking spices and one cooking spice raw material, namely overfermented tempeh flour, the conclusion is:

1. All variables TTA 01 and BM01, BM02, BM03, BM04, BM05, VC, BM07, BM08, BM09, MSG become variables from the first main component (F1), meaning all variables from overfermented tempeh flour (TTA01) and ten cooking spices have identical (similar) flavor characteristics.

2. Monosodium glutamate cooking spice, vetsin China, totole mushroom powder cooking spice, overfermented tempeh flour, balado flavored cooking spice, and curry flavored cooking spice have identical or similar flavor characteristics.

3. Overfermented tempeh flour can be used as the main raw material in the curry-flavored cooking seasoning industry (a typical Indonesian cooking spice).

## Author details

Mujianto Mujianto
Disciplines of Industrial Statistics and Process Engineering, Faculty of Engineering, Agriculture Industry Technology Department, Wijaya Kusuma Surabaya University, East Java, Indonesia

*Address all correspondence to: mujianto@uwks.ac.id

## References

[1] Clarke MW. Herbs and Spices, Handbook of Industrial Seasoning. Boston, MA: Springer; 1994

[2] Shylaja MR, Peter KV. The functional role of herbal spices [Internet]. In: Handbook of Herbs and Spices. Abington, Cambridge, England: Elsevier; 2004. pp. 11-21. DOI: 10.1533/9781855738355.1.11

[3] Malang B. Deskripsi Varietas Unggul Kacang-Kacangan dan Umbi-Umbian. ... dan Pengembangan Tanaman Pangan. Badan Litbang. 2008

[4] Balia RL, Utama GL. The occurrence of yeasts and functional properties of Indonesian ethnic fermented foods and beverages. Sci Pap D-Animal Sci. 2017;**60**:303-307

[5] Starzyńska-Janiszewska A, Duliński R, et al. Prolonged tempe-type fermentation in order to improve bioactive potential and nutritional parameters of quinoa seeds. Journal of Cereal Science [Internet]. Sep 2016;**71**: 116-121. Available from: https://www.sciencedirect.com/science/article/pii/S0733521016301515

[6] Yang Y, Kameda T, Aoki H, Nirmagustina DE, et al. The effects of tempe fermented with Rhizopus microsporus, Rhizopus oryzae, or Rhizopus stolonifer on the colonic luminal environment in rats. Journal of Functional Foods [Internet]. Oct 2018; **49**:162-167. Available from: https://www.sciencedirect.com/science/article/pii/S1756464618304389

[7] Ray RC, El Sheikha AF, Kumar S. Oriental fermented functional (probiotic) foods. In: Press C, editor. Microorganisms and Fermentation of Traditional Foods [Internet]. Boca Raton: CRC Press; 2014. pp. 283-311. Available from: https://books.google.com/books?hl=en&lr=&id=8ofSBQAAQBAJ&oi=fnd&pg=PA283&dq=over+fermented+tempe&ots=WIp3mHdrT-&sig=MnX7KqUe6kPyCCDAaifJDphKcww

[8] Nickavar B, Adeli A, Nickavar A. TLC-bioautography and GC-MS analyses for detection and identification of antioxidant constituents of Trachyspermum copticum essential oil [Internet]. Iranian Journal of Pharmaceutical. 2014;**13**(1):127-133. Available from: https://www.ncbi.nlm.nih.gov/pmc/articles/PMC3985236/

[9] Silva-Flores PG, Pérez-López LA, Rivas-Galindo VM, Paniagua-Vega D, Galindo-Rodríguez SA, Álvarez-Román R. Simultaneous GC-FID quantification of Main components of Rosmarinus officinalis L. and Lavandula dentata essential oils in polymeric Nanocapsules for antioxidant application. Journal of Analytical Methods in Chemistry. 2019;**2019**:1-9

[10] Mendes E, Duarte N. Mid-infrared spectroscopy as a valuable tool to tackle food. Food. 2021;**10**:477

[11] Pei Q, Liu Y, Peng S. Fatty acid profiling in kernels coupled with chemometric analyses as a feasible strategy for the discrimination of different walnuts. Food. 2022;**11**(4):1-12

[12] Alim A, Song H, Zou T. Analysis of meaty aroma and umami taste in thermally treated yeast extract by means of sensory-guided screening. European Food Research and Technology. 2020; **246**(10):2119-2133

[13] Sani Y, Nandiyanto ABD, Wardany OF, et al. Learning strategy of

production of tempe through various size of soybean particles for students with hearing impairments. Journal of Engineering [Internet]. 2020; **34**(Special issue):124-131. Available from: http://www.journaleet.org/index.php/jeet/article/view/157878

[14] Priatni S, Devi AF, LBS K, et al. Quality and sensory evaluations of Tempe prepared from various particle sizes of lupin beans [Evaluasi Sensorik dan Kualitas Tempe dari Kacang Lupin Berbagai]. Jurnal Teknologi dan Industri Pangan [Internet]. Dec 2013;**24**:209-214. Available from: https://jurnal.ipb.ac.id/index.php/jtip/article/view/7718

[15] Poppi RJ, Tizei PAG, Pereira AG. Raman spectroscopy and chemometrics for on-line control of glucose fermentation by Saccharomyces cerevisiae [Internet]. Biotechnology Progress. 2012;**28**(6):1598-1604. DOI: 10.1002/btpr.1615

[16] Bower A, Marquez S, de Mejia EG. The health benefits of selected culinary herbs and spices found in the traditional Mediterranean diet. Critical Reviews in Food Science and Nutrition [Internet]. 2016;**56**(16):2728-2746. Available from: https://www.tandfonline.com/doi/abs/10.1080/10408398.2013.805713

[17] Azah MAN, Ismail N, Mailina J, Taib MN, Rahiman MHF, Hafizi ZM. Chemometric study of selected agarwood oils by gas chromatography – Mass spectrometry. Journal of Tropical Forest Science. 2014;**26**(3):382-388

[18] Gıdık B. Antioxidant, antimicrobial activities and fatty acid compositions of wild berberis spp. by different techniques combined with chemometrics (PCA and HCA). Molecules. 2021;**26**(24):1-14

Chapter 3

# Reishi Mushroom and Spirulina Are Organic Whole Foods and Healthy Spices

*Maadh Abdulwahab Alfahad*


## Abstract

The most important cause of diseases that many do not know is increased acidity in the body! The body must remain moderate between acidic and alkaline in the normal ratio or lean slightly toward alkalinity, but now the human body has become more acidic! Due to the frequent consumption of sugars, meat, cheese, milk derivatives, pastries, canned food, preservatives, chemicals, etc., fungi and algae are microscopic living organisms rich in nutrients and compounds with therapeutic properties. With the increase in awareness of health and the search for sustainable food sources, interest has increased in its role in nutrition, spices, and treatment. In this chapter, we will discuss these concepts and the importance of the red or reishi mushrooms, *Ganoderma lucidium*, and the algae, *Spirulina platensis*.


**Keywords:** *Ganoderma*, reishi, *Spirulina*, organic, super food, anti-viruses, self-recovery

## 1. Introduction

Following a healthy diet throughout the life course helps prevent malnutrition in all its forms as well as a range of non-communicable diseases and related conditions. However, increased production of processed foods, rapid urbanization, and changing lifestyles have all led to a shift in dietary patterns. People are now consuming more foods that are higher in energy, fat, free sugars, and salt/sodium, and many people are not eating enough fruit, vegetables, and other dietary fiber such as whole grains.

The exact composition of a varied, balanced, and healthy diet will vary depending on individual characteristics (for example, age, gender, lifestyle, and degree of physical activity), cultural context, locally available foods, and dietary habits. However, the basic principles of what constitutes a healthy diet remain the same.

### 1.1 A healthy diet for adults includes

- Fruits, vegetables, legumes (such as lentils and beans), nuts, and whole grains (such as unprocessed corn, millet, oats, wheat, and brown rice).

- At least 400 grams (i.e. five servings) of fruit and vegetables daily, excluding potatoes, sweet potatoes, cassava, and other starchy roots.

- Less than 10% of total energy intake comes from free sugars, which is equivalent to 50 grams (or about 12 flat teaspoons) for a person who has a healthy body weight and consumes about 2000 calories per day, but the ideal percentage is less than 5% of total energy intake for additional health benefits. Free sugars are all sugars added to foods or drinks by the manufacturer, cook, or consumer, as well as sugars found naturally in honey, syrups, fruit juices, and fruit juice concentrates.

- Less than 30% of total energy intake comes from fat. Unsaturated fats (found in fish, avocados, and nuts, and in sunflower, soybean, canola, and olive oils) are better than saturated fats (found in fatty meats, butter, palm oils, coconut, cream, cheese, ghee, and lard) and trans fats of all kinds, including all of industrially produced trans fats (found in baked and fried foods, snacks, and pre-packaged foods, such as frozen pizza, pies, cakes, biscuits, chips, cooking oils, and spreads) and ruminant trans fats (found in meat and dairy products from ruminant animals, such as cattle, sheep, goats, and camels). It is suggested that saturated fat intake be reduced to less than 10% of total energy intake and trans fat intake to less than 1% of total energy intake. In particular, it is noted that industrially produced trans fats are not part of a healthy diet and should be avoided.

- Less than 5 grams of salt (equivalent to about one teaspoon) per day. The salt should be iodized.

Before I get into the explanation and concept of fungi and medical food algae, I would like to take you with me, dear intelligent reader, to the concept of acidity and alkalinity and the real causes of diseases!!

### 1.2 The effect of the body's pH on the occurrence of diseases

The most important cause of diseases that many do not know is increased acidity in the body! The body must remain moderate between acidity and alkalinity in the natural ratio, or slightly inclined toward alkalinity. The more the acidity and alkalinity are balanced, the farther the body is from disease. The human body was created by God with a certain degree of acidity and alkalinity, which is from 7.3 to 7.4. If this ratio is imbalanced toward acidity, the body is prepared for disease, because diseases prefer an acidic environment, and the human body needs foods that are alkaline at a rate of 80% and foods that are acidic at a rate of 20%, so the more If you maintain this percentage, your health and immunity to diseases will increase [1, 2].

But now the human body has become more acidic! Due to the large consumption of sugars, meat, cheese, milk derivatives, pastries, canned goods, preservatives, chemicals, etc., all of these foods make your body tend toward acidity more than alkalinity. This results in several diseases, including chronic diseases such as diabetes, high blood pressure, cancer, and various viral diseases.

Let us give an example of this fact with regard to viral diseases, where it was found that the building blocks of the viral protein envelope begin to disintegrate (**Figure 1**) and fail to build within the geometric shape of the virus in basic environments, and thus the viral nucleic acids, whether DNA or RNA, are destroyed by nucleic acid

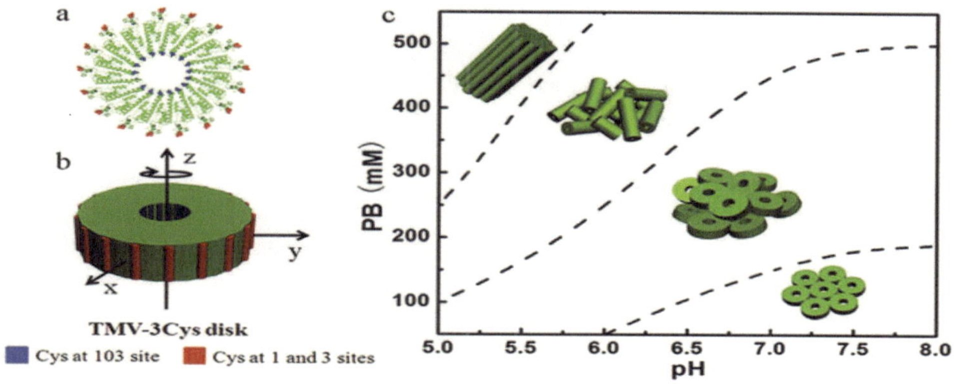


**Figure 1.**
*Disintegration of the subunits of the viral protein envelope with increasing pH value. From Ref. [3].*

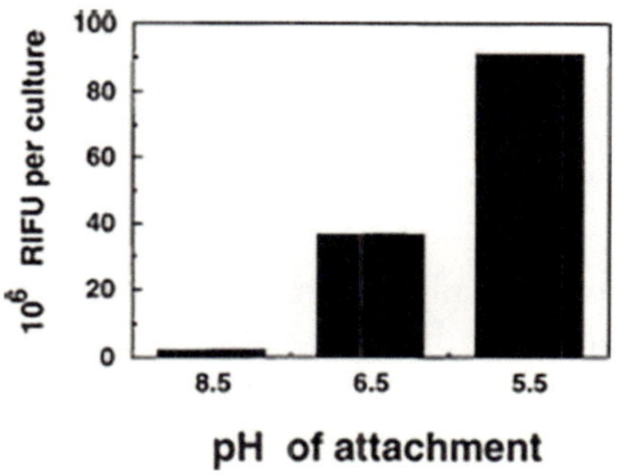


**Figure 2.**
*Effect of pH on the binding of hepatitis a virus (HAV) to host cells. From Ref. [4].*

enzymes such as DNase. And RNase, which in turn causes non-adhesion to the host cell (**Figure 2**) and a reduction in the replication process, thus preventing the disease from occurring or reducing the symptoms resulting from it, depending on the age, gender, and lifestyle of the patient.

Therefore, you must follow a healthy diet that relies mainly on vegetables, fruits, and whole grains, as well as other foods such as nuts and natural juices without sugar, honey, etc., and make the percentage of other foods such as meat, rice, bread, milk derivatives, and others not exceed 20%, and also stay away from them. Avoid refined sugar, soft drinks, fried foods, preservatives, dyes, artificial flavors, etc., from harmful substances. Thus, you will have maintained the alkalinity of your body, and it will become balanced and free from diseases, God willing.

But at this time, implementing this system has become difficult for most people!! What is the solution in light of this industrial chemical food chaos and this lifestyle? The solution is simply to take supplements or basics of natural organic foods that meet two conditions: that they are rich in alkaline elements and that they help expel toxins and free radicals from the cells.

The question that arises now is: Are there foods that are characterized by what we mentioned above? The other question is: how do these foods work inside the body?

These nutritional supplements or basics that we recommend as specialists and from my personal experiences must be from natural organic sources in general. They are foods that a person needs throughout his life and their consumption is not limited to a specific period. They are not medications targeting a specific disease but rather foods that nourish the body internally and externally. Yes, as well. While the body needs internal nutrition, it also needs external nutrition. You will find that some companies or institutions have produced products to nourish the body internally, such as nutritional supplements such as reishi mushrooms and *spirulina*, which we will talk about in detail, and the rest of the nutritional supplements... as well as natural juices, coffee, tea, and cocoa... and others that are grown under the organic farming system. It also produces products to nourish the body externally, such as body or skin care products such as shampoo, soap, toothpaste, creams, oils, and all kinds of cosmetic products.

## 2. The importance of using fungi and algae in nutrition and treatment

Fungi and algae are microscopic organisms rich in nutrients and compounds with therapeutic properties that restore the so-called self-recovery. With increasing health awareness and the search for sustainable food sources, interest in their role in nutrition and treatment has increased. Importance of fungi:

### 2.1 Importance of fungi

Some types of fungi contain a higher percentage of protein than meat and are a rich source of fiber that promotes digestive health. Fungi contain vitamins B, C, and D, and minerals such as potassium, magnesium, and phosphorus. Some types of fungi possess anti-inflammatory compounds, such as beta-glucan, which help reduce the risk of heart disease and cancer. It also contributes to enhancing the health of the immune system by stimulating the immune system and helping to fight infections.

### 2.2 Importance of algae

The importance of algae is that it contains a higher percentage of protein than meat and eggs, which makes it an excellent source of protein for vegetarians. It is rich in vitamins and minerals, such as vitamins A, B, C, and E, and minerals such as calcium, iron, magnesium, and zinc. It is a source of omega-3 fatty acids that promote heart and brain health. It possesses antioxidant compounds, such as carotenoids, that protect cells from damage. It also enhances the health of the digestive system by stimulating the growth of beneficial bacteria or probiotics in the intestines and enhancing the health of the digestive system.

### 2.3 Uses of fungi and algae

- As food: Fungi and algae can be used in many recipes, such as soups, stews, salads, pasta, juices, and coffee. Here, they can either be consumed directly or mixed in certain proportions with foods, fruit juices, and different types of coffee, to obtain the best results. It is required that the mixed materials be of natural and organic origins.

- As nutritional supplements: Fungi and algae are available as nutritional supplements in the form of capsules, powders, or liquids, and the best ones are called superfoods.

- In treatment: Some types of fungi and algae are used to treat some diseases, such as cancer, heart disease, digestive and nervous system diseases, allergies, asthma, diabetes, and various viral diseases. The higher ranks of these algae and fungi usually do not target a specific disease, but rather work to regulate acidity, raise immunity, remove toxicity, and restore the damaged ones. Thus, the cells return to their normal condition, in which God has placed an integrated protection and immunity system against diseases, and this is what we have personally observed through our research or other people who used these foods.

- Production of cosmetics and skin care products: Fungi and algae are involved in the manufacture of various types of cosmetics and skin care products, including creams, oils, shampoo, soap, body foam, lipstick, and oily serums, whether directly or by mixing with other natural materials.

- Production of medicines (pesticides), fertilizers, and plant stimulants: They have also been used directly in the manufacture of treatments against diseases that affect plants, as well as in the form of environmentally friendly fertilizers and stimulants that help raise the productivity of agricultural crops, which contribute to global food security [5–7].

- Sustainable source of food: They can be grown in small spaces, making them a sustainable source of food.

- Environmentally friendly: Cultivation of fungi and algae requires a small amount of water and energy, making it environmentally friendly.

Fungi and algae are living organisms that have many benefits for health and nutrition. It can be used to prepare many delicious dishes, and it has many medicinal properties.

## 3. Reishi mushroom *(Ganoderma lucidum)*

Also known as Lingzhi mushroom, it is a fungus belonging to the Ganodermataceae family. This mushroom is known for its bright red color and distinctive shape, consisting of a large fan-shaped cap and a short stem.

The taxonomic position of the fungus *G. lucidum*, according to what was approved by [8], is:

Kingdom: Fungi
Division: Basidiomycota
Class: Agaricomycetes
Order: Polyporales
Family: Ganodermataceae
Genus: *Ganoderma*
Species: *Lucidum*
Acronym: Reshi, Lingzhi mushroom

There are (38,000) mushroom species discovered in the world, only (2000) of which are edible (non-toxic). Among them, only (200) types of mushrooms have medicinal properties, whether preventive or therapeutic, of which only (6) types contain the highest medical value. Therapeutically, it is red in color and belongs to the *Ganoderma lucidum* family. Here it is worth noting the following:

1. Reishi (*Ganoderma*) does not treat diseases directly but rather helps the body to resist them.

2. Reishi is not a medical drug, but it is a nutritional element with a very high therapeutic value.

3. Reishi targets general health, not specific diseases.

For more than 2000 years, Reishi mushroom, or *Ganoderma lucidum*, has been considered by the Chinese to be a high-quality herbal medicine and an exceptional mushroom with wonderful therapeutic properties. From a botanical point of view, it is considered a very advanced species belonging to the kingdom of fungi. He has been immortalized in Chinese history and culture in paintings, statues, silk tapestries, illustrated woven fabrics, and on the clothing of emperors.

In traditional Chinese medical theory, Reishi or Lingzhi occupies the highest rank in the world of mushrooms thanks to its miraculous benefits, as it has a tremendous ability to improve overall health and increase the body's ability to recover, enhance, and prolong the life of cells without any side effects even when used for a long period of time. According to Dr. Shi Jin Li, the most famous doctor of the Ming Dynasty in China, long-term consumption of reishi will build a strong and healthy body and ensure a long life. Modern medical research has also proven that reishi contains many medicinal properties.

### 3.1 Reishi mushroom habitat

Reishi mushrooms grow naturally in East Asia, in areas with warm and humid climates.

### 3.2 Reishi mushroom life cycle

Its life cycle can be summarized in the following axes, the duration of which ranges from 70 to 90 days, depending on the available growth conditions and the type of fungal strain:

1. Spores (fungal seeds):

    - The life cycle of reishi mushrooms begins with spores found in the air.

    - These spores are very small and light, allowing them to be spread over long distances by wind.

    - When spores land on a suitable surface, such as the trunk of a dead tree, they begin to germinate.

2. Fungal network:

- The spores grow and form thin threads called the mycelial net.
- These hyphae branch and spread throughout the substrate, absorbing nutrients from the environment.
- The fungal network can live for many years.

3. Primitives:

- Some of the filaments in the fungal network begin to group together to form small mycelium called primordia.
- Protozoa look like small balls of cotton.
- The primordia grows rapidly and turns into a mature fungus.

4. Ripe mushrooms:

- A mature mushroom consists of a stem and a cap or mushroom hat.
- The stem is thick and strong, while the cap is large and flat.
- The surface of the cap is hard and shiny, and its color is brown, red, or black.
- Mature fungi produce new spores that begin a new life cycle.

5. Decomposition:

- After mature mushrooms produce new spores, they begin to decompose as the organic matter of the fungus breaks down and returns to the environment.

6. Factors affecting the life cycle of reishi mushrooms:

- Temperature: Reishi mushrooms grow best at temperatures between 15 and 25 degrees Celsius.
- Humidity: Reishi mushrooms grow best in a humid environment.
- Light: Reishi mushrooms do not need sunlight to grow.
- Substrate: Reishi mushrooms grow best on dead tree trunks.

The duration of each stage of the reishi mushroom life cycle can vary depending on the environmental conditions. Some factors, such as diseases and pests, can affect the life cycle of reishi mushrooms.

It can also be produced by humans under special conditions, as shown in **Figure 3**, where each stage of the life cycle needs a special temperature and lighting, as well as humidity.

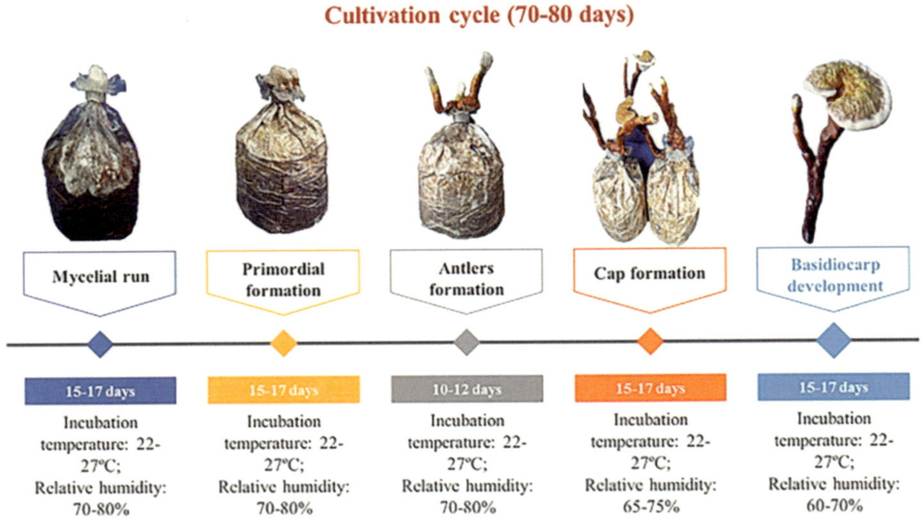


**Figure 3.**
*The life cycle of the Ganoderma lucidium mushroom produced in a private agricultural facility. From Ref. [9].*

### 3.3 Active ingredients in reishi mushrooms

Ganoderma species contain more than (200) active ingredients that have specific therapeutic properties. **Table 1** shows the main bioactive compounds of the reishi mushroom and their biological effects. The recent and extensive global research conducted on Ganoderma has revealed evidence of its ability to improve public health and prolong life (in health and wellness), God willing, in all pathological and normal conditions [11].

Its main function is to promote blood circulation, increase metabolic cell activities (cleansing), and also help in improving the functioning of all internal organs (balancing body functions). It also works to enhance natural healing processes, cleanses the body, neutralizes free radicals, rids it of toxins and metabolic waste, and improves the process of oxygen reaching the body.

Lingzhi can be taken daily for long periods without any harmful side effects. Long-term use of Lingzhi can help balance the body's system and strengthen the immune system to fight disease. It is not just a medicinal mushroom to treat a specific disease, but it is considered an adaptogen, meaning that it helps regulate and mobilize all bodily functions to function normally and properly and eliminates body disorders. The most important active ingredients in it are:

1. Polysaccharides:

Polysaccharides alert and control the work of the immune system by activating immune cells such as macrophages and helper cells and increasing the ability of aminoglobin (antibodies) to fight unwanted intruder cells such as bacteria, viruses, etc. They also help to get rid of toxic deposits present in the body and strengthen its ability to heal, transform abnormal cells into normal cells, strengthen the body's resistance in general, and thus improve overall health. Research conducted at the Ptotouma Drug Research Institute in Japan has proven that polysaccharides are tonic substances, and numerous medical reports have been issued in Japan confirming the effectiveness of reishi in treating patients suffering from kidney failure.

| Bioactive compounds | Biological effects |
|---|---|
| *Triterpenoids* | |
| Ganoderic acids, lucidumol, lucialdehyde, lucidenic acids, ganodermic, ganolucidic acids, ganoderals, ganoderiols | Anticancer |
| *Triterpenoids* | Antidiabetic |
| Ganoderic acids, ganodermin, ganoderic acid A, ganodermadiol, ganodermanondiol, lucidumol B, ganodermanontriol, ganoderic acid B, ganolucidic acid B | Antimicrobial |
| Triterpenoids, ganoderic acid, ganoderiol F, ganodermanontriol | Antiviral |
| *Polysaccharides:* | |
| 1 → 3, 1 → 4, and 1 → 6-linked _ and _-D (orL)-glucans, GLP-2B | Anticancer |
| Polysaccharides | Antidiabetic, Antioxidant, Antimicrobial |
| Polysaccharides (ganopoly) | Cardiovascular problems |
| *Proteins, Glycoproteins, and Peptidoglycans:* | |
| Glycopeptides and peptidoglycans | Anticancer |
| Protein Ling Zhi-8 (LZ-8), lectin, ribosome-inactivating proteins, antimicrobial proteins, glycopeptides/glycoproteins, peptidoglycans/proteoglycans, ganodermin A, ribonucleases, proteinases, metalloproteases, laccases | Immunomodulatory, anticancer, antitumor |
| Proteoglycans, proteins (LZ-8) | Antidiabetic |
| Polysaccharide–peptide complex | Antioxidant |
| *Phenolic compounds:* | |
| Phenolic components, phenolic extracts | Antioxidant |
| Saponins | Anticancer, antioxidant |
| Sterols, e.g., ergosterol | Provitamin D2 |
| Long-chain fatty acids | Antitumor |

**Table 1.**
*The main bioactive compounds of G. lucidium and their biological effects. From Ref. [10].*

2. Triterpenoids:

Triterpenoids are responsible for improving blood pressure and fat levels. Triterpenoids have a modifying effect on the body, especially on the immune and circulatory systems. It is what gives reishi the property of an adaptive agent that provides a person with protection from a wide range of biological, environmental, and social influences.

3. Indocin:

Indocin regulates and improves blood circulation and reduces the rate of platelet deposition. A previous study conducted by a group of researchers from the Beijing College of Traditional Chinese Herbs confirmed the ability of reishi to improve the performance of red blood cells in transporting oxygen. They attributed this result to indocin found in reishi. Indocin also helps reduce cholesterol levels and regulates food metabolism, thus enhancing strength and activity.

4. Organic germanium:

The amount of organic germanium found in reishi is considered very high when compared to the percentages found in any of the other herbs in the world, and it performs the following functions in our bodies:

- Increases the amount of oxygen entering the blood
- Activates cell tissue by increasing the amount of oxygen in the body's organs
- Increases the rates of vital reactions and food metabolism in the body
- It gets rid of the electrical energy around the abnormal cells and turns the abnormal cells into normal cells
- Helps get rid of the feeling of fatigue

5. Ganoderic extract:

Ganoderic extract is one of the ingredients that is rarely found in herbs, and it is the substance that gives reishi its shine property. It can also increase the beauty of the skin and help get rid of diseases that affect the skin. Due to the presence of ganoderic extract in reishi, it can be used externally to treat skin problems, wounds, and small boils. Studies conducted by many researchers have shown that using ganoderic extract regularly can delay aging.

### 3.4 The functions performed

Ridding the body of toxins deposited on cells and tissues and enhancing immunity by activating immune cells in the body, stimulating blood circulation, increasing the ability of blood cells to carry oxygen, activating the processes of food digestion and metabolism, renewing cells in general, giving skin cells freshness and vitality, and finally performing operations. Balance and adaptation to organ functions. This is the property due to Ganoderma's ability to help the body get rid of many health problems. Our latest contribution in this field is conducting research on the effect of reishi mushrooms on cancer and infertility cases [12].

But where is the secret behind this wonderful performance of this miraculous mushroom? The secret lies in the mechanism of action of reishi mushrooms, which I will summarize for you, dear reader, in the points below:

An important rule: There are two main causes of diseases: internal toxins in the body and malfunctions in the functions or organs of the body.

Ganoderma does not treat disease but rather helps restore the natural balance of the body's system and enhances immunity against diseases. Any reaction that occurs is the result of the body's system and not due to Ganoderma doses, meaning that the Ganoderma doses that are taken are not related to diseases.

In order for any food that contributes to self-healing or any natural medicine to rank highest, it must meet the following conditions: it should be non-toxic (free from any side effects), its effect should not be limited to a specific part or organ of the body's systems, and it should have the ability to normalize all body functions.

### 3.5 The mechanism of action of Ganoderma in the body's self-healing

This is done according to a biological-nutritional-medical theory developed by Dr. Lim Siow Jin through several studies and summarized in his research on the Coronavirus [13, 14]. It is called the stages of treatment with Ganoderma (or Ganotherapy) and includes the following sequential steps:

1. The stage of searching and identifying the diseased part of the body (1–30 days)

Ganoderma helps regulate body functions. After taking Ganoderma, the process of searching all cells of the body begins until reaching the diseased part of the body. This stage takes approximately 30 days.

2. Detoxify and cleanse the body (1–30 weeks)

Uric salts, high cholesterol levels, fats, calcium salt deposits, useless cells, and chemicals are considered harmful toxins accumulated within the body.

These toxins exit the body in several ways, such as through the process of sweating, through the blood circulation in the liver and kidneys, through urination and the excretion process (stool), and are also exited in secretions such as mucus and phlegm.

3. The process of rebalancing the body (1–12 months)

Ganoderma can restore balance to the body by regulating its functions, and the appearance of some body reactions during the balancing process is evidence that the body is healing and there is no need to worry about it. You must continue taking reichi, and if it is noticed that the reaction is strong, you must reduce the dose you take or take more. Janosylium, after the reaction disappears, continues taking your main dose again, and the reaction may return again, but you must continue taking your dose.

4. Construction process (6–24 months)

In this process, damaged tissues that were destroyed in the body are rebuilt, the body's immune system is strengthened, awareness and a sense of calm and comfort are improved, and resistance to diseases is increased.

Providing the body with important elements such as mineral salts and triterpenoids that are important to restore and improve the functioning of body functions.

5. The stage of regeneration and re-working the body's systems vigorously

These are the properties due to Ganoderma's ability to help the body get rid of many health problems. This is why this mushroom is a permanent protection against contemporary problems if consumed on a daily basis.

With this final stage (point No. 5), we have reached the goal of ganotherapy, which is to restore the vitality and strength of youth.

Here is a summary of what I detailed above: the reishi mushroom does

a. Ridding the body of toxins deposited in the tissues.

b. Strengthening immunity by activating immune cells in the body.

c. Stimulating blood circulation and increasing the ability of blood cells to carry oxygen.

d. Activating food catabolism (metabolism).

e. Cell regeneration in general and making skin cells fresh and vibrant.

f. Carrying out balance and adaptation processes for organ functions.

### 3.6 The mechanism of action of Ganoderma, with the addition of timing

Studies in this field have indicated that there are time periods for each stage of the action of reishi mushrooms after being eaten by humans, which can be reviewed in the following points:

1. Inspection and audit (SCANNNG) - 1 to 30 days

2. Detoxification - 1 to 30 weeks

3. REGULATION - 1 to 12 months

4. BUILDING - 6 months 24 months

5. Renewal and vitality (REGENERATION) - 1 to 3 years

An overview of the types of toxins that Ganoderma deals with and how to remove them!

Did you know that there are two types of toxins deposited in the body?

1. Toxins dissolve in water and are excreted in two ways:

   a. through the skin, and this can be observed by excessive sweating.

   b. through urine, and this can be known by the change in the color and smell of urine.

2. Toxins that are insoluble in water are excreted in the following ways:

   a. The skin. The person may feel itching with the appearance of discolored spots and the emergence of blisters - which are fatty toxins.

   b. Stool. This can be seen by a change in the color and smell of the stool.

c. Phlegm, and the person notices that he begins to cough and produce excess phlegm.

d. vomiting.

Here I would like to point out important notes when eating reishi mushrooms or even organic spirulina algae: all of these above-mentioned reactions in the body are good news for recovery!! - God willing - It is important to know that people differ on this matter. In some cases, reactions from the body are delayed, in others, they appear from the beginning of use of the mushroom product; and in some cases, they may only appear slightly, and the person may not feel them, and their use must be gradual.

Depending on the body's reaction and the increase in these reactions, we can know the amount of toxins deposited in the body. It is necessary to communicate during the reactions and not stop. Unless the reactions increase dramatically, in this case, the person must reduce the amount of the dose, and some may recommend leaving the dose for 2 days, and then starting to use it again. At that time, it is necessary to consult a specialist to find out the correct way to deal with these reactions.

The dose should not be dropped completely just because these reactions appear. All of this is natural, because Ganoderma contains about (400) natural compounds, and these natural compounds must work strongly in the body. Especially in sick bodies, merely stopping due to the appearance of these reactions is incorrect. Because it prevents the body from excreting toxins and strengthens weak immunity.

Ganoderma mushrooms have been used for thousands of years in all kinds of patients and unpatients, and very few side effects have ever been recorded. Reishi is classified as Class 1: Herbs that can be safely consumed when used appropriately.

On occasion, some mild digestive upset and skin rashes may occur, but these side effects seem to go away and are typically only a threat to sensitive people.

## 4. Spirulina super food!

Its scientific name is *Spirulina platensis*, a bluish-green algae with a spiral or filamentous shape. It has great importance in terms of its taxonomic position because it is located among the autotrophic common in eukaryotic cells in addition to bacteria. The genus Spirulina belongs to the blue–green algae known as cyanobacteria, which are non-eukaryotic spiral-shaped algae. Free-living animals are negative for the gram pigment [15]. Due to the characteristics they possess that overlap with different types of organisms, scientists have differed in their classification, as their chlorophyll content, as is the case in higher plants, has led botanists to classify them as microalgae belonging to... The Cyanophyceae class, but the structure of the nucleus is not true, has supported microbiologists in classifying it as a bacterial genera [16]. It has been classified in the past few years [17, 18] as one of the types of bacteria.

Kingdom (Domain): Eubacteria, Bacteria.
Phylum: Cyanobacteria.
Order: Spirulinales.
Family: Spirulinaceae.
Genus: *Spirulina.*
Species: *Platensis.*
Spirulina classification scheme [17].

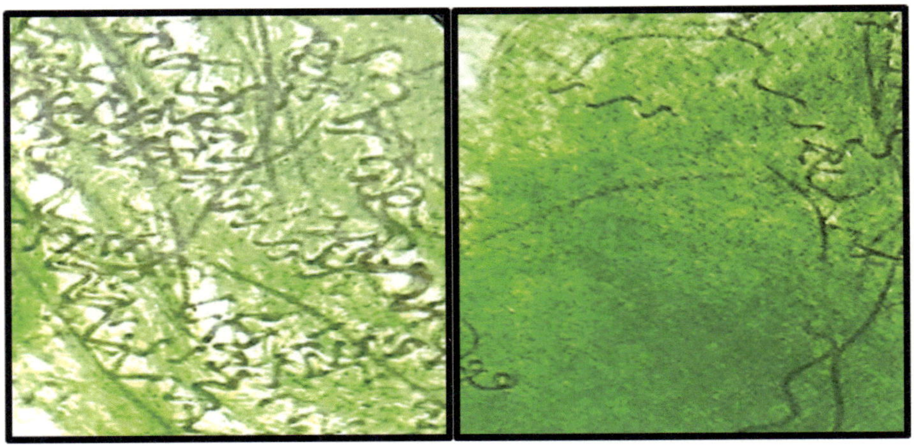

**Figure 4.**
*Spiral shapes of S. platensis algae under 40-100X optical microscope. From Ref. [19].*

Upon laboratory examination of samples containing them, they appear to be spiral-shaped or spiral-shaped cells under an optical microscope with magnification of 40X, **Figure 4**.

These green bacteria have received great attention from many organizations interested in health and healthy nutrition and have been described as (the savior of the world from hunger), (the fighter against malnutrition), and (the greatest food on earth)! It is rich in many important elements for the body: protein, beta-carotene, chlorophyll, antioxidants, mineral elements, vitamins, etc. It is one of the best and most useful alkaline foods for the body, and the American NASA uses it as food for astronauts! It also has the ability to fight viruses, enhance cellular immunity, and prevent cancer cells from growing.

God Almighty has deposited amazing secrets in spirulina! It contains more than 100 balanced nutrients, which makes it one of the best and most powerful foods for the body. As for why spirulina is called a superfood, it is because it is a food that is very rich in nutrients. By eating a small amount of spirulina, you get a benefit equivalent to hundreds of kilograms of vegetables and fruits.

It is recognized by the World Health Organization as a superfood and the best food for the future. It contains: amino acids (proteins), enzymes: more than 2000 balanced ones necessary for good health; vitamins such as vitamins A, D, and E; and the complete vitamin B complex group. Vitamin B12, anticancer vitamin E, important pigmented substances such as chlorophyll, beta-carotene, zeaxanthin, and cryptoxanthin, and a very high percentage of carotene. Mineral salts: They contain a large group of minerals such as potassium, calcium, chromium, copper, iron, and magnesium. Manganese, phosphorus, selenium, zinc, sodium, etc. They also contain many antioxidants.

### 4.1 Spirulina life cycle

Mature trichomes are divided into several pieces by the formation of specialized cells (**Figure 5**), called neuriocytes, which undergo degranulation, giving rise to biconcave separation discs. Segmentation of the trichome in the nochidia results in chains of short gliding cells (two to four cells) and hormones, which move away from the parental filament to give rise to the new trichome. Hormone cells lose the segments associated with necrotic cells and become rounded at the distal ends with little or no wall thickening. During this process, the cytoplasm appears less granular and the cells acquire a pale greenish-blue color like most cyanobacteria. Spirulina is an obligate photosynthetic organism and cannot grow in the dark in media containing organic sources of carbon [21].

### 4.2 The many benefits of spirulina

In addition to preserving the general health of the body, it has been used in serious diseases such as cancer, AIDS, bacterial or parasitic infections, anemia, poisoning, etc. All therapeutic indicators were positive and encouraging toward alleviating or stopping the disease.

Here we will mention some of its benefits very briefly: It is very beneficial in cases of malnutrition and weakness of the body. It is useful in the case of knee and joint pain for the elderly. Beneficial for breastfeeding women who suffer from dehydration and lack of milk. It is very useful for athletes and for all those interested in bodybuilding, because it is rich in protein and vitamin B complexes, such as vitamin B1, B2, B3, calcium, iron, and carotene, which is very useful for exercising. It strengthens the body's immunity. It is beneficial for many people suffering from osteoporosis. It is useful in cases of AIDS: As spirulina has the ability to prevent the virus from attaching to or penetrating cells, and thus works to prevent the spread of infection to the body's cells. It affects cancer cells in the body; it works to eliminate and stop important types of cancer. Beneficial for the growth of teenagers, children, and infants. It works to relieve the pain that precedes menstruation in women. It helps a lot in stopping the

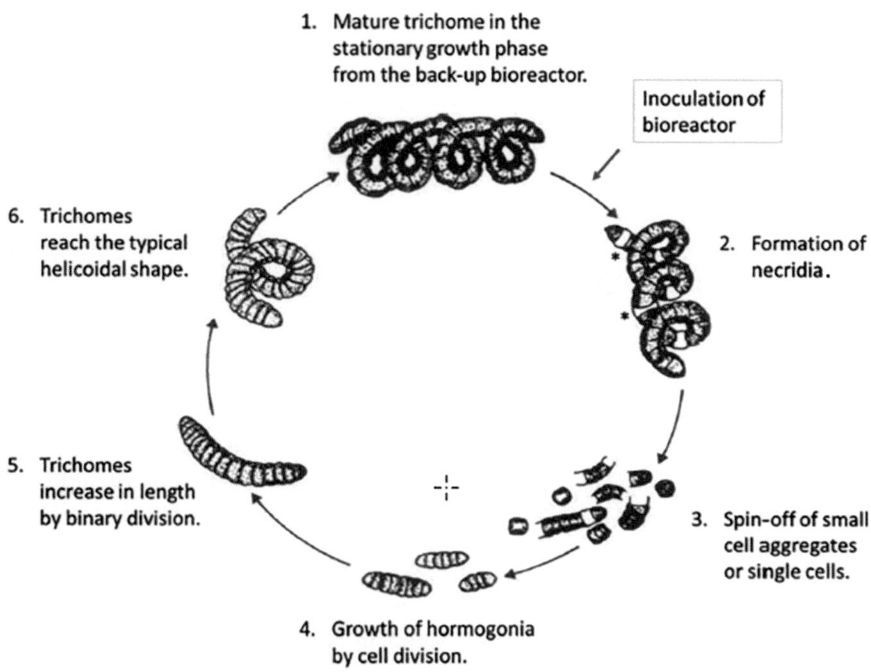


**Figure 5.**
*Spirulina life cycle. From Ref. [20].*

progression of the AIDS virus. Helps lower high blood pressure. It is beneficial for anemia patients and helps combat diabetes. Fights viral infections. It is beneficial for people who have stomach ulcers because it contains chlorophyll. Useful for eye allergies. Spirulina helps remove heavy metals such as lead and mercury from the body. It helps digestion because it is rich in enzymes. There are also many, many benefits of spirulina, and scientists are continuing to discover its many health secrets.

According to the University of Maryland Medical Center, spirulina is considered safe, even in high doses, with no side effects, and it must be from clean, uncontaminated organic sources.

### 4.3 The interest of international health organizations in spirulina

Many international organizations and bodies have paid attention to this algae, including the US Food and Drug Administration (USFDA), which says: (Spirulina is an excellent alkaline food) and also says: (It is an excellent and very powerful nutritional supplement). The Food and Agriculture Organization (FAO) said: It is the best food for tomorrow. Expo says: It is the greatest natural food in the world. The World Health Organization (WHO) says about spirulina: The greatest food on earth means the greatest food on the face of the earth. The American agency NASA gives spirulina as food to astronauts, and scientists say that every 1 kilo of spirulina is equivalent to 1000 kilos of mixed vegetables and fruits! The United Nations-affiliated IIMSAM Organization has become specialized in producing spirulina in several forms and supplying countries that suffer from malnutrition and infections with AIDS, Zika viruses, and other diseases. Some companies specialize in producing natural nutritional supplements, such as the Malaysian company DXN, and have contributed to this organization by donating and supplying organic spirulina to those countries.

### 4.4 Why do we need spirulina?

In this era and in this miserable state of health, in which water and food have become contaminated with all kinds of chemicals, and even the air that we inhale has become contaminated with smoke coming out of the exhausts of cars, factories, etc., this is no secret to anyone! From morning to night, we are surrounded by canned food, soft drinks, various sweets, harmful fats, etc., as we live in a near absence of healthy food that is beneficial to the human body. Food is available! But it does not contain the essential vitamins, proteins, and amino acids that the human body needs, and therefore it is an unhealthy food that does more harm than good.

Spirulina algae is widely used as food supplements for humans and as animal feed because it contains a high percentage of protein, reaching 60–70%, and very high levels of fat, in addition to amino acids and vitamins [22].

Blinkova and others [23] indicated that the compound sulfolipid, which was extracted from the spirulina algae, has a major and effective role in reducing infection and reducing the activity of the human immunodeficiency virus. It is also used to nourish plants, encourage their growth, and resist viral diseases that infect plants, such as TMV, CMV, BYDV, PVY, OYVMV, etc. [5, 24–26].

The latest studies indicate the importance of spirulina algae and consider it an important source as an antidote to genetic mutations [27]. It has also been used to treat many diseases, including allergies, diabetes, and obesity [28]. Spirulina algae helps lower blood cholesterol levels by stimulating the immune system [29].

In this world now, I call our food "food without health"!! Especially fast and processed foods that contain a lot of harmful chemicals such as flavorings, colorings, preservatives, etc.

Eating healthy foods, especially spirulina, which is very rich in vitamins, minerals, and essential amino acids, has become one of the most important things that help maintain your health and compensate for the deficiency in your food. According to reports received from the United Nations and the World Health Organization. The real challenge at this time is the challenge of fighting malnutrition more than famine!!, in addition to the lack of vitamins, minerals, and essential amino acids. **Table 2** shows the contents of 100 grams of spirulina algae, as indicated by [29, 30].

### 4.5 Nutritional value of spirulina and methods of taking it

Spirulina is rich in various nutrients. NASA has proposed growing it in space for use by astronauts. The standard dose for spirulina is (1–3) grams, as one tablespoon (7 grams) of dried spirulina powder contains:

Protein: 4 grams.

- Vitamin B1 (thiamine): 11% of the daily allowance per person.
- Vitamin B2 (riboflavin): 15% of the daily allowance per person.
- Vitamin B3 (niacin): 4% of the daily allowance per person.

| Percentage% | Fatty Acids | Quantity/mg | Metals | Quantity/mg | Vitamins |
|---|---|---|---|---|---|
| 27.85 | Identified | 468 | Calcium Ca | 350 | A |
| 27.04 | Octadecana | 87.4 | Phosphorus Ph | 1090 | K |
| 4.23 | Heptadecane | 461 | Iodin I | 0.5 | B1 |
| 10.01 | Triacontane | 142 | Magnesium Mg | 4.35 | B2 |
| 14.04 | Pentacosane | 1.45 | Zinc Zn | 14.9 | B3 |
| 2.12 | Phytol | 25.5 | Selenium Se | 6.96 | B6 |
| 9.61 | Hetacosane | 3.26 | Manganese Mn | 162 | B12 |
| 12.36 | Hexacosane | 400 | Chromium Cr | — | |
| 6.66 | Hetdetane | 1.660 | Potassium K | — | |
| 6.88 | Henocosane | 6.41 | Sodium Na | — | |
| 4.51 | Nonacosane | 1.55 | Iron Fe | — | |
| 5.24 | Nexadecane | — | | — | |
| 14.72 | Dictroponten | — | | — | |
| 24.41 | Hanecosane | — | | — | |
| 10.75 | Nabtacosane | — | | — | |
| 10.74 | Hetacosane | — | | — | |
| 10.28 | Hetadecane | — | | — | |

**Table 2.**
*The most important substances found in 100 grams of spirulina algae.*

- Copper: 21% of the daily quota per person.
- Iron: 11% of the daily quota per person.
- 20 calories.
- 1.7 grams of digestible carbohydrates.
- 1 gram of fat, including omega-6 and omega-3 fatty acids.
- 14 g of magnesium, 95 mg of potassium, plus small amounts of every other nutrient we need.

**4.6 Fun ways to eat spirulina for lunch**

1. Mix a tablespoon of spirulina with two cups of unsweetened natural fruit juice, and add a banana and a few cubes of ice to make a smoothie for a healthy breakfast.
2. Place in the blender two peeled avocados, lemon juice, a chopped tomato, salt, pepper, and garlic powder as desired, then add a spoon or two of spirulina powder and mix the mixture well to obtain guacamole, which is a type of dipping sauce.
3. Mix two teaspoons each of cumin, paprika, and chili powder, add two tablespoons of spirulina, and sprinkle it over a Mexican salad or oven-roasted potatoes.
4. Sprinkle a tablespoon of spirulina powder on fresh popcorn.

The crucial question that arises is: Should you incorporate spirulina into your diet as a nutrient or supplement? Is it recommended to take spirulina if you have a genetic predisposition to cancer linked to the gene? What if your genetic risk stems from the ALK gene? Is it beneficial to include spirulina in your diet if you have been diagnosed with primary urethral squamous cell carcinoma, or if your diagnosis is a primary solitary fibrous tumor?

Furthermore, how should your spirulina intake be adjusted if you are undergoing Avastin treatment or if your treatment plan switches from Avastin to radiation? It is necessary to realize that simplistic assertions such as “Spirulina is natural, so it is always beneficial” or “Spirulina enhances immunity” are not sufficient for informed food/supplement choices, despite their great and practically tangible benefits published in several international research and studies. We recommend taking it in gradual doses and monitoring the condition. And conduct periodic examinations.

In addition, it is necessary to re-evaluate the appropriateness of including spirulina in your diet if there are changes in your treatment regimen. In short, when making decisions about incorporating foods or nutritional supplements like spirulina into your diet to gain their benefits, you should take into account the overall biochemical effects of all ingredients, taking into account factors such as the type of cancer, the specific treatments you are undergoing, genetic predisposition, and lifestyle choices.

The two most important things to remember are that cancer treatments and nutrition are never the same for everyone. Nutrition, including food and supplements like spirulina, is a powerful tool you can control while fighting cancer. "What should I eat?" It is the most common question among cancer patients and those at risk of cancer. The correct answer is that it depends on factors such as the type of cancer, the genetics of the tumor, current treatments, allergies, lifestyle, and body mass index. Cancer changes over time, so the nutritional pattern and programs must be changed from time to time depending on the nature of the condition, and you must choose a reliable product. From a solid company supported by certificates of organic agriculture, nature, and good manufacturing.

### 4.7 Biological properties of spirulina components

Here I will highlight the biological effectiveness of the components of spirulina and the possibility of using it to balance immunity, self-heal from diseases, and live a healthy, upscale lifestyle.

Below is a summary of the biologically active properties of spirulina as approved by many studies, including [31]:

Bioactive properties of Spirulina:

Properties \ Activity

1. As food supplement (in cookies, ice cream, and soft cheese enriched with Spirulina) Increases:

   - The protein content
   - The minerals
   - The phycobiliproteins
   - The fatty acids (myristic acid, palmitic acid, palmitoleic acid, oleic acid, gamma-linolenic acid)
   - The diversity of amino acids
   - The antioxidant activity
   - The vitamins
   - The shelf-life

2. Anti-nephrotoxicity - Induced by cisplatin, cyclophosphamide, cadmium, and chromium

   - Anti-genotoxicity
   - Elevates the DNA repair procedure
   - Repairs process of cell nucleus enzymes
   - Protective effects against cyclophosphamide, cisplatin, and urethane
   - Decreases the percentage of DNA fragmentation

3. Antiviral activity

    - Inhibitory activity against viruses such as HIV-1, HSV-1, HSV-2, HCMV, and influenza type A
    - Minimizes HIV-I replication

4. Anti-inflammatory property

    - Lowers the extent of beta-glucuronidase
    - Preserves against rheumatoid arthritis
    - Raises the activity of antioxidant enzyme

5. Anti-obesity and Weight loss

    - Inhibits NADPH oxidase
    - Induces insulin resistance
    - Suppresses adipocyte oxidative stress

6. Hypoglycemic and hypolipidemic properties

    - Reduces the level of blood glucose
    - Manages cholesterol and triglycerides
    - Enhances insulin resistance

7. Neuroprotective property

    - Decreases the level of ROS, nitric oxide, and lipid peroxidation
    - Improves locomotor activity

8. Anticancer Effects

    - Enhances cell nucleus enzyme activity
    - Enhances DNA repair synthesis
    - Inhibits the growth of human colon carcinoma cells
    - Inhibits the hepatocellular carcinoma cells (HCC)
    - Inhibits the growth of breast cancer cells

- Anti-proliferative activity against the cancer cells
- Inhibits the growth of liver cancer cells, hepatoma cells, and lung cancer cells
- Activates apoptosis enzymes

9. Anti-anemic activity

- Increases the concentration of mean capsular hemoglobin (MCH)
- Increases the mean concentration of hemoglobin
- Increases mean corpuscular volume (MCV)
- Increases mean corpuscular hemoglobin (MCH)
- Increases mean corpuscular hemoglobin concentration (MCHC)

10. Probiotic property

- Enhances the growth of Lactobacillus
- Extension of vitamin B1

11. Spirulina for eyesight

- Increases the serum zeaxanthin level
- Inhibits the corneal neovascularization

12. Immunological Applications

- Improves anti-allergic effects
- Inhibits the release of histamine

## 5. Conclusion

In our journey through the pages of this chapter, we can say with confidence that reishi mushrooms and spirulina algae represent natural treasures rich in nutrients and various health benefits because they are nature's masterpieces in spices, nutrition, and human health.

Scientific studies have shown that reishi mushrooms and spirulina algae possess anti-inflammatory and antioxidant properties, making them effective in strengthening the immune system and protecting the body from various diseases. These two natural ingredients are also rich in essential vitamins and minerals, making them essential for maintaining the body's health and vital functions. In addition, reishi mushrooms and spirulina can be used to enhance the taste of food and increase its nutritional value, making them an ideal alternative to traditional spices.

In conclusion, we hope that this chapter has provided you with valuable information about reishi mushrooms and spirulina, and encouraged you to include them in your diet to improve your health and well-being. Feel free to experiment with the recipes presented in this chapter, and explore your creativity in using these two wonderful ingredients. We wish you good health and a happy life!

## Acknowledgements

All appreciation to Dr. Lim Siow Jin, the Goodwill Ambassador to the United Nations and the founder and Chairman of the Board of Directors of DXN Company, for the science he gave to humanity in the field of organic nutrition and self-healing, the discovery of the mechanism of action of the reishi mushroom according to the theory of ganotherapy, and for paving the way for us in the company's laboratories and factories and delivering the reishi mushroom. Spirulina is ours, and thanks go to Mr. Raed Abbas, his work team, and the trained professors in Iraq for opening centers for the sale and delivery of organic products derived from the mushrooms and algae above, with a free training service. I also thank the Ministry of Higher Education and Scientific Research, Tikrit University, the College of Agriculture, the Department of Plant Protection, and the Virus Research Laboratory.

## Thanks

I extend my sincere thanks to God Almighty, who helped me to complete this chapter, and then to my family, who bore the burden of my preoccupation. My appreciation and thanks to the team working with me in the Virus Research Laboratory, who are Dr. Basma Dhabab, Janan Al-Azzawi, Ashwaq Talib, Yasmine Al-Nasri, Raghda Jamal, Engineer Muhammad Ghazi, Ali Kadhim, Amjad Khalaf, Abdul Majeed Al-Ghathwan, and Zuhair Al-Qaisi, for their assistance in conducting the procedure. The research, some of the results of which we presented in this chapter, and to everyone who contributed to this achievement.

## Author details

Maadh Abdulwahab Alfahad
Tikrit University, Salahdin, Iraq

*Address all correspondence to: maadh.alfahd@tu.edu.iq

## References

[1] Seo K, Lee JE, Lim MY, Ko G. Effect of temperature, pH, and NaCl on the inactivation kinetics of murine norovirus. Journal of Food Protection. 2012;**75**(3):533-540. DOI: 10.4315/0362-028X.JFP-11-199

[2] David SC, Vadas O, Glas I, Schaub A, Luo B, D'angelo G, et al. Inactivation mechanisms of influenza a virus under pH conditions encountered in aerosol particles as revealed by whole-virus HDX-MS. mSphere. 2023;**8**(5):e0022623. DOI: 10.1128/msphere.00226-23. Epub 2023 Aug 18

[3] Zhang J, Zhou K, Wang Q. Tailoring the self-assembly behaviors of recombinant tobacco mosaic virus by rationally introducing covalent bonding at the protein–protein Interface. Small. 2016;**12**(36):4955-4959. DOI: 10.1002/smll.201503487

[4] Bishop NE, Anderson DA. Early interactions of hepatitis a virus with cultured cells: Viral elution and the effect of pH and calcium ions. Archives of Virology. 1997;**142**:2161-2178. DOI: 10.1007/s007050050233

[5] Alfahad MA, Abd Al-Wahab M, Mohammad AT. Biological control to tobamovirus tobacco mosaic virus (isolated from some plants and desert soil gypsiferous desert soil) using DXN-ganoderma and DXN-spirulina food supplements in Saladin gov. Iraqi Journal of Desert Studies. 2018;**8**(2):167-175

[6] Alsamarray SMA, Al-Fahed MA, Anis AHA. Molecular diagnosis of maize dwarf mosaic virus (MDMV) and bio-control with food supplements and Bacillus subtilis on several maize genotypes. Arab Journal of Plant Protection. 2019;**37**(1):49-58

[7] Al-Fahd MA, Abdelmagid MG, Abboud ON. Molecular diagnosis of a cucumber mosaic virus and its biological control using the algae Spirulina platensis, bacterium pseudomonas fluorescens and some herbal extracts under field conditions. Arab Journal of Plant Protection. 2020;**38**(2):137-148

[8] Bhosle S, Ranadive K, Bapat G, Garad S, Deshpande G, Vaidya J. Taxonomy and diversity of Ganoderma from the Western parts of Maharashtra (India). Mycosphere. 2010;**1**(3):249-262

[9] Seethapathy P, Sankaralingam S, Muniraj IK, Perumal M, Pandurangan N. Mass multiplication, economic analysis, and marketing of Ganoderma sp. (Reishi Mushroom). In: Food Microbiology Based Entrepreneurship, Making Money From Microbes. Singapore: Springer; 2023. pp. 89-113. DOI: 10.1007/978-981-19-5041-4_6

[10] El Sheikha AF. Nutritional profile and health benefits of Ganoderma lucidum "Lingzhi, Reishi, or Mannentake" as functional foods: Current scenario and future perspectives. Food. 2022;**11**:1030. DOI: 10.3390/foods11071030

[11] Frljak J, Mulabećirović AH, Isaković S, Karahmet E, Toroman A. Biological active components of selected medical fungi. Open Journal of Preventive Medicine. 2021;**11**(01):9-22. DOI: 10.4236/ojpm.2021.111002

[12] Abdullah BA, Alfahad MA, Hdree DH. Effects of red reishi mushroom (Ganoderma lucidum) on the reproductive system in female and male rats. Iraqi Journal of Veterinary Sciences. 2019;**33**(1):137-141. DOI: 10.33899/IJVS.2019.125547.1067

[13] Karuppiah S, Jin LS. Ganotherapy and holistic human system is the pathway of holistic health for immediate relief for COVID19. Open Journal of Preventive Medicine. 2020;**10**:45-61. DOI: 10.4236/ojpm.2020.103003

[14] Sonar S, Gondkar S, Saudagar RB. Ganotherapy. Journal of Drug Delivery and Therapeutics. 2019;**9**(3-s):828-831. DOI: 10.22270/jddt.v9i3-s.2791

[15] Gabr GA, El-Sayed SM, Hikal MS. Antioxidant activities of phycocyanin: A bioactive compound from Spirulina platensis. Journal of Pharmaceutical Research International. 2020;**32**(2):73-85

[16] Sudhakar K, Premalatha M. Characterization of micro algal biomass through FTIR/TGA/CHN analysis: Application to Scenedesmus sp. Energy Sources, Part A: Recovery, Utilization, and Environmental Effects. 2015;**37**(21):2330-2337. DOI: 10.1080/15567036.2013.825661

[17] Kewi C, David L, Keith K, William D, Brett AN. Molecular classification of commercial spirulina strains and identification of their sulfolipid biosynthesis genes. Journal of Microbiology and Biotechnology. 2011;**21**(4):359-365. DOI: 10.4014/jmb.1008.08016

[18] Ali SK, Saleh AM. Spirulina—An overview. International Journal of Pharmacy and Pharmaceutical Sciences. 2012;**4**(3):9-15

[19] Al-Dulami AKE, Maadh A-F, Wahab A. Bio-mass production of spirulina platensis and its use in inducing hostility resistance against Potyvirus potato virus y. IOP Conference Series: Earth and Environmental Science. 2023;**1158**:072018. DOI: 10.1088/1755-1315/1158/7/072018

[20] Jung CHG, Braune S, Waldeck P, Küpper J-H, Petrick I, Jung F. Morphology and growth of arthrospira platensis during cultivation in a flat-type bioreactor. Life. 2021;**11**(6):536. DOI: 10.3390/life11060536

[21] Ciferri O. Spirulina, the edible microorganism. Microbiological Reviews. 1983;**47**(4):551-578

[22] Juarez-Oropeza MA, Mascher D, Torres-Duran PV, Farias JM, Paredes-Carbajal MC. Effects of spirulina on vascular reactivity. Journal of Medicinal Food. 2009;**12**:15-20

[23] Blinkova LP, Gorobets OB, Baturo AP. Biological activity of spirulina. Zhurnal Mikrobiologii, Epidemiologii, i Immunobiologii. 2001;**2**:114-118

[24] Al-ajeele BD, Fahad A, Al-Wahab MA. Efficiency of algae Spirulina platensis the and the aqueous extract of Moringa oleifera seeds and leaves to induce ten cultivars of cucumber plants against the Cucumovirus cucumber mosaic virus (CMV). Euphrates Journal of Agricultural Science. 2023;**15**(4):217-229

[25] Jadoua JA, Al-Fahd MAW. Biological resistance to okra yellow vein mosaic virus using three biological agents on three okra cultivars. Arab Journal of Plant Protection. 2022;**40**(2):148-157. DOI: 10.22268/AJPP-40.2.148157

[26] Alfahad MA. The activity of new bio-agent to control cucumovirus cucumber mosaic virus (CMV). In: Cucumber Economic Values and Its Cultivation and Breeding. IntechOpen; 2021. pp. 145-177. DOI: 10.5772/intechopen.96587

[27] Aly FM, Kotb AM, Hammad S. Effects of spirulina platensis on DNA damage and chromosomal aberration against cadmium chloride-induced

genotoxicity in rats. Environmental Science and Pollution Research International. 2018;**25**(11):10829-10836. DOI: 10.1007/s11356-018-1329-3 Epub 2018 Feb 3

[28] Khan Z, Bhadouria P, Bisen PS. Nutritional and therapeutic potential of spirulina. Current Pharmaceutical Biotechnology. 2005;**6**(5):373-379

[29] Amamoor VJ, Pieme CA, Biapa PCN, Matip MEN, Moukette BM, Nzufo FT, et al. Chemical composition spirulina platensis of Nomayos-Yaounde (Cameroon). Annals Food Science and Technology. 2016;**17**(2):524-528

[30] Joventino IPHGR, Alves LCNF, Pinheiro-Joventino L, Kalyne MAML, Neves SA, Ferreira FV, et al. The microalga Spirulina platensis presents anti-inflammatory action as well as hypoglycemic and hypolipidemic properties in diabetic rats. Journal of Complementary and Integrative Medicine. 2012;**9**(1). DOI: 10.1515/1553-3840.1534

[31] Anvar AA, Nowruzi B. Bioactive properties of spirulina: A review. Microbial Bioactives. 2021;**4**(1):134-142

Chapter 4

# The Possible Use of Herbs and Spices in the Food Processing and Food Safety

*Melaku Tafese Awulachew*


## Abstract

Spices and herbs are utilised in many different cuisines and food items, but they are consumed in small amounts, making them a special market within the food industry. These plants are rich sources of valuable bioactive components that are utilised for a range of functions, such as flavouring, colouring, medicinal, and preservation. The food industry recognises the potential of plant extracts, and not just because of their nutritional benefits. Antimicrobials and antioxidants, such as various essential oils with antibacterial qualities, are mostly found in plants. Herbs and spices including rosemary, sage, basil, oregano, thyme, cardamom, and clove contain a variety of essential oils. To increase product value and shelf life, plant essential oils are also included as additives to edible or biodegradable films and coatings for active food packaging. However, it is important to consider and thoroughly research any potential harmful consequences of specific phyto-compounds. However, this chapter's findings are more comprehensive, represent significant and timely advances towards potential applications in food sector as well as the food safety aspects of commodities containing spices and herbs by effectively protecting consumers from potential risks resulting from unintentional or intentional contamination of spices and herbs.


**Keywords:** herbs and spices, safety, potential use, food, nutrition, preservative effect, health effect

## 1. Introduction

One key factor in combating bacteria' resistance to antibiotics is the antimicrobial activity of plant extracts. Because bacteria can genetically transfer and acquire drug resistance, the last three decades have seen a rapid expansion in the pharmaceutical industry, which has led to an increase in drug resistance among microorganisms. As a result, new illnesses have emerged that have led to a high death rate, particularly in individuals with impaired immune systems. Thus, limiting antibiotic use and researching the pathways of genetic resistance in bacteria to produce new medications, particularly those of natural origin, are critical to addressing this issue [1]. To learn more about the characteristics of plant derivatives that prevent bacterial growth, it has also been demonstrated that spice and herb extracts have health-promoting qualities [2].

Foodborne disease outbreaks caused by pathogenic microbes have been associated with fresh foods that have short shelf life, such as those from livestock, shellfish, and horticulture. Much work has been put into finding natural antimicrobials that can stop bacteria, viruses, and fungi from growing in food in recent years. The food industry has simultaneously tried to replace or enhance more recent, less intrusive methods like pulsed light, high-pressure, pulsed electric field, ultrasound, oscillating magnetic field, and UV treatments with more conventional methods like heat treatment, acidification, salting, drying, and chemical preservation [3, 4]. This is because some consumers choose meals that have been lightly processed over those that have been processed, preferring those that include less additives and/or natural components [3].

The use of food additives produced from plants has grown significantly in popularity in recent years. Synthetic antioxidants, such as butyl hydroxyl anisole or butyl hydroxytoluene, are frequently added to food products in an effort to block the processes that cause food quality to deteriorate (such as the breakdown of proteins, lipids, and carbohydrates). These antioxidants do, however, quickly degrade at high temperatures and are volatile. Moreover, it is still unknown if long-term use of these antioxidants poses any health hazards. Phenolic chemicals, which have been shown to have good antioxidant activity and can be employed as natural food preservatives, are abundant in many herbs and spices [2, 5].

Natural antimicrobials are defined as substances that develop naturally from biological systems without being altered or modified in a lab. Numerous living organisms, including fungi, bacteria, plants, and algae, can provide these. The fact that plant extracts have been consumed by humans for thousands of years is a plus. In addition to being used as antimicrobials, many plants are also used in functional foods, nutritional supplements, traditional medicine, and the creation of recombinant proteins. Their ability to regulate natural deterioration processes (food preservation) and to prevent the growth of bacteria, particularly hazardous bacteria (food safety), gives them their position as antimicrobials [6].

As previously indicated, in response to consumer demand for safer and more natural foods as well as their rejection of synthetic antioxidants, which are being reevaluated for the potential toxicity and carcinogenicity of the components formed during their degradation, food companies are becoming more and more interested in incorporating natural antioxidants and antimicrobials into food. The goal of this research is to provide an overview of existing understanding and recent advancements in the use of plant-derived compounds obtained from herbs and spices as food preservatives and additives, as well as their potential, limitations, and food safety aspect.

## 2. Nutritive value and possible side of herbs and spices

Effects of herbs and spices are essential components in many recipes. They enhance flavour, aroma, colour, texture, and even nutritional value. The common assumption about herbal medicines, on the other hand, is that "because they are natural and have fewer side effects than prescription drugs, they are safe." Combinations of herbs are widely utilised because they are more effective and may have fewer adverse effects. However, additional research has found that they may have detrimental effects if taken seldom, excessively, or in conjunction with certain drugs [7]. Interactions between medicines and herbs can have unintended consequences. **Table 1** shows the applications and negative effects of several herbs and spices that can be used safely.

| Herbs/ Spices | Botanical name/ Scientific name | Benefits | Side effect |
|---|---|---|---|
| Parsley | *Petroselinum crispum* | Parsley can be used to treat urinary tract infections, kidney stones (nephrolithiasis), gastrointestinal disorders, constipation, jaundice, intestinal gas (flatulence), indigestion, colic, diabetes, cough, asthma, fluid retention (oedema), osteoarthritis, "tired blood" (anaemia), high blood pressure, prostate problems and spleen problems. It is high in vitamin C, iron, and antioxidants, and it protects against rheumatoid arthritis. | Parsley can trigger allergic skin responses in certain people. It contains a lot of oxalic acid, 1.70 mg per 100 g. Gouty arthritis, kidney stones, and mineral nutrient deficits may develop from consuming oxalate-rich foods over an extended period of time. |
| Aloe vera | *Aloe barbadensis* miller | Anti-inflammatory, antiproliferative, anti-ageing properties, wound healing, burn damage recovery, cell growth, and immunological modulation | Abdominal spasms, pain, allergic responses, cramping, and kidney damage are all symptoms of hepatotoxicity. |
| Dill | *Anethum graveolens* | Used to treat digestive ailments such as indigestion, flatulence, and liver difficulties. It is also used to treat urinary tract issues like kidney disease and painful or difficult urination. Treatment for fever and colds, cough, haemorrhoids, infections, nerve soreness, genital ulcers, and period cramps. | Some people may have skin sensitivity when exposed to dill. Fresh dill juice can also make your skin more sun sensitive. |
| Rosemary | *Rosmarinus officinalis* | Accepted as a highly effective antioxidant, anti-inflammatory, antiviral, and antibacterial agents. Carnosic acid in rosemary has also been demonstrated in studies to protect against damaging carcinogens and Alzheimer's disease. | Undiluted oil is dangerous to consume; high amounts can result in vomiting, uterine bleeding, renal irritation, increased sun sensitivity, skin redness, and allergic responses. |
| Peppermint | *Mentha x piperita L Lamiaceae* | Colds, coughs, sinus infections, and respiratory infections are all treated with this medication. Additionally, it is used to treat digestive issues such as gas, upset stomach, diarrhoea, small intestine bacterial overgrowth, heartburn, nausea, vomiting, morning sickness, irritable bowel syndrome, and cramping in the upper gastrointestinal tract and bile ducts. Peppermint oil is used topically to treat headaches, muscle discomfort, nerve pain, toothaches, oral inflammation, joint issues, itching, allergic rash, bacterial and viral infections, and to relax the colon during barium enemas. | Peppermint can produce heartburn as well as allergic symptoms such as flushing, headache, and mouth sores. Large doses of peppermint oil may be harmful to the kidneys. |

| Herbs/ Spices | Botanical name/ Scientific name | Benefits | Side effect |
|---|---|---|---|
| Dandelion | *Taraxacumofficinale* | Fresh dandelion herb is one of the richest sources of vitamin A among culinary herbs, with 10,161 IU per 100 g. Vitamin A is an antioxidant that is necessary for maintaining healthy mucus membranes and skin. A good source of minerals and vitamins, such as folic acid, riboflavin, pyridoxine, niacin, and vitamins E and C, all of which are necessary for optimal health. It is high in vitamin K, which helps to increase bone mass and has been shown to play a role in Alzheimer's disease by reducing neuronal damage in the brain. | It may increase potassium toxicity in people receiving potassium sparing diuretic treatment. Dandelion plant can potentially cause allergic contact dermatitis in some people. |
| Celery | *Apium graveolens* | Celery is an adaptable vegetable. Flavonoids, which include antioxidant, cancer-preventive, and immune-boosting qualities, are abundant in the leaves of this plant. It also contains high levels of many essential vitamins, such as riboflavin, niacin, folic acid, vitamins A and C, which are needed for healthy metabolism, and vitamin K, which helps build strong bones and has been connected to a lower risk of Alzheimer's disease by protecting brain neurons. It is used to alleviate anxiety, osteoarthritis, and gouty arthritis. It is also a rich source of important volatile oils and minerals. | It has a lot of soluble and insoluble fibre. Eating meals high in fibre may induce stomach pain, indigestion, and bloating, and it frequently worsens pre-existing constipation. |
| Cinnamon | *Cinnamomum zeylanicum* | Cinnamon possesses anti-inflammatory, anti-oxidant, anti-diabetic, warming, relaxing, and carminative qualities. It contains antibacterial essential oils such as eugenol, which is effective in tooth and gum care. The active ingredients in this spice may boost intestinal motility and digestive capacity by enhancing gastrointestinal enzyme secretions. Minerals, vitamin A, and flavonoid phenolic antioxidants such as carotenes are abundant. | Excessive use of the cinnamon stick may result in taste bud inflammation, gum swelling, and mouth ulcers. Excessive consumption might result in trouble breathing, dilation of blood vessels, tiredness, and sadness (Healthy herbal nutrition). |

| Herbs/ Spices | Botanical name/ Scientific name | Benefits | Side effect |
|---|---|---|---|
| Ginse | *Zingiber officinale* | Used to address issues with the heart and blood vessels. In addition, it is used to treat rheumatoid arthritis, diabetes, Alzheimer's disease, attention deficit hyperactivity disorder, renal disease, colds, flu, chronic bronchitis, and tuberculosis. It is also used to relieve the side effects of cancer treatment. | Drowsiness, changes in heart rate, and muscle spasms can occur in certain patients. It may also result in skin rashes, asthma attacks, high blood pressure, diarrhoea, exhilaration, and anxiousness. |
| Thyme | *Thymus vulgaris* | Thyme is used to treat whooping cough, sore throat, colic, arthritis, upset stomach, stomach pain (gastritis), diarrhoea, intestinal gas (flatulence), parasitic worm infections, and skin conditions. It is also used to enhance urine flow (as a diuretic), disinfect urine, and stimulate hunger. Thyme essential oil contains 20–54% thymol, which has antibacterial and antifungal qualities and is utilised in a range of products, including the treatment of respiratory infections. | It has the potential to disturb the digestive system. Applying the oil to the skin can cause irritation in some persons. There is insufficient evidence to determine whether thyme oil is safe to ingest orally in medical doses. |
| Ginger | *Allium sativum L* | Useful for motion sickness, morning sickness, colic, upset stomach, gas, diarrhoea, cancer-related nausea, arthritis or muscular discomfort, menstruation pain, upper respiratory tract infections, cough and bronchitis, chest pain and stomach pain. | Heartburn, diarrhoea, and general stomach discomfort are some of the side effects. |
| Nutmeg | *Myristica fragrans* | Flavouring agent used in baked products, sweets, and some beverages. Used to treat rheumatoid arthritis and as a digestive aid. Diarrhoea, nausea, stomach spasms and pain, and intestinal gas are all treated with this medication. They are also used to treat cancer, kidney illness, and sleep disorders (insomnia). | Nausea, vomiting, difficulty urinating, dizziness, constipation, and dry mouth are some of the side effects. Seizures, hallucinations, and death are all possible side effects. |
| Cardamom | *Elettaria cardamomum* | Heartburn, intestinal spasms, irritable bowel syndrome, intestinal gas, constipation, liver and gallbladder complaints, and loss of appetite are all symptoms of this medication. It is also used to treat the common cold, cough, bronchitis, sore mouth and throat, and infection susceptibility. | Contact dermatitis caused by allergies. Cardamom seed can cause gallstone colic. |

| Herbs/ Spices | Botanical name/ Scientific name | Benefits | Side effect |
|---|---|---|---|
| Turmeric | *Curcuma longa* | Antibacterial, anticancer, antifungal, antioxidant, hypoglycemic, colourant, antiseptic, and wound healer are some of the properties of this plant. | Increase the risk of bleeding or the effects of warfarin medication. |
| Fennel | *Foeniculum Vulgare* | Aromatic, carminative, diuretic, and flavouring agent | Allergic responses, occupational rhinitis, asthma, conjunctivitis, and estrogenic activity are all possible side effects. |
| Cloves | *Eugenia aromaticum* | Used to treat diarrhoea, gas, bloating, intestinal spasms, nausea, and as an antioxidant and pain reliever in toothaches. | Pneumonia, bronchitis, hemoptysis, central nervous system depression, and occupational allergic contact dermatitis |

**Table 1.**
*Benefits and possible side effects of some important herbs and spices [8].*

## 3. Application herbs and spices in food and natural medicine

Herbs are derived from fresh or dried plant leaves or flowering parts, whereas spices are derived from other plant components such as roots, stems, bark, seeds, and bulbs [9, 10]. They are usually also dried before use. Herbs and spices may come from the same plant but from separate portions in some circumstances. Herbs and spices have long been used for both food and medicine (**Table 2**). They are used in cooking as preservatives (antioxidants or antimicrobials), flavour enhancers, colourants, and salt and sugar substitutes. Attempts to quantify dietary intake of spices and herbs are challenging since they are used in a variety of ways and are ingested in trace amounts alongside other foods.

### 3.1 Food application

#### *3.1.1 Deodorising/masking effect of spices and herbs*

Garlic and onions are commonly used in the preparation of meats [12]. These can mask or deodorise undesirable meat odours. In general, spices have three types of deodorising properties. Chemical deodorization is the process of converting odorous materials into odourless compounds through a chemical reaction with nonvolatile or other odourless molecules. A masking effect occurs when a strong flavour conceals a disagreeable aroma, or when two distinct odours combine to generate an odourless combination [12]. There is deodorization of locations where odorous molecules have been absorbed by porous materials. It is unusual to accomplish the latter with spices. The timing and method of adding a spice to food influence whether or not it will deodorise. Garlic has little value when added near the end of food preparation, but it has the desired effect when used right at the start.

#### *3.1.2 Preservative effect*

To preserve food, there is a rising interest in developing novel, safe, and effective natural antibacterial compounds, such as extracts from herbs and spices.

| Spices/ Herbs | Scientific name | Parts used | Food applications | Therapeutically properties |
|---|---|---|---|---|
| Saffron | *Crocus sativus* | Stigma | Colourant | Metabolic disorders, macular degeneration |
| Mustard Seed | *Brassica nigra, Brassica juncea, Brassica hirta* | Seed | Flavouring, Antimicrobial | Cancer, metabolic disorders |
| Turmeric | *Curcuma longa* | Rhizome | Colourant, Antimicrobial L | Cancer, cardiovascular disease, neurodegenerative disorders, skin inflammation, osteoarthritis metabolic |
| Paprika | *Capsicum annum* | Fruit | Flavouring, colourant | Cardiovascular disease, headache |
| Vanilla | *Vanilla tahitensis* | | Flavouring | — |
| Pepper, red | *Capsicum frutescens* | Fruit | Flavouring | Metabolic disorders, osteoporosis |
| Star anise | *Illicium verum* | Fruit | Antioxidant | Metabolic disorders, neurodegenerative disordersb |
| Onion | *Allium cepa* | Bulb | Flavouring | Cardiovascular disease, metabolic disorders |
| Horseradish | *Armoracia lapathifolia* | Root | Flavouring | Metabolic disorders |
| Anise | *Pimpinella anisum* | Fruit | Flavouring | Metabolic disordersa |
| Ginger | *Zingiber officinale* | Rhizome | Flavouring, Antioxidant | Cardiovascular disease, metabolic disorders, anxiety, psoriasis |
| Cinnamon | *Cinnamomum zeylanicum* | Bark | Flavouring, antioxidant, antimicrobial | Cardiovascular diseasea, metabolic disorders, neurodegenerative disorders, dental caries |
| Garlic | *Allium sativum* | Bulb | Flavouring | Cancer, cardiovascular disease, metabolic disorders |
| Coriander | *Coriandrum sativum* | Fruit | Flavouring, Antioxidant | Cardiovascular disease, metabolic Disorders |
| Pepper, black/white | *Piper nigrum* | Fruit | Flavouring, antioxidant | Cardiovascular disease, metabolic disorders |
| Celery | *Apium graveolens* | Fruit | Flavouring | Cardiovascular disease, Neurodegenerative disordersb |
| Cloves | *Eugenia aromaticum* | Flower Bud | Flavouring, antioxidant, Antimicrobial | Cardiovascular disease, metabolic disorders, cancer, dental caries |
| Cardamom | *Elettaria cardamomum* | Fruit | Flavouring, antioxidant | Cancer, cardiovascular disease |
| Caraway | *Carum carvi* | Fruit | Antioxidant | Metabolic disorders, back pain |
| Cumin | *Cuminum cyminum* | Fruit | Flavouring, antioxidant, antimicrobial | Metabolic disorders, fungal infection |

| Spices/ Herbs | Scientific name | Parts used | Food applications | Therapeutically properties |
|---|---|---|---|---|
| Fenugreek | *Trigonella Foenumgraecum* | Fruit | Flavouring | Metabolic disorders, hypogonadism |
| Fennel | *Foeniculum vulgare* | Fruit | Flavouring, antioxidant | Chronic constipation |
| Dill | *Anethum graveolens* | Fruit | Flavouring | Cancer, kidney injury |
| Sage | *Salvia officinalis* | Leaves | Flavouring, antioxidant, antimicrobial | Metabolic disorders, cardiovascular diseases |
| Basil | *Ocimum basilicum L.* | Leaves | Flavouring, antimicrobial | Cancer, metabolic disorders |
| Tea | *Camellia sinensis L* | Leaves | Antioxidant, Beverage | Metabolic disorders, cancer, cardiovascular diseases, neurodegenerative diseases, urinary disorders |
| Thyme | *Thymus vulgaris* | Leaves | Antioxidant, antimicrobial | |
| Senna plant | *Cassia angustifolia* | Leaves, Flowers | Beverage | Metabolic disorders, respiratory disorders |
| Bay | *Laurus nobilis* | Leaves | Flavouring, colourant | Metabolic disorders |
| Borage | *Borago officinalis L.* | Leaves | Antioxidant | Cardiovascular diseases, metabolic disorders |
| Calendula | *Calendula officinalis L* | Flowers | Flavouring, Colourant | Cardiovascular diseases, neurodegenerative disordersb |
| Chamomile | *Matricaria chamomilla L.* | Flowers | Colourant | Cardiovascular diseases |
| Lemon grass | *Cymbopogon citratus* | Leaves | Flavouring, beverage | Cancer |
| Lemon verbena | *Aloysia citrodora* | Leaves | Beverage | Neuromuscular diseases |
| Lemon Balm | *Melissa officinalis L.* | Leaves | Flavouring, Antioxidant | Neurodegenerative disorders, metabolic disorders, cancer, cardiovascular diseasesb |
| Lovage | *Levisticum officinale* | Leaves | Beverage | Urologic diseases |
| Feverfew | *Tanacetum parthenium L.* | Leaves | Beverage | Migraine |
| Linden | *Tilia americana L.* | Leaves | Beverage | Neurodegenerative disorders |
| Marjoram | *Origanum majorana* | Leaves | Flavouring, antioxidant | Respiratory disorders |
| Oregano | *Origanum vulgare* | Leaves | Flavouring, antioxidant, Antimicrobial | Metabolic disorders, cardiovascular diseases, sleep disorders, respiratory disorders |
| Parsley | *Petroselinum crispum* | Leaves | Flavouring | Metabolic disorders, urinary disorders |

| Spices/ Herbs | Scientific name | Parts used | Food applications | Therapeutically properties |
|---|---|---|---|---|
| Rosemary | *Rosmarinus officinalis* | Leaves | Flavouring, antioxidant, Antimicrobial | Cancer, metabolic disorders, cardiovascular diseases, respiratory disorders, urinary disorders |

**Table 2.**
*Food applications and therapeutically properties of the most consumed herbs [11].*

Furthermore, due to their antioxidative properties, herbs and spices stabilise lipids in diet [13]. Fat is a necessary ingredient in all cuisine. One of the reasons food deteriorates is the oxidation of fat in the food. Rancidity is caused when food lipids combine with airborne oxygen to form peroxides, which oxidise and break down into molecules of low molecular alcohol and aldehyde [12]. Synthetic antioxidants are routinely used to mitigate this. The majority of antioxidants found in spices work by interacting with free radicals produced during the autoxidation beginning stage or by building complexes with metal ions [14]. **Table 3** lists some of the most common antioxidants found in herbs and spices. Because of their antibacterial and antioxidant actions, herbs and spices can be used effectively in the food business to extend the shelf life of food goods.

#### *3.1.2.1 Antioxidants*

Tea, borage, balm, marjoram, rosemary, oregano, sage, and thyme are a few herbaceous plants that naturally contain antioxidants and free radical scavengers [15]. Kumar et al. [18] looked into the use of herbs in meat and meat-based products. Pork batters were mixed with extracts of oregano and rosemary in order to determine the primary antioxidant components and their effect on colour and oxidation. The antioxidant activity of rosemary extracts was shown to be higher than that of the phenol components separately. These extracts also had the highest antioxidant capacity, which could be attributed to the presence of large levels of carnosic acid and carnosol, among other chemicals.

However, ethanol oregano extracts strong in phenols, primarily rosmarinic acid, effectively prevented colour degradation [19]. Additionally, fermented sausages have been known to contain borage as an antioxidant [20], while pork products have been known to use balm [21]. Sage proved effective in lowering the prooxidant effects of salt, cooking, and storage in chicken flesh, as well as lipid and cholesterol oxidation [22]. Additionally, green tea extract was used instead of butylated hydroxytoluene (BHT) to prevent oxidation of Turkish dry-fermented sausage (sucuk) during the ripening process. Green tea extract has been found to be capable of scavenging oxygen radicals and chelating metal ions [23].

Rosemary extract (E-392) is the only herb commercially available for its use as an antioxidant and has been classified as a food additive in the EU and the US. The European Commission also uses and regulates carnosic acid and its derivative carnosol as major antioxidant components in rosemary extracts [24]. Rosemary leaf extract antioxidant can be extracted using a solvent (ethanol, acetone, or ethanol followed by hexane) or supercritical carbon dioxide extraction, which is then deodorised, decoloured, and standardised. Only deodorised rosemary extracts containing carnosic acid and carnosol are considered additives according to EU regulations. Food matrices

| Spice/Herb | Scientific name | Antioxidant compounds | Mode of action |
|---|---|---|---|
| Mustards | *Brassica nigra, alba, juncea* | Carotenes, glucosinolates | Free radical scavenger |
| Coriander | *Coriandrum sativum L apiaceae* | p-hydroxy-benzoic acid, protocatechuic acid, quercetin, rhamnetin, rutin, scopoletin, tannin, terpinen-4-ol, trans-anethole, vanillic acid, beta-carotene, beta-sitosterol, camphene, gammaterpinene, isoquercitrin, myricene, and myristicin | Antioxidant of free radicals and metal chelator |
| Turmeric | *Curcuma domestica, Valeton, Zingiberaceae* | Ascorbic acid, carotenes, caffeic acid, curcumin, p-cumaric acid | Oxygen scavenger, metal chelator, and free radical scavenger |
| Laurel or bay leaf | *Laurus nobilis L Lauraceae* | Beta-carotene, tocopherols, ascorbic acid, methyl-eugenol, eudesmol, kaempferol, kaempferol-3-rhamnopyranoside, kaempferol-3, 7-dirhamnopyranoside, 8-cineole, α-terpinyl acetate, terpinen-4-ol, catechin, and cinnamtannin B1 may be present. | oxygen and free radical scavengers |
| Onion | *Allium cepa L Amaryllidaceae* | Allicin, taxifolin, cyanidin glucosides, peonidin glucosides, quercetin, and kaempferol | Free radical scavenge |
| Oregano | *Origanum vulgaris L Lamiaceae* | Protocatechuic acid, rosmarinic acid, 2-caff eoyloxy-3-[2-(4-hydroxybenzyl)-4,5-dihydroxy], caffeic acid phenylpropionic acid; flavonoids: carvacrol, thymol, camphene, gamma-terpinene, terpinen-4-ol, myricene, linalyl-acetate; apigen, eriodictyol, dihydroquercetin, dihydrokaempherol | Free radical scavenger |
| Sage | *Salvia officinali L Lamiaceae* | Methyl and ethyl esters of carnosol, rosmarinic acid, ascorbic acid, beta-carotene, beta-sitosterol, camphene, gamma-terpinene, hispidulin, labiatic acid, oleanolic acid, terpinen-4-ol, ursolic acid, selenium, salvigenin, nevadensin, apigenin, cirsileol, and cirsimariti | Free radical scavenger |
| Rosemary | *Rosmarinus officinalis L Lamiaceae* | Diterpenes (epirosmanol, isorosmanol, rosmaridiphenol), rosmariquinone, rosmarinic acid, carnosol, carnosic acid, rosmanol, rosmadial, and | Scavenge superoxide radicals, lipid antioxidant, and metal chelator |
| Thyme | *Thymus vulgaris L Lamiaceae* | Ascorbic acid, beta-carotene, isochlorogenic acid, labiatic acid, p-coumaric acid, rosmarinic acid, thymol, carvacrol, p-cumene-2,3-diol, phenolic acids (gallic acid, caffeic acid, rosmarinic acid), phenolic diterpenes, flavonoids | Free radical scavenger, oxygen scavenger |
| Ginger | *Allium sativum L Amaryllidaceae* | Ascorbic acid, beta-carotene, camphene, gamma-terpinene, p-coumaric acid, gingerol, shogaol, zingerone, and caffeic acid | Oxygen and free radical scavengers |
| Turmeric | *Curcuma domestica Valeton Zingiberaceae* | Ascorbic acid, caffeic acid, curcumins, p-cumaric acid, carotenes, and 4-hydroxycinnamoyl methane | Free radical scavenger, oxygen scavenger, free radical scavenger, metal chelator |

| Spice/Herb | Scientific name | Antioxidant compounds | Mode of action |
|---|---|---|---|
| Black pepper | *Piper nigrum L Piperaceae* | Ubiquinone, piperine, quercetin, eugenol, beta-carotene, camphene, carvacrol, methyl eugenol, kaempferol, ascorbic acid, rhamnetin, gamma-terpinene | Free radical scavenger |
| Chilli pepper | *Capsicum frutescence L Solanaceae* | Capsaicinol, capsaicin | Free radical scavenger |
| Clove | *Eugenia caryophyllata Thunb Myrtaceae* | Flavonol glucosides, phenolic acids (gallic acid), tannins, and phenolic volatile oils (eugenol, isoeugenol, and acetyl eugenol) | Metal chelator and scavenger of free radicals |
| Marjoram | *Majorana hortensis Moench Lamiaceae* | Phenol, ursolic acid, beta-sitosterol, linalyl-acetate plant, caffeic acid, tannin, myricene, beta-carotene, caffeic acid, carvacrol, hydroquinone, beta-carotene, trans-anethole, oleanolic acid, terpinen4-ol, myricene, rosmarinic acid, eugenol | Free radical scavenger |
| Cumin | *Cumimum cyminum L Apiaceae* | Cuminaldehyde, apigenin, carotol, γ-terpinene, β-pinene, luteolin, 1-methyl-2-(1-methylethyl) benzene, linalool, pinocarveol, cuminic alcohol, and luteolin | Metal chelator, free radical scavenger |
| Garlic | *Allium sativum L Amaryllidaceae* | Vanillic, allicin, p-hydroxybenzoic, p-coumaric acids, caffeic | Metal chelator, free radical scavenger |
| Peppermint | *Mentha x piperita L Lamiaceae* | Luteolin 7-O-β-glucoside, menthoside, diosmin, hesperidin, luteolin 7-O-rutinoside, rosmarinic acid, caffeic acid, piperitoside, lithospermic acid, ascorbic acid, eriodictyol, narirutin, eriodictyol 7-O-β-eriocitrin, and beta-carotene | Free radical and oxygen scavenger |

**Table 3.**
*Antioxidants isolated from herbs and spices (Source: Refs. [15–17]) and the botanical names of the plants were verified using the world flora online (http://www.worldfloraonline.org/)).*

such as oils, animal fats, sauces, baked goods, and meat and fish items are among the application areas [24]. Rosemary is a superoxide dismutase. Rosemary is known to be a superoxide radical scavenger, lipid antioxidant, and metal chelator [15].

Coriander has been shown to be effective as an antioxidant in the inhibition of lipid oxidation in frozen white hake fish meatballs [25]. The effects of the Chinese five-spice ingredients—cinnamon, cloves, fennel, pepper, and star anise—both separately and in combination with cooked ground beef were examined by Dwivedi et al. [26]. The findings showed that every spice and combination decreased the flavour and odour of rancidity in cooked ground beef. However, the spices provided antioxidant effects rather than masking rotten off-flavours [26].

#### *3.1.2.2 Antimicrobials*

The antimicrobial qualities of herbs and the products made from them are gaining popularity as a means of lowering food contamination from pathogenic microorganisms such as *Salmonella typhimurium*, *Bacillus cereus*, *Listeria monocytogenes*, *Escherichia coli* O157:H7, and Staphylococcus aureus. In addition to extending the shelf life of food, antimicrobial compounds and extracts can lessen microbial illnesses and toxins [11]. However, these chemicals or their extracts operate as antimicrobials *in vitro*, and a higher concentration is required to produce the same effect in foods. There are now just a few commercially available food preservatives with antioxidant and antibacterial qualities that contain essential oils from rosemary, sage, thyme, oregano, and basil. They have been employed successfully either alone or in conjunction with other methods of preservation.

The antibacterial activity of rosemary hydroalcoholic extract was shown in all but one test against standard strains of *Lactobacillus casei*, *Streptococcus mitis*, *Streptococcus sanguinis*, *Streptococcus mutans*, and *Streptococcus sobrinus* [27]. When it comes to multiresistant bacteria (such as *E. Coli*, *Listeria monocytogenes*, *Bacillus cereus*, *Staphylococcus aureus*, *Enterococcus faecalis*, and *Saccharomyces cerevisiae*), Costa et al. [28] looked into the antibacterial activity of oregano essential oils. Although they were more effective against gram-positive bacteria, the essential oils of oregano, thyme, marjoram, and basil were similarly effective against strains of *Salmonella enteritidis* and *Listeria monocytogenes* in meat products [29].

Studies on the preservation properties of spices have been conducted on a range of foods, including rice, fruit, dairy products, meat, fish, and animal feeds. Spices have antimicrobial activity in two ways: they can hinder the growth of harmful microorganisms (food safety) and stop the growth of rotting bacteria (food preservation) [30]. Because of the presence of significant quantities of phenolic chemicals, many spice extracts have antibacterial activity against a wide spectrum of bacteria, yeast, moulds, and viruses. Cumin and clove essential oils suppressed overall bacterial growth on meat samples for 15 days at 2 degrees Celsius [31].

Furthermore, treatment of raw chicken meat with extracts of clove, oregano, cinnamon, and black mustard inhibited microbial development [32]. Turmeric extract (1.5%) added to entire gutted rainbow trout can also slow microbiological development, postpone chemical changes, maintain sensory qualities (texture, odour, colour, and overall), and increase shelf life during refrigerated storage [33]. Although spices and their derivatives have been studied for antibacterial activity against a wide variety of microbes throughout the years, their mode of action is still unknown. In truth, spices and essential oils can contain a wide range of bioactive chemicals in varying concentrations.

Spices' antibacterial effect is attributed to bioactive substances such as terpenes, terpenoids (thymol and carvacrol), and phenylpropenes (eugenol and cinnamaldehyde). Terpenes have less antibacterial action than other substances. Terpenoids' antimicrobial activity is mediated by their functional groups (hydroxyl groups and delocalized electrons), which impair permeability or interrupt critical energy-generating mechanisms, resulting in cell death. Eugenol's antibacterial activity occurs at the membrane and protein levels, where it interacts and cross-links with DNA and proteins. Overall, the antibacterial action of spices is a synergistic impact of all the bioactive chemicals found in them [30]. The fundamental restriction of spices as antibacterial agents is the requirement for a large amount of natural components that make up the organoleptic profile of meals. As a result, combining spices or their pure natural constituents with other technologies is a possible alternative for reducing the amount of spices utilised and solving this problem.

### *3.1.3 Flavour*

Herbs and spices include flavouring compounds that range in strength from moderate to intense [12], owing to the presence of volatile essential oils.

A culinary herb is an edible plant that gives significant flavour and scent when taken in little amounts. Many different herbs (**Table 1**), either fresh or dried, are commonly used as a food element in salads, spaghetti, sausages, soups, marinades, meat, eggs, vinegar, and even sweets and biscuits, as well as in some alcoholic beverages such as liqueurs and wine [10, 34]. Terpene- or phenolic-based aromatic chemicals are found in herbs. Carvacrol, thymol (in oregano and thyme), and linalool (in sage and rosemary) are three key chemical components for herb flavouring [35].

One of the most prevalent applications for spices is to flavour meals. There is a traditional classification of spices depending on their level of flavour [35]: Spices that are hot include ginger, mustard, black, red, and white pepper; moderate spices are paprika and coriander. Aromatic spices include nutmeg, onion, garlic, fenugreek, dill, cardamom, celery, cinnamon, clove, cumin, and dill. Eugenol (allspice, cinnamon, and clove), piperine (black pepper), gingerol (ginger), myristicin (nutmeg), and turmerone (turmeric) are the most important flavour compounds found in culinary spices [35].

### *3.1.4 Pungency*

Pungency is caused by essential oils stimulating taste senses. The concentration of essential oils in a spice gives it a sharper sting. Some spice components are highly aromatic even in little doses. The volatile components, according to Hirasa and Takemasa [36], are carbonyl, thioether, acid-amide, or isothiocyanate compounds. **Table 4** outlines some of the toxic chemicals found in the species. The mucous membranes of the oral cavity and nose are stimulated, and a burning sensation spreads throughout the mouth. Sharp sensations are generally perceived with volatile drugs, whereas hot sensations are mostly felt with nonvolatile compounds.

### *3.1.5 Herbs and spices as food colourants*

Colourants are food additives that change or add colour to foods [37]. They are used to enhance the appearance of meals while also preserving their natural colour

| Spice | Scientific name | Pungent compound | Basic structure | Sensation |
|---|---|---|---|---|
| Red pepper | *Capsicum frutescens* | Capsaicin, Dihydrocapsaicin | Acid amide group | Hot Sharp |
| Ginger | *Allium sativum* | Zingerrol, Shogaol | Carbonyl group | |
| Onion | *Allium cepa L* | Diallyl sulfide | Thioether group | |
| Garlic | *Allium sativum L* | Diallyl sulfide | | |
| Mustard | *Brassica carinata A. Braun brassicaceae* | Allyl isothiocyanate, Phydroxybenzyl isothiocyanate | Isothiocyanate group | |
| Horseradish | *Armoracia lapathifolia* | Allyl-isothiocyanate | Isothiocyanate group | |

**Table 4.**
*Pungent compounds in spices [36].*

throughout preparation and storage [38]. Green chlorophylls, yellow-orange-red carotenoids, redblue-purple anthocyanins, and red betanin are the four families of plant-derived colours. Colourants derived from natural sources are becoming increasingly popular, particularly in the food industry. Colour and freshness are the primary factors favoured by the social trend towards the use of natural products rather than manufactured ones due to their side effects, toxicity, and allergic reactions [39]. Flavonoids are the primary chemicals responsible for colour in herbs, ranging from mild yellow (isoflavones) to deep yellow (chalcones, flavones, flavonols, aurones), orange (aurones), and reds and blues (anthocyanins) [35].

Because of the presence of anthocyanins, bay leaf has been used as a food colourant [40]. Calendula flower is commonly used as a culinary colourant to impart a yellow hue because of the presence of carotenoids. If the final goods are to be appealing and acceptable, the stability of these chemicals during commercial shelf life is critical [41]. Furthermore, essential oils extracted from fresh or dried chamomile flower heads have colouring capabilities [42]. Spices, on the other hand, are more widely employed as a source of natural colourants than herbs.

Paprika (E160c) and curcumin (E100) are two natural colourants derived from spices that are permitted as food additives in the EU and the United States [11].

The main colouring ingredients capsanthin and capsorubin are found in paprika extract (E160c), a naturally occurring dark red colour made by solvent extraction from crushed fruit pods of *Capsicum annuum*, with or without seeds [43]. Both the EU and the US have allowed the use of paprika extract (E 160c) in a quantum satis manner, meaning that no maximum level is set and that it must be used in compliance with good manufacturing practices, at a level no higher than is necessary to achieve the intended purpose, and that the consumer is not misled [44]. However, the European Commission has set a limit of 10 mg/kg product for meat preparations and processed meat [45].

Curcumin (E100) is extracted from turmeric, which is the pulverised rhizomes of *Curcuma longa*. Curcumin, which accounts for around 2–8% of most turmeric preparations, is responsible for turmeric's characteristic colour and flavour. Crystallisation produces a concentrated curcumin powder. Curcumins, the colouring ingredient (1,7-bis(4-hydroxy-3-methoxyphenyl)hepta 1,6-dien-3,5-dione), and its two desmethoxy derivatives are present in variable amounts in the orange-yellow powder [43]. The EU has defined a curcumin daily intake of 3 mg/kg; however, the United States limits the amount of turmeric used in foods by good manufacturing practices (GMP), which require producers to use only the amount of an addition required to achieve the desired outcome [45]. **Table 5** shows common spice bio-colours found in herbs and spices, as well as their applications.

### 3.2 Cooking and roasting with herbs and spices

Spices have an essential role in the creation of food compositions in order to improve flavour and acceptance. As a result, spices can be added to food in a variety of ways, including slicing, roasting, grinding, frying, boiling, or heating, to release their characteristic aromatic components. Spices add flavour, aroma, texture, and colour to food and beverages while cooking. Furthermore, some herbs and spices operate as an antibacterial agent, extending the shelf life of packaged foods [52], while others act as an antioxidant, preventing the oxidation of lipids and fats [53]. When combined with food, many herbs and spices provide health advantages. The method of cooking, depending on the duration and temperature, is reported to help stabilise and improve

| Crop | Scientific name | Parts used | Pigment | Colour shade | Applications | References |
|---|---|---|---|---|---|---|
| Sea fennel | *Foeniculum vulgare* | Leaves | Phycocyanin | Green | Colourant in food | [46] |
| Saffron | *Crocus sativus* | Stigma | Crocin | Yellow-Orange | Colourant in sweet preparations | [47] |
| Turmeric | *Curcuma longa* | Rhizome | Curcumin | Yellow | Food colourant | [48] |
| Tamarind | *Tamarindus indica* | Pulp Seed | Leucoanthocynidine Tannin | Red Brown | Textiles Dye | [49] |
| Chilli | *Capsicum annuum L.* | Pericarp | Capsanthin | Red | Culinary preparation | [50] |
| Kokum | *Brindonia indica* | Rind | Anthocyanin | Red | Beverages, textiles | [51] |
| Fenugreek | *Trigonella foenumgraecum* | Seed | Carotenoid | Yellow | Dye | [48] |

**Table 5.**
*Bio-colours from spices and their applications.*

the quality of stored meals, maintaining the particular scent, flavour of the spice, texture, and nutritional content of the food [54].

Herbs and spices lose some of their flavouring or essential component as they cook. The material swells when it is submerged in water during cooking because the water permeates the herb's or spice's cell membrane from the outside in. This is followed by a rupture on the cell's surface; ultimately, because of the concentration gradient, the essential component emerges as cell breakage takes place and gives the food a delicious flavour. A similar mechanism occurs during the extraction and distillation processes of herbs and spices to produce the essential component [55]. The active component of several spices is lost after cooking. Tur dal (split red gramme) was created in one study by adding a known amount of curry powder, turmeric powder, red pepper powder, and black pepper powder. The effect of cooking on the active principle component was investigated using the thin-layer chromatographic (TLC) method. Cooking time and pH were found to have a substantial effect on the active components of spices [56].

The objectives should be obvious while utilising spices in cookery. Four categories can be used to classify the main use of spices in cooking: flavour, pungency, colouring, and masking/deodorising. The first three are the direct uses of spices, and any overlaps between deodorising and masking suggest more complex activities [57]. When certain foods, such as freshwater fish, have an unpleasant odour, spices play a significant role in disguising the odour during cooking. Ginger's volatile compound and garlic's sulphur-containing compound have a high capacity to minimise off-odour [58]. Combining caraway, cassia, clove, ginger, laurel, mace, nutmeg, onion, pepper, sage, and thyme helps mask the off flavour or odour of raw materials during cooking, according to numerous studies [59–61].

Vanilla is used to increase the sweetness of ice creams and pastries. In other words, adding a few drops of vanilla essence to the same amount of sugar boosts the sweetness significantly. Similarly, adding cinnamon to cakes has the same effect. Consider the flavour of peppered ice cream. There are numerous applications for a single spice across different regions of the globe. Whereas exotic spices are more frequently utilised in Europe and the Americas, locally available spices are more frequently employed in Asia and Africa. Most of the time, using spices in cooking comes down to personal taste and cultural customs. Pepper and black pepper have equal heat and pungency levels. Furthermore, when spices are combined, they can have either suppressive or synergistic effects [12].

### 3.3 Herbs and spices in natural medicine

#### *3.3.1 Metabolic disorders*

Several *in vitro* studies demonstrate that herbs and spices have anti-diabetic characteristics, but just a few clinical trials studying the effect of herbs and spices on metabolic disorders have been done. Cinnamon has the ability to reduce blood glucose levels in both animal models and humans. Several randomised controlled trials have been conducted to date to investigate the effect of cinnamon on T2DM in people. Cinnamon's influence on glycosylated haemoglobin, fasting plasma glucose, total cholesterol, LDL cholesterol, and triglycerides has been studied [62]. The research's brief duration is problematic, though, as the design makes it challenging to assess the information that is now available.

The first randomised double-blind placebo-controlled scientific research assessing cinnamon's potential advantages for individuals with type 2 diabetes was conducted in 2003. Thirty men and thirty women with diabetes were given a placebo or one of three doses of cinnamon powder (1, 3, or 6 g/day, respectively) for a period of 40 days. Research has demonstrated that cinnamon reduces total cholesterol, LDL cholesterol, triglycerides, and fasting blood glucose [63]. Another clinical investigation found that cinnamon reduced glycosylated haemoglobin (HbA1c) in 109 T2DM patients [64].

*Fenugreek seeds:* Fenugreek seeds have been shown in various clinical trials over the last several years to reduce hyperglycaemia in both healthy and diabetic people [65]. Diabetic patients had considerably lower fasting blood glucose, 24-hour urinary sugar excretion, serum cholesterol, and triglyceride levels. Clinical symptoms such as polyuria, polyphagia, and polydipsia improved as well. The gum fibre in fenugreek seeds appears to be responsible for these effects. Incorporating 25–50 g of fenugreek into the daily diet can be a useful supportive medication in the control of diabetes. According to a long-term animal study [65], fenugreek is completely safe to consume.

*Ginger* supplementation may be an effective natural therapy for lowering triglycerides and LDL cholesterol. The effect of 2 g of ginger per day on triglycerides and total cholesterol was larger [66]. At standard dosages, ginger is a safe, inexpensive, and widely available traditional treatment with negligible adverse effects. However, additional long-term research are required to corroborate these findings.

*Turmeric:* The roots, also known as rhizomes and bulbs, are utilised in both medicinal and culinary preparations. Turmeric is a popular spice that has been shown in a small number of clinical investigations to have favourable hypoglycemic effects and to increase glucose tolerance. Moreover, turmeric and metformin had a synergistic effect on lowering fasting blood glucose in people with type 2 diabetes. Patients with type 2 diabetes who received turmeric in addition to metformin had significantly lower fasting blood glucose and HbA1c readings than those who received metformin alone [67]. In the treatment of type 2 diabetes, cumin supplementation was found to be more beneficial than glibenclamide. Cumin's antihyperglycaemic impact could be attributed to its ability to protect surviving pancreatic cells, enhance insulin production, and increase glycogen storage [68].

*Clove extracts:* In diabetics, clove extracts also improved insulin function and decreased glucose, LDL, triglycerides, and total cholesterol. Thirty-six type 2 diabetic individuals were given capsules containing 0, 1, 2, or 3 g of cloves daily for 30 days, after which there was a 10-day washout period. After 30 days, the diabetic patients who took a clove supplement showed a reduction in their serum levels of LDL, triglycerides, total cholesterol, and glucose [69].

*Lemon balm leaf powder:* The ingestion of 3 g of lemon balm leaf powder per day for 2 months was related to a substantial decrease in mean blood LDL cholesterol in a parallel randomised doubleblind clinical trial with 82 patients with borderline hyperlipidaemia [70]. Increased insulin sensitivity, increased insulin secretion, decreased absorption of carbohydrates, increased peripheral glucose uptake, inhibition of hepatic glycogenolysis, antioxidant effects, and potentiation of endogenous incretins are some of the anti-diabetic characteristics of these spices [71].

*Bay leaf:* In comparison with the placebo group, men and women over the age of 40 who consumed bay leaf for 30 days had lower fasting serum glucose, total cholesterol, LDL, and triglycerides [51]. Similarly, after consuming 3 g of bay leaf daily for 4 weeks, both male and female individuals over 40 with controlled type I diabetes showed greater serum HDL and decreased levels of total cholesterol, LDL, triglycerides, and fasting glucose [72]. These findings imply that eating bay leaves may help manage hyperlipidaemia.

*Oregano extract:* Humans have also been tested for the potential hypolipidaemic impact of oregano extract consumption. Forty-five nonsmoking men were divided into three groups: placebo (mango-orange juice, n = 15); low phenolic (mango-orange juice enriched with oregano extract, n = 15); and high phenolic (mango-orange juice enriched with oregano extract, n = 15). The low phenolic group demonstrated a significant reduction in LDL oxidation after 90 minutes of juice drinking [73].

*Sage:* Sage extract is also advantageous to the human blood lipid profile. Patients with newly diagnosed primary hyperlipidaemia took a pill containing 500 mg of sage extract every 8 hours for 2 months in a randomised doubleblind placebo-controlled clinical research. Subjects had reduced serum total cholesterol, triglycerides, LDL, and VLDL and higher HDL at the end of the study compared to both the baseline and placebo groups [74]. Another study on healthy female individuals aged 40 to 50 found that drinking sage aqueous infusion twice a day for 2 weeks was related to reduced plasma LDL, total cholesterol, and higher plasma HDL [75]. In a double-blind, placebo-controlled study, 56 obese, hypertensive individuals were randomly assigned to receive 1 capsule daily containing either 379 mg of green tea extract or a comparable placebo for 3 months. In individuals with obesity-related hypertension, daily supplementation with 379 mg of green tea extract improved blood pressure, insulin resistance, inflammation, oxidative stress, and lipid profile [76].

### *3.3.2 Cardiovascular diseases*

A randomised crossover study discovered that by blocking the small intestine's lipid-digesting enzymes, a culinary spice blend (black pepper, cinnamon, cloves, garlic, ginger, oregano, paprika, rosemary, and turmeric) decreased the post-meal triglyceride response [77]. Triglycerides after meals are an important sign of cardiovascular risk and a possible therapeutic target. These findings imply that including spices in the diet on a regular basis may assist to mitigate the effect of high fat loading on cardiovascular risk. Congestive heart failure, systolic hypertension, angina pectoris, atherosclerosis, cerebral insufficiency, venous insufficiency, and arrhythmia have all been treated with herbs [78]. Chamomile, rosemary, sage, and thyme have significant flavonoid concentrations and so play an essential role in dietary flavonoid consumption; however, aside from few epidemiological researches, there is little evidence to support a direct cardiovascular health benefit from these herbs [10].

*Green tea:* According to research, green tea may help treat hypertension and hyperlipidaemia in both medication-treated patients and healthy individuals [79]. It may also lower blood pressure and cholesterol levels. A 12-week, double-blind, randomised, placebo-controlled study was carried out on patients with mild-to-moderate hypercholesterolemia who were already following a low-fat diet. The capsules included 75 mg theaflavin, 150 mg green tea catechins, and 150 mg other tea polyphenols. When compared to the placebo group, the extract decreased mean serum total cholesterol and LDL by 11.3 and 16.4%, respectively, in the treated people [80].

*Lemon balm:* In a 30-day study involving 55 radiology staff members, it was discovered that consuming an infusion of lemon balm twice a day enhanced the activity of guttation peroxidase (GPx), catalase (CAT), and superoxide dismutase (SOD), while decreasing the levels of lipid peroxidation (LPO), 8-hydroxy-2′-deoxyguanosine (8-OHdG), and myeloperoxidase (MPO). Because oxidative stress is associated with cardiovascular diseases (CVDs), lemon balm may be useful in the prevention or treatment of these conditions as a result of its reduction of radiation-induced oxidative stress indicators [81].

Its phenolic components, including phenolic acids (rosmarinic acid, gallic acid, and ferulic acid), flavonoids (luteolin 7-O-glucoside, quercetin 3-rutinoside, and quercetin 3-O galactoside), and antioxidant capacity, may be responsible for the putative antioxidant properties of lemon balm infusion [82]. There was no control group in another clinical trial, but patients with hypercholesterolemia who took 140 mg of lemon grass oil daily observed a significant decrease in mean cholesterol concentrations of up to 38 mg/dL [83].

*Hawthorn:* Hawthorn is a herb that is used to improve blood flow. Hawthorn leaves, fruit, and flowers are frequently used in Europe to improve heart pumping capacity and treat angina. Hawthorn's main activity is assumed to be mediated *via* flavonoids. Patients with congestive heart failure (NY Heart Association class II) who received 600 mg/day of a hawthorn extract had significantly reduced heart rates, blood pressure, and dyspnoea during exercise as compared to non-hawthorn participants [84].

### *3.3.3 Anticancer effects*

*Garlic:* Garlic has also been researched in a few clinical trials to look into its anticancer properties. Several randomised clinical trials have been conducted to assess the effect of garlic consumption on the risk of stomach cancer. In one trial, patients who took garlic extract had a 33% lower risk of all tumours combined and a 52% lower risk of stomach cancer when compared to the placebo group [57]. However, a different randomised experiment involving individuals with precancerous stomach lesions found that taking a supplement containing 800 mg of garlic extract and 4 mg of steam-distilled garlic oil daily did not lower the incidence of gastric cancer or the prevalence of precancerous gastric lesions [85].

Many mechanisms, including as the suppression of mutagenesis, the stimulation of phase II detoxification enzymes, the inhibition of DNA adduct formation, the impact on the intrinsic pathway for apoptotic cell death, and the cell cycle machinery, can allow garlic compounds to exhibit anticancer activity [86]. Future study in the clinical evaluation of these substances for the prevention or treatment of cancer in humans is required.

*Green tea* is the only herb with clinical data confirming its anticancer properties [87]. Green tea catechins were found to be safe and beneficial for treating premalignant prostate cancer in a double-blind placebo-controlled research [88]. However, several studies [89, 90] discovered that green tea has little therapeutic benefit in the fight against prostate cancer. It has been determined that the main active components of green tea, known as catechins (phenolic compounds), are mostly responsible for the biological activities of the tea.

Among the bioactive substances found in herbs with cancer-preventive qualities are terpenes (basil, marjoram, mint, rosemary, oregano, sage, and thyme), polyphenols, mainly flavonoids (basil, marjoram, mint, rosemary, oregano, sage, parsley, and thyme), and epigallocatechin gallate and other catechins (green tea) [91, 92].

*Turmeric:* Turmeric is one of the most researched spices for its potential use in cancer treatment. Curcumin derived from turmeric has been shown in preclinical and clinical experiments to have antioxidant, anti-inflammatory, and anticancer effects. Antitumor action is thought to be caused *via* apoptosis activation and suppression of inflammation, angiogenesis, and metastasis in the tumour microenvironment [93].

Numerous clinical trials are being conducted to confirm curcumin's *in vitro* anticancer activity and usefulness as a therapeutic agent in many tumour types such as colon, gastric, cervical, endometrial, breast, pancreatic, prostate, lung, and lymphoma [93]. As a result, multiple patents for the administration and use of curcumin against various forms of cancer have been generated.

#### *3.3.4 Neurodegenerative disorders*

*Lemon balm:* Lemon balm has traditionally been used to treat dementia and amnesia, two illnesses that are closely related to Alzheimer's disease. A 16-week clinical trial of 42 patients revealed that a hydroalcoholic extract of lemon balm (60 drops/day) standardised to contain 500 g citral/mL (terpenoid component) reduced agitation and improved cognitive and behavioural functioning [94, 95]. A possible mechanism for lemon balm's memory-enhancing effects is the inhibition of acetylcholinesterase activity, stimulation of acetylcholine (nicotinic and muscarinic receptors), stimulation of -aminobutyric acid (GABAA) receptors, and inhibition of matrix metalloproteinase-2 (MMP-2).

*Curcumin:* The majority of the work on nutraceuticals derived from spices focuses on curcumin, which has the potential to fight neurodegenerative illnesses due to its powerful anti-inflammatory and antioxidant properties. Nevertheless, it is also necessary to look into the possibilities of numerous other spices. Therefore, to fully explore the potential of spice-derived nutraceuticals as neuroprotective agents, more preclinical and clinical research is desperately needed.

## 4. Food safety of spices and herbs along global food systems

### 4.1 Possible risks from herbs and spices

Spices and herbs are used and consumed in little amounts, but because they are added to so many different recipes, particularly ready-to-eat (RTE) foods, keeping track of one's consumption has become more important. Spices' physiological effects [36], health advantages [94–98], anti-inflammatory properties, and pharmaceutical applications [10, 11], including cancer therapies [99–101], are only a few of the many applications.

In fact, recent large-scale cohort studies have demonstrated a negative link between frequent consumption of spicy meals and overall mortality as well as major causes of death (cancer, ischemic heart disease, and respiratory disorders) [102]. Herbs and spices, on the other hand, have been linked to some illnesses or dysfunction syndromes, particularly in the digestive system [103], as well as food-borne infections [104]. While this is definitely a risk to human health in and of itself, the principal health concerns have been linked to a range of pollutants that periodically surface in spice goods owing to technical or other causes, rather than to the natural ingredients of spices. Through numerous flaws in the production and supply systems, microbiological and chemical agents can cause severe harm to people. Spices and herbs, despite being relatively insignificant as food additives, have a high potential to contaminate a wide range of foods due to their widespread use and distribution.

However, because outbreak investigations usually concentrate on the main food ingredients and the cause of infection in food-borne outbreaks is typically only proven in less than 20% of cases, contaminated spices and herbs would be challenging to identify as the source of a foodborne infection or intoxication [105]. Furthermore, many commonly used detection methods are unsuitable for matrices containing a wide range of herbs and spices.

### 4.2 General considerations, sampling

Contaminants in spices and herbs must be evaluated in terms of the risk they cause to human (or environmental) health as well as the anticipated exposure to these substances. A risk matrix built from the severity of the adverse effects based on available toxicological reference levels and the probability of exposure based on historical data from available monitoring surveys and the frequency of notifications in the EU's Rapid Alert System for Food and Feed (RASFF) allows for the risk ranking of chemical contaminants in spices [106]. The concept of traceability holds great importance in the realm of food safety. Nevertheless, the intricate manufacturing and trade chains of spices and food products containing them pose a challenge to traceability in the global networks of dried culinary herbs and spices. As previously indicated, in the field, adequate and standardised documentations, as well as harmonised information technology solutions, are vital. The selection of proper sample techniques is crucial to identifying contamination effectively [107]. Sample is recognised to have the biggest experimental random error in the analytical process, and stratified random sample designs may provide enhanced efficacy at higher predicted contaminated levels, and network analysis might help with spice contamination monitoring [107].

### 4.3 Chemical contamination

Three sources account for the majority of organic micro-contaminants detected in spices: chemicals used in food adulteration, pesticide residues, and phytopathogenic fungal mycotoxins. Mycotoxins are mainly created during harvesting and the storage that follows. They are a serious problem since these secondary metabolic chemicals have the potential to induce endocrine disruption or cancer in humans or other exposed species. Although numerous decontamination techniques for mycotoxinogenic fungi are available, they are only effective if employed before mycotoxin contamination has occurred [107]. Pesticide residues are primarily the consequence of agrochemical treatments used to protect farmed spice plants from various weeds, fungi, or pest animals (mostly insects). Compliance with technology specifications in cultivation, harvesting, and storage is crucial to avoiding mycotoxin and pesticide contamination [108], and prudent pesticide use is required to avoid MRL violations [109, 110]. Unwanted substances such as polycyclic aromatic hydrocarbons (PAHs), anizoles, benzidines, unsubstituted, halogenated biphenyls, and other molecules may exist in addition to these two types of major chemical pollutants. In contrast to the aforementioned chemical contaminants, which are often present in spices due to accidental contamination, other chemicals may appear as deliberate contamination for financial reasons (spice adulteration) or other malicious purposes. As can be seen, the "from farm to fork" concept solidifies environmental safety as a component of food safety evaluation by including Environmental Risk evaluation (ERA).

Furthermore, as recently reported, pesticide toxicity is linked to a number of additives, including agricultural surfactants used, for example, in the formulation of plant protection products [109]. As a result, the probable presence of these co-formulates, which were previously assumed to be "inert", is expected to be monitored in the future.

### 4.4 Microbial contamination

When it comes to dried spices and herbs, microbial food safety is likely to be the most concerning [111, 112]. Spices and herbs are potential sources of foodborne disease outbreaks due to their low water activity, which allows microbial contamination to remain [104]. For example, the US Food and Drug Administration developed risk profiles for spices in order to assess and limit the risk associated with microbial concerns [113, 114].

Decontamination methods are widely accepted for reducing or eliminating microbial contamination in spices [115]. Appropriate quality standards and indices have also been identified [36, 116]. Despite the fact that many dried spices and herbs can act as media or sources of microbial contamination, they have been connected to their own natural antibacterial effect. Numerous investigations have shown that spices have antibacterial effects [6, 36]. Some spices and spice combinations (such as clove, cinnamon, oregano, and rosemary) have been recommended as potential alternatives for standard food preservation additives due to their qualities [117, 118].

Given the importance of the issue, the majority of the pieces in this Special Issue focus on the microbiological safety of spices. Three studies have found and confirmed the presence of *Salmonella* spp., *Bacillus cereus*, and *Staphylococcus aurous* in dried spices and herbs. *B. cereus* species' phylogenetic and toxicogenic properties have been identified [107]. Molecular biological approaches for the detection and quantification of microorganisms from spices were used to determine the potential origin of spice paprika samples and to detect hazardous bacteria in spices and herbs at high throughput [107]. Among the nine condiments examined, oregano and cinnamon were found to have strong antibacterial action against a strain of *Salmonella* known as Lins. Various cleaning approaches, including steaming and radiation, microwave heating independently and in conjunction with rewetting, and radiofrequency heat treatment, were tested for effectiveness and effects on the composition of spice paprika.

## 5. Conclusions and future outlook

For millennia, herbs and spices have been employed in culinary practices and traditional medicine. The controversy and uncertainty surrounding chemical additives have intensified the search for natural substitutes that are palatable to customers. Combining the fads of safe, unprocessed, nutritious eating, exercise, and health-conscious eating, there is a growing push to reduce or eliminate food additive usage or swap them out with special natural additions. However, their contribution to maintaining and/or raising standards for food quality justifies their usage in particular manufacturing and storage processes, and hence the search for creative and practical solutions.

Herbal medicine is a component of alternative treatment approaches that incorporates the use of various plants and their extracts because of their numerous benefits. Practitioners of conventional medicine are now acknowledging it as one of the safest and most effective therapy options. Conversely, herbal supplements may be advantageous to users but can also have serious adverse effects and even be fatal. The FDA can do nothing to safeguard consumers from these health hazards under a regulatory environment. Food contamination, on the other hand, is a major public health issue that could be mitigated by the use of natural preservatives such as essential oils derived from spices. Numerous studies in recent years have demonstrated that many essential oils have antibacterial action. The product and the species of bacteria or fungi that the essential oil is intended to combat dictate the type of essential oil and the optimal concentration. The use of naturally derived preservatives may alter food taste or exceed established flavour thresholds; hence, consideration must be given to the sensory impact if essential oils are to be widely employed as antibacterial and antifungal agents. To obtain efficient antibacterial action at adequately low doses, research in this area should focus on improving essential oil blends and applications. It could be inferred that consumers and physicians should be aware of the negative effects of herbal medicines as well as their interactions with other prescription drugs. Herbal treatments are known to have numerous benefits for people under conventional therapy, which is true, but one should be completely aware of their adverse effects at normal and high doses.

## Author details

Melaku Tafese Awulachew
Food Science and Nutrition Research Sector, Kulumsa Agricultural Research Center, Ethiopian Institute of Agricultural Research, Addis Ababa, Ethiopia

*Address all correspondence to: melakutafese12@gmail.com; melaku.tafese@eiar.gov.et

## References


[1] Gislene GFNN, Locatelli J, Freitas PC, Silva GL. Antibacterial activity of plant extracts and phytochemicals on antibiotic-resistant bacteria. Brazilian Journal of Microbiology. 2000;**31**(4):247-256

[2] Hinneburg I, Dorman HJD, Hiltunen R. Antioxidant activities of extracts from selected culinary herbs and spices. Food Chemistry. 2006;**97**:122-129

[3] Carocho M et al. Adding molecules to food, pros and cons: A review on synthetic and natural food additives. Comprehensive Reviews in Food Science and Food Safety. 2014;**13**:377-399

[4] Pinto VF, Patriarca A, Pose G. Plant extracts as natural antifungals: Alternative strategies for mold control in foods. In: Novel Technologies in Food Science. New York: Springer; 2012. pp. 205-218

[5] Leja KB, Czaczyk K. The industrial potential of herbs and spices – A mini review. Acta Scientiarum Polonorum. Technologia Alimentaria. 2016;**15**(4):353-365

[6] Tajkarimi M, Ibrahim SA, Cliver D. Antimicrobial herb and spice compounds in food. Food Control. 2010;**21**:1199-1218

[7] Stickel F, Patsenker E, Schppan D. Herbal hepatotoxicity. Journal of Hepatology. 2005;**43**:901-910

[8] Abdel-Aziz SM, Aeron A, Kahil TA. Health benefits and possible risks of herbal medicine. Microbes in Food and Health. 2016:97-116. DOI: 10.1007/978-3-319-25277-3_6

[9] JOINT FAO/WHO FOOD STANDARDS. Revision of the classification of food and feed: Class A: Primary Food Commodities of Plant Origin Type 05: Herbs and Spices (CODEX (CX/PR 18/50/7)), 50th Session. Haikou, PR. China; 2017. pp. 1-23

[10] Tapsell LC, Scientific TC, Sullivan DR. Health benefits of herbs and spices: The past, the present, the future. The Medical Journal of Australia. 2006;**185**(Suppl. 4):S4-S24

[11] Teresa H, Amaia ID, Jaime U, María DC. Beneficial herbs and spices. Journal of Nutrition, Fitness, and Mindfulness Nutrition and Health. 2020:1-22. DOI: 10.1007/978-3-030-30892-6_6

[12] Divakara Sastry EV. Understanding spices for processing. Ethiopian Journal of Science and Technology. (Special Issue No.1). 2013:81-89

[13] Yanishlieva NV, Marinova E, Pokorný J. Natural antioxidants from herbs and spices. European Journal of Lipid Science and Technology. 2006;**108**:776-793

[14] Zachariah TJ, Leela NK, Shamina A. Methods of analysis of herbs and spices. In: Peter KV, editor. Handbook of Herbs and Spices. Philadelphia: Woodhead Publishing; 2012. pp. 89-116

[15] Embuscado ME. Spices and herbs: Natural sources of antioxidants – A mini review. Journal of Functional Foods. 2015;**18**:811-819. DOI: 10.1016/j.jff.2015.03.005

[16] Pokorný J, Pánek J. The effect of natural antioxidants in herbs and spices on food shelf-life. In: Peter KV, editor. Handbook of Herbs and Spices. Vol. 2.

Oxford: Woodhead Publishing; 2012. pp. 51-57

[17] Cazzola R, Cestaro B. Antioxidant spices and herbs used in diabetes. In: Preedy V, editor. Diabetes: Oxidative Stress and Dietary Antioxidants. Elsevier; 2014. pp. 89-97. DOI: 10.1016/B978-0-12-405885-9.00009-7

[18] Kumar Y et al. Recent trends in the use of natural antioxidants for meat and meat products. Comprehensive Reviews in Food Science and Food Safety. 2015;**14**(6):796-812. DOI: 10.1111/1541-4337.12156

[19] Hernández-Hernández E et al. Antioxidant effect rosemary (*Rosmarinus officinalis* L.) and oregano (*Origanum vulgare* L.) extracts on TBARS and colour of model raw pork batters. Meat Science. 2009;**81**(2):410-417. DOI: 10.1016/J.MEATSCI.2008.09.004

[20] Ciriano MG et al. Use of natural antioxidants from lyophilized water extracts of Borago officinalis in dry fermented sausages enriched in ω-3 PUFA. Meat Science. 2009;**83**(2):271-277. DOI: 10.1016/J. MEATSCI.2009.05.009

[21] de Ciriano MG-I et al. Effect of lyophilized water extracts of Melissa officinalis on the stability of algae and linseed oil-in-water emulsion to be used as a functional ingredient in meat products. Meat Science. 2010;**85**(2):373-377. DOI: 10.1016/j.meatsci.2010.01.007

[22] Mariutti LRB, Nogueira GC, Bragagnolo N. Lipid and cholesterol oxidation in chicken meat are inhibited by sage but not by garlic. Journal of Food Science. 2011;**76**(6):C909-C915. DOI: 10.1111/j.1750-3841.2011.02274.x

[23] Bozkurt H. Utilization of natural antioxidants: Green tea extract and thymbra spicata oil in Turkish dry-fermented sausage. Meat Science. 2006;**73**(3):442-450. DOI: 10.1016/J.MEATSCI.2006.01.005

[24] Birtić S et al. Carnosic acid. Phytochemistry. 2015;**115**:9-19. DOI: 10.1016/J.PHYTOCHEM.2014.12.026

[25] Sancho RAS et al. Effect of annatto seed and coriander leaves as natural antioxidants in fish meatballs during frozen storage. Journal of Food Science. 2011;**76**(6):C838-C845. DOI: 10.1111/j.1750-3841.2011.02224.x

[26] Dwivedi S, Vasavada MN, Cornforth D. Evaluation of antioxidant effects and sensory attributes of Chinese 5-spice ingredients in cooked ground beef. Journal of Food Science. 2006;**71**(1):C12-C17. DOI: 10.1111/j.1365-2621.2006. tb12381.x

[27] Silva IF et al. Antimicrobial screening of some medicinal plants from Mato Grosso Cerrado. Brazilian Journal of Pharmacognosy. 2009;**19**(1 B):242-248. DOI: 10.1590/S0102-695X2009000200011

[28] Costa A et al. Antibacterial activity of the essential oil of *Origanum vulgare* L. (Lamiaceae) against bacterial multiresistant strains isolated from nosocomial patients. Revista Brasileira de Farmacognosia. 2009;**19**(1B):2

[29] Silva N, Fernandes JA. Biological properties of medicinal plants: A review of their antimicrobial activity. Journal of Venomous Animals and Toxins including Tropical Diseases. 2010;**16**(3):402-413. DOI: 10.1590/S1678-91992010000300006

[30] Gottardi D et al. Beneficial effects of spices in food preservation and safety. Frontiers in Microbiology. 2016;**7**:1-20. DOI: 10.3389/fmicb.2016.01394

[31] Hernández-Ochoa L et al. Use of essential oils and extracts from spices in meat protection. Journal of Food Science and Technology. 2014;**51**(5):957-963. DOI: 10.1007/s13197-011-0598-3

[32] Radha Krishnan K et al. Antimicrobial and antioxidant effects of spice extracts on the shelf life extension of raw chicken meat. International Journal of Food Microbiology. 2014;**171**:32-40. DOI: 10.1016/j.ijfoodmicro.2013.11.011

[33] Pezeshk S, Rezaei M, Hosseini H. Effects of turmeric, shallot extracts, and their combination on quality characteristics of vacuum-packaged rainbow trout stored at 4 ± 1°C. Journal of Food Science. 2011;**76**(6):387-391. DOI: 10.1111/j.1750-3841.2011.02242.x

[34] Bown D. New Encyclopedia of Herbs and their Uses: The Definitive Guide to the Identification, Cultivation and Uses of Herbs. New York: Dorling Ki; 2001

[35] Peter KV, Shylaja MR. In: Peter KV, editor. Introduction to Herbs and Spices: Definitions, Trade and Applications. Cambridge: Woodhead Publishing; 2012. pp. 1-24

[36] Hirasa K, Takemasa M. Spice Science and Technology. Food Science and Technology. Vol. 89. New York, USA: Marcel Dekker; 1998

[37] Nations, F. and A. O. of the U. Codex Alimentarius: General standard for food additives. In: Codex Alimentarius: General Standard for Food Additives. Rome: Food and Agriculture Organization of the United Nations (FAO); 2017

[38] Llamas NE et al. Second order advantage in the determination of amaranth, sunset yellow FCF and tartrazine by UV–vis and multivariate curve resolution-alternating least squares. Analytica Chimica Acta. 2009;**655**(1-2):38-42. DOI: 10.1016/J.ACA.2009.10.001

[39] Santos-Buelga C, Mateus N, De Freitas V. Anthocyanins. Plant pigments and beyond. Journal of Agricultural and Food Chemistry. 2014;**62**(29):6879-6884. DOI: 10.1021/jf501950s

[40] Longo L, Vasapollo G. Anthocyanins from bay (*Laurus nobilis* L.) berries. Journal of Agricultural and Food Chemistry. 2005;**53**(20):8063-8067. DOI: 10.1021/jf051400e

[41] Rjzk EM, El-Gharably AM, Tolba KH. Carotenoid pigments composition of calendula flower and its potential uses as antioxidant and natural colorant in manufacturing of hard candy. Arab Universities Journal of Agricultural Sciences. 2008;**16**(2):407-417

[42] El-Agbar ZA et al. Comparative antioxidant activity of some edible plants. Turkish Journal of Biology. 2007;**32**:(2008):193-196

[43] European Commission. Annex II commission regulation (EU) No 231/2012 of 9 March 2012 laying down specifications for food additives listed in annexes II and III to regulation (EC) No 1333/2008 of the European Parliament and of the council. Official Journal of the European Union. 2012:1-295. DOI: 10.1201/9781315152752

[44] European Commission. Regulation (EC) No 1333/2008 of the European Parliament and of the council of 16 December 2008 on food additives. Official Journal of the European Communities. 2008:1-342. DOI: 2004R0726-v.7.of.05.06.2013

[45] Lehto S et al. Comparison of food colour regulations in the EU and the US:

A review of current provisions. Food Additives & Contaminants. Part A, Chemistry, Analysis, Control, Exposure & Risk Assessment. 2017;**34**(3):335-355. DOI: 10.1080/19440049.2016.1274431

[46] Rennan M, Gonnella M. The use of the sea fennel as a new spice-colorant in culinary preparations. International Journal of Gastronomy and Food Science. 2012;**1**:111-115

[47] Raja SM, Pareek PK, Shakyawar DB, Sarfaraz AW, Nehvi FA, Asif H. Extraction of natural dye from saffron flower waste and its application on pashmina fabric. Advances in Applied Science Research. 2012;**3**(1):156-161

[48] Selvam MR, Athinarayanana G, Nanthinib RUA, Singh AJA, Kalirajana KP, Selvakumard M. Extraction of natural dyes from Curcuma longa, Trigonella foenum graecum and Nerium oleander plants and their application in antimicrobial fabric. Industrial Crops and Products. 2015;**70**:84-90

[49] Mayavel A, Singh GB, Durai A, Murugeasan S. Evaluation of colour and stability of anthocyanin in red tamarind (*Tamarindus indica* L.). International Journal of Advanced Life Sciences. 2012;**5**(2):137-144

[50] Arimboor R, Babu R, Menon NKR, Lekshmi P, Chandrasekhar MV. Red pepper (*Capsicum annuum*) carotenoids as a source of natural food colors: Analysis and stability - A review. Journal of Food Science and Technology. 2015;**52**(3):1258-1271

[51] Vasundhara M, Radhika B, Thara BS, Priyanka R, Jayaram A. Organic colours for ayurvastra from kokum fruits and rinds. Journal of Medicinal Plants Studies. 2016;**4**(6):104-107

[52] Billing J, Sherman PW. Antimicrobial functions of spices: Why some like it hot. The Quarterly Review of Biology. 1998;**73**(1):3-49

[53] Tiwari V, Shanker R, Srivastava J, Vankar PS. Change in antioxidant activity of spices-turmeric and ginger on heat treatment. Electronic Journal of Environmental, Agricultural and Food Chemistry. 2006;**5**(2):1313-1317

[54] James SJ, James C. Minimal processing of ready meals. In: Emerging Technologies for Food Processing. London: Elsevier Academic; 2005. pp. 717-732

[55] Douglas M, Heyes J, Smallfield B. Herbs, Spices and Essential Oils: Post-Harvest Operations in Developing Countries, 61. First published 2005. Mar 5, 2006. Vienna, Austria: UNIDO and FAO; 2005:**91153**(300):1-61

[56] Srinivasan K, Sambaiah K, Chandrasekhara N. Loss of active principles of common spices during domestic cooking. Food Chemistry. 1992;**43**(4):271-274

[57] Hirasa K, Takemasa M. In: Höferl M, editor. Spice Science and Technology. Food chemistry. Boca Raton: CRC Press; 1998:1-232. DOI: 10.1201/9780367800451

[58] Li J, Tu Z, Zhang L, et al. The effect of ginger and garlic addition during cooking on the volatile profile of grass carp (*Ctenopharyngodon idella*) soup. Journal of Food Science and Technology. 2016;**53**(8):3253-3270

[59] Kikuchi T, Hirai K, Sudarso AS. Suppressing effect of spice for fishy odor: A model experiment with the mixture of trimethylamine and spice. Eiyo to Shokuryo. 1968;**21**:253-256

[60] Yoshida A, Sasaki K, Ohshiba K. Suppressing experiments for fishy odor. II. Seikatsu Esei. Journal of Urban Living and Health Association. 1984;**28**(4):211-218

[61] Kasahara K, Osawa C. Combination effects of spices on masking of odor in boiled sardine. Fisheries Science. 1998;**64**(3):415-418

[62] Medagama AB. The glycaemic outcomes of cinnamon, a review of the experimental evidence and clinical trials. Nutrition Journal. 2015;**14**(1):1-12. DOI: 10.1186/s12937-015-0098-9

[63] Khan A et al. Cinnamon improves glucose and lipids of people with type 2 diabetes. Diabetes Care. 2003;**26**(12):3215-3218

[64] Crawford P. Effectiveness of cinnamon for lowering hemoglobin A1C in patients with type 2 diabetes: A randomized, controlled trial. Journal of American Board of Family Medicine. 2009;**22**(5):507-512. DOI: 10.3122/jabfm.2009.05.080093

[65] Srinivasan K. Plant foods in the management of diabetes mellitus: Spices as beneficial antidiabetic food adjuncts. International Journal of Food Sciences and Nutrition. 2005;**56**(6):399-414. DOI: 10.1080/09637480500512872

[66] Pourmasoumi M et al. The effect of ginger supplementation on lipid profile: A systematic review and metaanalysis of clinical trials. Phytomedicine. 2018;**43**:28-36. DOI: 10.1016/j.phymed.2018.03.043

[67] Maithili Karpaga Selvi N et al. Efficacy of turmeric as adjuvant therapy in type 2 diabetic patients. Indian Journal of Clinical Biochemistry: IJCB. 2015;**30**(2):180-186. DOI: 10.1007/s12291-014-0436-2

[68] Bi X, Lim J, Henry CJ. Spices in the management of diabetes mellitus. Food Chemistry. 2017;**217**:281-293. DOI: 10.1016/j.foodchem.2016.08.111

[69] Kahn SE, Hull RL, Utzschneider KM. Mechanisms linking obesity to insulin resistance and type 2 diabetes. Nature. 2006;**444**:840-846

[70] Jandaghi P et al. Lemon balm: A promising herbal therapy for patients with borderline hyperlipidemia—A randomized double-blind placebo-controlled clinical trial. Complementary Therapies in Medicine. 2016;**26**:136-140. DOI: 10.1016/j.ctim.2016.03.012

[71] Beidokhti MN, Jäger AK. Review of antidiabetic fruits, vegetables, beverages, oils and spices commonly consumed in the diet. Journal of Ethnopharmacology. 2017;**201**:26-41. DOI: 10.1016/j.jep.2017.02.031

[72] Aljamal A. Effects of bay leaves on blood glucose and lipid profiles on the patients with type 1 diabetes. World Academy of Science, Engineering and Technology. 2010;**4**(9):409-412

[73] Nurmi A et al. Consumption of juice fortified with oregano extract markedly increases excretion of phenolic acids but lacks short- and long-term effects on lipid peroxidation in healthy nonsmoking men. Journal of Agricultural and Food Chemistry. 2006;**54**(16):5790-5796. DOI: 10.1021/jf0608928

[74] Kianbakht S et al. Antihyperlipidemic effects of *Salvia officinalis* L. leaf extract in patients with hyperlipidemia: A randomized double-blind placebo-controlled clinical trial. Phytotherapy Research. 2011;**25**(12):1849-1853. DOI: 10.1002/ptr.3506

[75] Sá C et al. Sage tea drinking improves lipid profile and antioxidant defences in humans. International Journal of Molecular Sciences. 2009;**10**(9):3937-3950. DOI: 10.3390/ijms10093937

[76] Bogdanski P et al. Green tea extract reduces blood pressure, inflammatory biomarkers, and oxidative stress and improves parameters associated with insulin resistance in obese, hypertensive patients. Nutrition Research. 2012;**32**(6):421-427. DOI: 10.1016/j.nutres.2012.05.007

[77] McCrea CE et al. Effects of culinary spices and psychological stress on postprandial lipemia and lipase activity: Results of a randomized crossover study and in vitro experiments. Journal of Translational Medicine. 2015;**13**(1):1-12. DOI: 10.1186/s12967-014-0360-5

[78] Rastogi S, Pandey MM, Rawat AKS. Traditional herbs: A remedy for cardiovascular disorders. Phytomedicine. 2016;**23**(11):1082-1089. DOI: 10.1016/J.PHYMED.2015.10.012

[79] Momose Y, Maeda-Yamamoto M, Nabetani H. Systematic review of green tea epigallocatechin gallate in reducing low-density lipoprotein cholesterol levels of humans. International Journal of Food Sciences and Nutrition. 2016;**67**(6):606-613. DOI: 10.1080/09637486.2016.1196655

[80] Maron DJ et al. Cholesterol-lowering effect of a theaflavin-enriched green tea extract. Archives of Internal Medicine. 2003;**163**(12):1448. DOI: 10.1001/archinte.163.12.1448

[81] Zeraatpishe A et al. Effects of *Melissa officinalis* L. on oxidative status and DNA damage in subjects exposed to long-term low-dose ionizing radiation. Toxicology and Industrial Health. 2011;**27**(3):205-212. DOI: 10.1177/0748233710383889

[82] Kulisic-Bilusic T et al. Antioxidant and acetylcholinesterase inhibiting activity of several aqueous tea infusions in vitro. Food Technology and Biotechnology. 2008;**46**(4):368-375

[83] Elson CE et al. Impact of lemongrass oil, an essential oil, on serum cholesterol. Lipids. 1989;**24**(8):677-679. DOI: 10.1007/BF02535203

[84] Schmidt U et al. Efficacy of the hawthorn (Crataegus) preparation LI 132 in 78 patients with chronic congestive heart failure defined as NYHA functional class II. Phytomedicine. 1994;**1**(1):17-24. DOI: 10.1016/S0944-7113(11)80018-8

[85] You WC et al. Randomized double-blind factorial trial of three treatments to reduce the prevalence of precancerous gastric lesions. Journal of the National Cancer Institute. 2006;**98**(14):974-983. DOI: 10.1093/jnci/djj26

[86] Srinivasan K. Antimutagenic and cancer preventive potential of culinary spices and their bioactive compounds. PharmaNutrition. 2017;**5**(3):89-102. DOI: 10.1016/j.phanu.2017.06.001

[87] Hosseini A, Ghorbani A. Cancer therapy with phytochemicals: Evidence from clinical studies. Avicenna Journal of Phytomedicine. 2015;**5**(2):84-97

[88] Bettuzzi S et al. Chemoprevention of human prostate cancer by oral administration of green tea catechins in volunteers with high-grade prostate intraepithelial neoplasia: A preliminary report from a one-year proof-of-principle study. Cancer Research. 2006;**66**(2):1234-1240. DOI: 10.1158/0008-5472.CAN-05-1145

[89] Choan E et al. A prospective clinical trial of green tea for hormone refractory prostate cancer: An evaluation of the complementary/alternative therapy approach. Urologic Oncology. 2005;**23**(2):108-113. DOI: 10.1016/j.urolonc.2004.10.008

[90] Jatoi A et al. A phase II trial of green tea in the treatment of

patients with androgen independent metastatic prostate carcinoma. Cancer. 2003;**97**(6):1442-1446. DOI: 10.1002/cncr.11200

[91] Huang W-Y, Cai Y-Z, Zhang Y. Natural phenolic compounds from medicinal herbs and dietary plants: Potential use for cancer prevention. Nutrition and Cancer. 2009;**62**(1):1-20. DOI: 10.1080/01635580903191585

[92] Yashin A et al. Antioxidant activity of spices and their impact on human health: A review. Antioxidants. 2017;**6**(3):70. DOI: 10.3390/antiox6030070

[93] Doello K et al. Latest in vitro and in vivo assay, clinical trials and patents in cancer treatment using curcumin: A literature review. Nutrition and Cancer. 2018;**70**(4):569-578. DOI: 10.1080/01635581.2018.1464347

[94] Akhondzadeh S et al. Melissa officinalis extract in the treatment of patients with mild to moderate Alzheimer's disease: A double blind, randomised, placebo controlled trial. Journal of Neurology, Neurosurgery, and Psychiatry. 2003;**74**(7):863-866. DOI: 10.1136/JNNP.74.7.863

[95] Tapsell LC, Hemphill I, Cobiac L, Patch CS, Sullivan DR, Fenech M, et al. Health benefits of herbs and spices: The past, the present, the future. The Medical Journal of Australia. 2006;**185**(4):4-24

[96] Williams PG. Health benefits of herbs and spices: Public health. Medical Journal of Australia. 2006;**185**(4):17-18

[97] Peter KV. Handbook of Herbs and Spices. 2nd ed. Vol. 2. Cambridge, UK: Woodhead Publishing; 2012b

[98] Duke JA. CRC Handbook of Medicinal Spices. Boca Raton, USA: CRC Press; 2002

[99] Harborne JB. Food Phytochemicals for Cancer Prevention, Volume 1, Fruits and Vegetables. Volume 2, Teas, Spices and Herbsedited by Huang MT, Osawa T, Ho CP, Rosen RT, 1994. ACS Symposium Series 546 & 547. Phytochemistry. Washington: American Chemical Society; 1995;**38**(2):559-559. DOI: 10.1016/0031-9422(95)90059-4

[100] Kaefer CM, Milner JA. The role of herbs and spices in cancer prevention. Journal of Nutritional Biochemistry. 2008;**19**(6):347-361

[101] Ferrucci LM, Daniel CR, Kapur K, Chadha P, Shetty H, Graubard BI, et al. Measurement of spices and seasonings in India: Opportunities for cancer epidemiology and prevention. Asian Pacific Journal of Cancer Prevention. 2010;**11**(6):1621-1629

[102] Chopan M, Littenberg B. The association of hot red chili pepper consumption and mortality: A large population-based cohort study. PLoS One. 2017;**12**(1):e0169876

[103] Esmaillzadeh A, Keshteli AH, Hajishafiee M, Feizi A, Feinle-Bisset C, Adibi P. Consumption of spicy foods and the prevalence of irritable bowel syndrome. World Journal of Gastroenterology. 2013;**19**(38):6465-6471

[104] Zweifel C, Stephan R. Spices and herbs as source of salmonella-related foodborne diseases. Food Research International. 2012;**45**(2):765-769

[105] EFSA. The European Union summary report on trends and sources of zoonoses, zoonotic agents and food-borne outbreaks in 2015. EFSA Journal. 2016;**14**(12):4634

[106] RASFF. Rapid Alert System for Food and Feed. Brussels, Belgium: European Commission; 2017. Available

from: https://webgate.ec.europa.eu/rasff-window/portal/

[107] Székács A, Wilkinson MG, Appel B. Environmental and food safety of spices and herbs along global food chains. Food Control. 2018;**83**:1-6. DOI: 10.1016/j. foodcont.2017.06.033

[108] Pathirana R. Peppers: Vegetable and Spice Capsicums. In: Bosland PW, Votava EJ, editors. 2nd edition. Vol. 41. New Zealand Journal of Crop and Horticultural Science. 2013. pp. 2102-2103. DOI: 10.1080/01140671.2012.745161

[109] Klátyik Sz, Darvas B, Mörtl M, Ottucsák M, Takács E, Bánáti H. et al. Food safety aspects of pesticide residues in spice paprika. International Journal of Biological, Biomolecular, Agricultural, Food and Biotechological Engineering. 2016;**10**(3):156-159

[110] Klátyik SZ, Darvas B, Oláh M, Mörtl M, Takács E, Székács A. Pesticide residues in spice paprika and their effects on environmental and food safety. Journal of Food and Nutrition Research. 2017;**56**(3):201-2018

[111] Sagoo SK, Little CL, Greenwood M, Mithani V, Grant KA, McLauchlin J, et al. Assessment of the microbiological safety of dried spices and herbs from production and retail premises in the United Kingdom. Food Microbiology. 2009;**26**(1):39-43

[112] Gurtler JB, Doyle MP, Kornacki JL, editors. The Microbiological safety of spices and low-water activity foods: Correcting Historic Misassumptions. New York, USA: Springer; 2014:3-13. DOI: 10.1007/978-1-4939-2062-4_1

[113] EFSA. Scientific opinion on the risk posed by pathogens in food of non-animal origin. Part 1. Outbreak data analysis and risk ranking of food/pathogen combinations. EFSA Journal. 2013;**11**(1):3025

[114] USFDA. Draft Risk Profile: Pathogens and Filth in Spices. Washington DC, USA: US FDA (US Food and Drug Administration); 2013

[115] Chakraverty A, Mujumdar AS, Vijaya Raghavan GS, Ramaswamy HS. Handbook of Postharvest Technology. Cereals, Fruits, Vegetables, Tea, and Spices. New York, USA: Marcel Dekker; 2003

[116] Kónya É, Szabó E, Bata-Vidács I, Deák T, Ottucsák M, Adányi N, et al. Quality management in spice paprika production as a synergy of internal and external quality measures. International Journal of Biological, Biomolecular, Agricultural, Food and Biotechological Engineering. 2016;**10**(3):160-166

[117] Mytle N, Anderson GL, Doyle MP, Smith MA. Antimicrobial activity of clove (*Syugium aromaticum*) oil in inhibiting listeria monocytogenes on chicken frankfurters. Food Control. 2006;**17**(2):102-107

[118] Moore-Neibel K, Gerber C, Patel J, Friedman M, Jaroni D, Ravishankar S. Antimicrobial activity of oregano oil against antibiotic-resistant salmonella enterica on organic leafy greens at varying exposure times and storage temperatures. Food Microbiology. 2013;**34**(1):123-129

Section 2

# Healthy Lifestyle

Chapter 5

# Herbs and Spices-Based Value Addition for Nutritional and Healthy Living

*Ogu Eneojo and Emeje Martins*

## Abstract

The demand for herbs and spices has been rising recently due to their application or use in human diet and medicine. The use of herbs and spices, their bi-products and value-added products as sources of food, medicine, wealth and great power has historical importance. Despite being inexpensive commodities, their value cannot be underestimated in the health and food industries, serving as therapeutic agents, fortifying food, adding flavour and colour, and acting as preservatives. Due to the abundance of nutritional and health benefits offered by herbs and spices, it is imperative to ensure a sustainable supply through value addition. This process involves elevating the raw form or creating added value through processing, packaging, grading, branding and marketing. The advantages of value addition in herbs and spices include portability, reduced bacteria contamination, prolonged flavour, affordability and availability, among others. Forms of value addition in herbs and spices include powder, tea, juice, chocolate and sweets, essential oils, oleoresin functional foods, etc.

**Keywords:** natural, antioxidants, bioactive, free radicals, food

## 1. Introduction

Recently, there has been a growing demand for herbs and spices because of their widespread use in human diets and medicinal treatments. These natural ingredients have held historical significance in providing food, medicine and wealth with great power. As far back as 1555 BCE, thyme, juniper, garlic, coriander, fennel and cumin have been reported to have been used for cooking and treating diseases [1]. Despite being inexpensive commodities, their value cannot be underestimated in the health and food industries, serving as therapeutic agents, fortifying food, adding flavour and colour, and acting as preservatives [2]. Worthy of note, is the application of herbs and spices in the production of functional foods. Functional foods provide not just the basic nutrients but are capable of meeting health requirements in terms of providing bioactive compounds which are required to maintain normal bodily functions, prevent deficiency, support health and prevent abnormalities and susceptibility to diseases. Irrespective of the numerous advantages presented by herbs and spices, care must be taken to avoid excess use as well as adulterated products, as such excess and adulteration may result in undesirable side effects on the consumers.

## 2. Herbs and spices

Different parts of plants produced herbs and spices. Herbs are obtained from the leaves of medicinal and aromatic plants while spices are obtained from parts of plants such as the flowers, seeds, buds, rhizomes, stigma, stem, bark, root, fruit, pods and bulb [2–4]. Herbs can be referred to as fresh or dried leaves of aromatic plants often used to add aroma, colour and flavour to foods [5]. Due to the presence of phytochemicals in herbs, their application is not only limited to culinary use as condiments. They are also consumed as essential oils and beverages in the form of herbal infusion or tea to boost the immune system or as medicine to cure ailments. Nigeria, because of its rich biodiversity, is blessed with varieties of herbs which finds application as condiments, beverages or essential oils. Among the herbs that are used as condiments are moringa (which is an excellent source of vitamins A, C and E, mineral compounds and amino acids), bitter leaf (which adds taste and flavour to food, and an excellent source of vitamins and mineral compounds), Aloe vera, scent leaf (which add taste and flavour to food), wild lettuce, water leaf, thyme (a fragrant herb used for food seasoning), curry leaves (also fragrant herbs for food seasoning), clove, basil (which its fresh leaves are used in a variety of ways in a different part of Nigeria. Some parts like Delta region use it as an ingredient for pottages and pepper soups, in Kwara as a sauce for egusi while among the Igala's in Kogi State, it serves as a vegetable), spring onions (like the common onion but lack a fully developed root bulb. It is a fragrant herb; the flavour is a cross between an onion and garlic but milder to taste. It is used as a vegetable, either raw or cooked), garlic chives (this have a slight garlic flavour and are delicious), roselle or sorrel (this herb is popular with the Hausas who use it to prepare different kinds of soups including groundnut soup) [6]. Herbs that find applications in beverages (tea and infusion) include eucalyptus, tea plant, lemon grass, moringa, basil, bitter leaf, coriander, parsley, ginger and roselle, among others. Those with essential oils include basil, fennel, lemon grass, marjoram, oregano, peppermint, rosemary, sage, eucalyptus, thyme, camomile, etc. Herbs can be consumed fresh where there is a need to preserve their highly valued aromatic attributes; however, the herbs trade is basically on dried herbs. Thus, they are used in the food industries as natural colourants, flavouring and preservatives [3].

Spices on the other hand are "aromatic vegetable substances, in the whole, broken, or ground form, whose significant function in food is seasoning rather than nutrition" and from which "no portion of any volatile oil or other flavouring principle has been removed" [3, 7]. They are regarded as natural plant products derived from whole or plant parts such as bud, flower, fruit, leaves, seed, root, stem, bark, berry, bulb, pod, latex and rhizome used to improve, alter or enhance the flavour, aroma, taste and colour of food products. Spices gotten from the different parts of the plant include bud (carper and clover), flower (marjoram and saffron), fruit (cardamom and chilli), leaf (curry leaves, bay leaves and mint), seed (coriander, anise, nutmeg, fennel, pepper, star anise, tamarind, cumin, cardamom, etc.) [8], root (horseradish and angelica), bark (cinnamon and cassia), bulb (onions and garlic), rhizome (ginger and turmeric), pod (tamarind and vanilla), berry (black pepper) and latex (hing spice) [3, 5, 9]. Spices like herbs also contain bioactive substances, which make them useful as nutritional and therapeutic agents. The introduction of spices in food has many beneficial effects such as stimulation of appetite, promotion of digestion, reduction of vomiting and nausea, prevention of cold and influenza, and change in the physical appearance and taste of food. Due to their preservative characteristics, spices found application in food industry, beverages, liquor and pharmaceuticals [7].

## 3. Value addition in herbs and spices

Due to the nutritional and health importance of herbs and spices, their all-year-round availability, affordability and accessibility are vital following the fact that some of these horticultural plants are seasonal. To ensure all-year-round availability, there is a need to create value addition in herbs and spices to ensure that humanity continues to derive the benefit attached to these natural products of importance.

Value addition refers to the process of transforming or changing a product from its raw or original form to a more valuable form or product. This process involves elevating the raw form or creating added value through processing, packaging, grading, branding and marketing. The advantages of value addition in herbs and spices include portability, reduced bacteria contamination, prolonged flavour, affordability and availability, among others. Value addition in herbs and spices can be in any of the following forms.

a. Powders: This involves pulverising the dried herbs and spices, sieving using a fine mesh to remove the chaff. Examples include rose, garlic, turmeric, thyme, curry, ginger, mint, basil, vanilla powders, etc. These powders are in turn used in the food and beverages industry as condiments and for producing functional food products.

b. Teas: dried herbs and spices can be pulverised to a specific mesh size and packed in tea sachets or bags. Examples of such tea include lemongrass base tea, eucalyptus base tea, etc.

c. Juice: herbs and spices can be transformed into juice and preserved to ensure their availability in the season. Leaves, seeds, fruits and flowers of some herbs and spices can be processed into juice such as zobo a local drink from rose consumed in Nigeria, garlic juice, ginger juice, turmeric juice, etc.

d. Chocolates and sweets: chocolate diets, sugar-free sweets, etc., can be gotten from herbs and spices like mint and rose.

e. Essential oils: herbs and spices contain essential oils which can be extracted from either the bark, leaf, stem, bud, etc., through steam distillation or hydro-distillation. Examples of such oil include ginger oil, garlic oil, eucalyptus oil, clove oil, cinnamon oil, etc.

f. Oleoresin: this is another value-added product from herbs and spices. The product is obtained through organic solvent extraction of buds or stems, which the solvent is then removed through evaporation to have concentrated products which contain all the flavouring ingredients soluble in the solvent used for the extraction, such that it is much closer to the original herbs and spices odour and flavour.

g. Functional foods: these are value-added food products which perform physiological functions beyond nutritional function. Examples include turmeric bread, diabetic flour, etc.

## 4. Nutritional benefits of herbs and spices value-added products

Herbs and spices used for food are found to be good sources of vital nutrients such as carbohydrates, protein and fat. They also play an important role as good source of micro and macro-nutrients in nutrition. The authors [10] noted that the results of some Nigerian herbs and spices analysed showed to be fairly rich in nutrients. Many herbs and spices contribute vital mineral such as calcium, iron, magnesium, phosphorus, potassium, sodium and zinc to nutrition. They also contain vitamins such as thiamine, riboflavin, niacin, pantothenic acid, folate and vitamin C. Examples of herbs and spices with the above nutrients include garlic and ginger; others like cinnamon are rich in fibres, clove contains vitamins A and C, cumin are rich in thiamine, vitamin E and niacin with manganese, African cardamon contains copper, amino acids, threonine, serine, valine, proline, glutamic acid, lysine, leucine and glycine while saffron are rich in selenium among others [7, 10].

### 4.1 Health benefits of herbs and spices value-added products

Herbs and spices are consumed as a whole or processed products in the form of powder, juice, oil and tea, among others. The presence of bioactive compounds in herbs and spices accounts for the strong connection between increased consumption of these natural products and human disease prevention. Nowadays, people are more concerned about herbs and spices due to their nutritive and therapeutic properties. Many herbs and spices are believed to possess therapeutic properties; thus, consuming a diet rich in herbs and spices will provide an environment of bioactive substances and non-nutritive compounds in plants that have health and protective benefits. Diets rich in herbs and spices such as liquorice, cabbage, soybeans and garlic are said to have the highest anticancer activity [7]. The addition of herbs and spice in food, even in a little quantity could boost the body immune system, fight against cancer, and decrease the risk of high blood pressure and cardiovascular disease. Herbs and spices used in the food industry for food seasoning, colouring, flavouring and preservation contain a variety of phytochemicals and phytonutrients which contribute to human health and well-being.

The need for a more streamlined approach to nutrition has been created due to changes arising from an increasing desire for improvement or maintenance of healthy living by intake of natural products in combination with changing lifestyle. With the rise in interest and awareness of healthy foods among countries around the globe, the value of herbs and spices as medicinal plants has increased among today's foods and has caused them to be among high-quality food products. Herbs and spices in food constitute a useful diet which has an impact on human health. The concept of medicinal foods entails food items that exercise a valuable effect on health and/or decrease the risk of chronic disease beyond just the dietetic functions. Herbs and spices like other medicinal plants, enter into the daily human food directly, such as turmeric, ginger, clove and garlic rich in mineral compounds and vitamins.

Many herbs and spices are used in the beverage industry for the preparation of daily drinks such as cocoa, coffee and tea [11]. Herbal teas have been in use in many countries of the world before drinks such as black tea and coffee started gaining wider acceptance today. Herbal teas are preferred because of their pleasing taste and their healing properties for some health challenges [12]. On this note, tea can be produced from many herbal materials such as sage, linden, mint, fennel, chamomile, coneflower, rosehip, apple, mountain tea, lemon balm, rosemary, cassia, thyme, nettle, tarragon, raspberry, basil and anise [13]. It is stated that herbal teas, whose composition can change with various factors, have antioxidant, anti-inflammatory, antimicrobial, anti-carcinogenic, anti-atherogenic, antiaging, cardio-protective and functional properties [14]. Herbal teas with this type of functional properties can be used to relieve problems such as psychosomatic diseases, colds and congestion, gastrointestinal diseases, urinary system diseases, diarrhoea, constipation, local use as a mouthwash or, as a taste and odour corrector, menstrual complaints and physical and mental fatigue [14, 15].

### 4.2 Application of herbs and spices in the food industries

Herbs and spices present vital opportunities to food industries following the fact that there is a growing concern for natural food products of plant origin, which provide not only nutritional benefits but also contribute to human health. Food industries are faced with the prevailing challenges of preventing foods from physical, chemical and microbial spoilage [4]. Currently, there is rising consumer dissatisfaction/rejection of food preserved with chemical/synthetic preservatives due to the possible negative health effects. Reports from studies [7] show that herbs and spices, either raw or processed (powder, essential oils, etc.), are natural products of plants. Applications in food have the tendency to extend its shelf life and improve aroma, flavour, colour and quality, thereby making it generally acceptable without side effects. Thus, herbs and spices find their application in the food industry as.

#### *4.2.1 Preservatives*

The quality of food and its acceptability for consumption can be affected or compromised as a result of spoilage. Thus, food spoilage due to lipid oxidation, microbial growth and pathogenic organisms constitutes major challenges in food industry, thereby making the food industry use synthetic antioxidants such as propyl gallate (PG), tert-butyl hydroquinone (TBHQ), butylated hydroxyanisole (BHA) and butylated hydroxytoluene to prevent spoilage [16, 17]. However, the safety of these synthetic antioxidants used as preservatives is in doubt. The literature revealed that there exists a relationship between the long-term intake of synthetic antioxidants and health-related issues, such as cases of increased risk of cancer, skin allergies, DNA damage, induced premature senescence, adverse effects and gastrointestinal tract [17]. The need for the replacement of these synthetic preservatives with natural ones by consumers is on the increase. Herbs and spices present potent tools for the food industry as preservatives due to the presence of bioactive compounds such as phenolic compounds, which have both antioxidant and antimicrobial properties.

Phenolic compounds constitute the most abundant groups of plant secondary metabolites, whose strong antioxidant properties are due to the possession of aromatic rings with a hydroxyl group. The antioxidant potential is due to the ability of the compounds to inhibit free radical formation as well as high radical scavenging activity in relation to the hydrogen-donating ability of phenolic compounds. Among the compound that exhibit antioxidant characteristics in herbs and spices are gallic acid, phenolic acid oleoresin, essential oils, eugenol, curcumin, gingerol, carvacrol, thymol, capsaicin, pimento, oleuropein, ligstroside, verbascosides, catechins, rosmarinic acid, caffeic acid, kaempferol etc. [18]. Among the most important herbs and spices with antioxidant possibilities are ginger, thyme, rosemary, sage, black pepper, clove, basil, red pepper, garlic, nutmeg which if added to food containing lipids has the ability to prevent oxidative rancidity by stabilising lipids in food products thereby, prevent food deterioration and extends the shelf life [19].

More so, herbs and spices not only possess antioxidant properties but also possess antimicrobial characteristics, which make them useful in the food industry as natural preservatives for extending the shelf life of foodstuffs due to their capacity to inhibit the growth of food spoilage microorganisms and pathogenic bacteria. The authors [20] conducted a study on the use of neem and garlic dried powders for controlling some stored grain pests and found that dried powders of neem and garlic were effective in the control of storage insect pests of grains such as beans, rice and maize. The application of these herbs and spice in prolonging the shelf life of these food stuff provides cheap, safe and eco-friendly alternatives for the treatment of stored food products and reduce financial wastage thereby increase the profit that will be generated by the food industry.

Among the herbs and spices-based antimicrobial for inhibiting microorganisms and pathogenic bacteria and for increasing the overall quality of food products are lemon grass, cloves, garlic, rosemary, cinnamon, coriander and sage [3]. The most common chemical compounds that provide antimicrobial activity to herbs and spices are phenolic acids, caffeic acid, romarinic acid, gingerol, cinnamaldehyde, eugenol, glucosinolates, isothiocyanates, quercetin, cuminaldehyde, β-caryophyllene, aspiperine, carnosic acid and coumarin [18].

#### *4.2.2 Flavouring*

There is a high demand for herbs and spices in the food industry due to their application in flavouring food. Herbs and spices are used as flavours either in whole or ground form or as extractives. Value-added herbs and spices in ground form or extractive release their flavours more readily because the cells containing the flavour have been broken during grinding. The presence of metabolites in herbs and spices enables them to impart flavour to food, thereby improving it characteristics such as fragrance, aroma and taste. Food industry needs ground herbs and spices in food processing in order to provide sensual appeal. The sensual qualities of each herbs and spices is as a result of the predominant chemical compounds it contains. These chemical compounds can contribute mild to strong flavours. The flavours are a result of certain families of chemicals such as phenol, monoterpenes and phenylpropanoids compound. The characteristics of the flavour profile are given by the balance of these chemical compounds [7, 10].

The flavouring potentials of herbs and spices are due to the presence of key volatile compounds such thymol and carvacrol (marjoram, thyme and oregano), α-pinene, rosmarinic acid, caryophyllene (rosemary), α-thujone, β-caryophyllene, viridiflorol (sage), safranal (saffron), aromatic-turmerone (turmeric), zingiberene, gingerol (ginger), anethole, estragole, fenchone (fennel), estragole, linalool (basil), cinnamaldehyde, cinnamyl alcohol, eugenol, coumarins and cinnamyl acetate (cinnamon), capsaicin (chilli peppers), myristicin, safrole, elemicin (nutmeg), sabinene (oregano), α-pinene (parsley) and santene (peppermint) [21, 22]. Depending on the region, different herbs and spices used for flavouring foods add distinguished flavour to each food style that even gives culinary identity. In Nigeria, unique flavours are added to food such as bread, cakes, zobo, kunu, etc., using herbs and spices like curry leaves, ginger, garlic, cloves and nutmegs.

#### *4.2.3 As colourant*

Certain herbs and spices such as turmeric, ginger and saffron can be utilised as colouring materials in some food. Value-added products of spices such as dehydrated or freeze-dried forms in canned or brine forms like white and green pepper are sources of natural colourant in food. Natural colourants are considered safer than synthetic colourants. The natural colourants are components of the plant's primary and secondary metabolites. The primary and secondary metabolites in herbs and spices responsible for the colouring of foods include chlorophylls (tea plants, fresh spinach, turkey berries and winter-sweet), carotenoids which are a derivative of tetraterpenes and are responsible for the yellow, red and orange colour (chilli, parsley, coriander, carrot, marjoram, sage, turmeric, pumpkin and ripe red bell peppers) and anthocyanins a derivative of flavonoids (blueberry, hibiscus, blackberry and blue barberry) [23]. Various processed products utilises spices such as dried chilli, red pepper oleoresin and paprika oleoresin.

#### *4.2.4 Food supplement/functional foods*

Another important area, where herbs and spices find its use in the food industry, is for the manufacturing of food supplement products/functional food. For any food product to be considered a "food supplement," it must contain essential nutrients such as vitamins, mineral compounds, amino acids and one or more chemical substances in the plant, which must be purified. Products that have been standardised in this way and turned into tablets, capsules or syrups are considered food supplements [13]. Considering herbs and spices as supplements requires focusing on their function in the diet rather than their use as medicines.

Similarly, examining herbs and spices from the perspective of functional food involves the consideration of how herbs and spices are used in the diet rather than their role. Thus, for any food to be considered functional food, according to [13], it has to provide benefits beyond basic nutrition. This implies that besides the vitamins and minerals (nutrients) required for the maintenance of normal bodily functions, the food components have the ability to actively interact with the body to prevent abnormality and overt disease, thereby supporting healthy livelihood.

From the perspectives of function and use, herbs and spices are found to occupy an excellent position as raw materials in the food industry for the production of food supplements as well as a functional food in that they possess both nutritional and medicinal attributes required for the maintenance of normal bodily functions as well as boasting the human immunity, supply natural antioxidant which has the ability to scavenge free radicals and other reactive oxygen species capable of causing degenerative diseases thereby ensures healthy living.

#### *4.2.5 Edible coating or films for food packaging*

Herbs and spices also find applications in the food industry for the production of protective active barriers. The protective active barriers are applied directly on food product's surface as edible coating for food packaging purposes. The detrimental impacts of oxygen on food can be resolved using this new approach known as edible coating or active barriers [17]. Apart from the antimicrobial activity, most herbs and spices contain natural antioxidants, which serve as active barriers designed to extend food products' shelf lives as they are effective against oxidation reactions such as oxidative rancidity or enzymatic browning but also against the proliferation of microbes. The antioxidant effect of the bioactive compounds in herbs and spices is primarily due to their redox properties and is the effects of various practical chemical processes such as metal chelating conversion activity, singlet oxygen quenching capacity and free radical scavenging activity [24]. The incorporation of edible films and coatings directly into bulk food presents important merits over the artificial antimicrobial or antioxidant agents, “the possibility of active compound diffusion control at the surface of the food and the reduction of the amount of preservatives added in the food” [17]. The use of natural biopolymers for the production of these active barriers is gaining increasing attention as an alternative to non-biodegradable, synthetic plastic packaging.

The potential of extracts from agricultural by-products incorporated as edible films and coating in packaging materials in the food industry has shown promising results. The authors [17] reported that chitosan films prepared with ethanolic mango leaf extracts have antioxidant properties. The report revealed that 5% mango leaf extract film, when applied as packaging materials for cashew nuts storage, reduced fatty acid oxidation by 56% compared to synthetic commercial polyamide/polyethylene films. The application of edible coating has the tendency to reduce water vapour and gas exchange, especially in fruits and vegetables with the aim of increasing their shelf lives and ensuring that their firmness levels, sensory properties, antioxidant activities and colours are maintained. The use of edible coating helps to delay fruit ripening rate, control weight loss and reduce loss of ascorbic acid, polyphenols and titratable acidity.

### 4.3 Application of herbs and spices in human health

The use of herbs and spices by human for the treatment/management and prevention of ailment is as ancient as man himself. Various parts of plants such as leaves, roots, bark, seeds and the whole plants have been used significantly in the indigenous healing arts following the fact that they exhibit medicinal properties. The presence of phytoconstituents in herbs and spices explained the strong association between the increased consumption of these natural products and human disease prevention. The health effects especially the suppression of active oxygen species of phytoconstituents in some herbs and spices have been studied [10]. Herbs and spices can offer protection from peroxidative damage owing to the presence of dietary antioxidants.

Dietary antioxidants are found to have several health benefits, among which is the maintenance of balance between the activity of free highly reactive compounds known as reactive oxygen species (ROS) and antioxidant potentials by regulating oxidative stress. Oxidative stress occurs as an imbalance between the produced reactive oxygen species and defence systems, thus resulting in oxidative damage. Oxidative damage and free radicals constitute a significant cause of human disease [25]. There has been a correlation between oxidative damage and diseases such as arthritis, atherosclerosis, ocular diseases, diabetes and cancer.

Reactive oxygen species and other free radicals, which are responsible for several diseases, are products of metabolic processes in the human body as well as a result of exposure to industrial chemicals, ozone, radiation, air pollution, etc. [25, 26]. For healthy livelihood, a Recommended Dietary Allowance (RDA) for specific antioxidants is required. Herbs and spices constitute an excellent source of natural antioxidants, and their fresh consumption in the diet contributes to the daily antioxidant intake. Examples of natural dietary antioxidants includes lipoic acid, vitamin C, vitamin E and beta-carotene found in rosemary, clove, garlic and ginger whose biological function is to protect against oxidative damage, hence preventing neurological diseases, delaying chronic health problems such as cataracts and glaucoma, carcinogenetic and cardiovascular diseases [27]. Through direct scavenging, antioxidants can inhibit reactive oxygen species production, limit the propagation of oxidants, and decrease the amount of oxidants within and around the cells, thereby preventing reactive oxygen species from reaching their target organs. Antioxidants perform other biological roles such as stabilising or deactivating free radicals before attacking cells.

Besides the consumption of herbs and spices in the diet, they are used as ingredients in local medicine for the treatments of diseases such as tonsils, inflammation of throat, tapeworms, post-partum pain, fever, dysentery, gastrointestinal troubles and as carminative such as Alligator pepper or grains of paradise. Value-added products of clove like clove oil are used during surgery as an antiseptic, nutmeg for cough and black pepper for promoting healing in small wounds. Spices like garlic have been established to possess diuretic and antipyretic properties [10].

## 5. Conclusion

Herbs and spices possess both nutritional and therapeutic properties. The nutritional properties are due to the presence of vitamins and trace mineral compounds, whereas the therapeutic properties are due to the presence of phytochemicals or bioactive compounds in plants. They are commonly used in the food industry for the production of functional foods, food preservation, flavour addition, natural colourant as well as edible coating or films for food packaging. While in health, they are used in the treatment of some ailments and their consumption as condiments in food helps to build or boost human immunity against several diseases. Applying herbs and spices in food provides both nutritional and therapeutic benefits, thereby contributing to healthy living. Despite the numerous benefits presented by herbs and spices, care must be taken in their application in food as excess and/or irregular use or application is capable of producing negative side effects such as heartburn, abdominal spasms, vomiting, stomach pain and indigestion, among others.

## Author details

Ogu Eneojo* and Emeje Martins
Nigeria Natural Medicine Development Agency (NNMDA), Victoria Island, Lagos, Nigeria

*Address all correspondence to: eneojo.ogu@nnmda.gov.ng

## References

[1] Tapsell LC, Hemphill I, Cobiac L, Sullivan FP. Health benefits of herbs and spices: The past, the present, the Furture. The Medical Journal of Australia. 2006;**185**(4):s1-s24

[2] El-Sayed SM, Youssef AM. Potential application of herbs and spices and their effects in functional dairy products. Heliyon [Internet]. 2019;**5**(6):e01989. DOI: 10.1016/j.heliyon.2019.e01989

[3] Herrera T, Iriondo-DeHond A, Uribarri J, del Castillo MD. Beneficial herbs and spices. In: Uribarri J, Vassalotti J, editors. Nutrition, Fitness, and Mindfulness. Nutrition and Health. 1st ed. Chapter 6. Cham: Humana; 2020. pp. 65-85. DOI: 10.1007/978-3-030-30892-6

[4] Leykun ST. Systematic review on spices and herbs used in food industry. American Journal of Ethnomedicine [Internet]. 2020;7(5):20 Available from: http://www.imedpub.com/ethnomedicine/

[5] Madhusoodanan KJ, Rao YS. Cardamom (large). In: Peter KV, editor. Handbook of Herbs and Spices. 1st ed. Vol. 1, No. 1. Chapter 10. Oxford, UK: Woodhead; 2001. pp. 134-142

[6] Collins W. 10 Common Nigerian Herbs and their Health Benefits. [Internet]. 2020. Available from: https://www.healthguide.ng/nigerian-herbs-their-health-benefits/ [Accessed: November 15, 2023]

[7] Sunday RJ. National Open University of Nigeria Advanced Macroeconmics. Abuja: NOUN; 2018. 2-3 pp

[8] United States Department of Agriculture (USDA). Spices and Herbs. Available from: http://www.fs.usda.gov

[9] Das C, Sahana N, Mandal S. Spices Based Value Added Products for Livelihood Security. Patna: ICAR; 2021. 1-6 pp

[10] Fasoyiro S. The Value of Spices; Uses, Nutritional and Health Benefits. [Internet]. 2015. Available from: https://www.researchgate.net/publication/282353976_The_Value_of_Spices_Uses_Nutritional_and_Health_Benefits [Accessed: November 16, 2023]

[11] Samiha HA, Rasha AA, Khuzama IB, Noura SA, Mai AA. A review of medicinal plants, their definition, uses, active ingredients and prevalence in the Kingdom of Saudi Arabia. International Journal of Science and Research (IJSR). 2022;**11**(12):1183-1188

[12] Igwillo UC, Ola-Adedoyin AT, Abdullahi MM, Chukwuemeka AE. A review of opportunities and challenges in conservation and use of medicinal and aromatic plants in Nigeria. International Journal of Advanced Research. 2019;7(4):770-778

[13] Tuğba G. D 2023 use of medicinal and aromatic plants in food. Eurasian Clinical and Analytical Medicine. 2023;**11**(1):6-10

[14] Gölükcü M, Toker R, Tokgöz H. Farklı Sıcaklık ve Sürelerde Demlemenin Dağ Çayının (Sideritis congesta) Bazı Kalite Özellikleri Üzerine Etkisi (the effect of brewing at different temperatures and times on some quality traits of mountain tea (Sideritis congesta)). GIDA. 2014;**39**(3):155-162

[15] Zhao J, Deng JW, Chen YW, Lİ SP. Advanced phytochemicals analysis of herbal tea in China. Journal of Chromatography A. 2013;**1323**:2-23

[16] Gottardi D, Bukvicki D, Prasad S, Tyagi AK. Beneficial effects of spices in food preservation and safety. Frontiers in Microbiology. 2016;7(SEP):1-20

[17] Lourenço SC, Moldão-Martins M, Alves VD. Antioxidants of natural plant origins: From sources to food industry applications. Molecules. 2019;**24**(22):14-16

[18] Jessica Elizabeth DLT, Gassara F, Kouassi AP, Brar SK, Belkacemi K. Spice use in food: Properties and benefits. Critical Reviews in Food Science and Nutrition. 2017;**57**(6):1078-1088

[19] Leja KB, Czaczyk K. The industrial potential of herbs and spices - a mini review. Acta Scientiarum Polonorum. Technologia Alimentaria. 2016;**15**(4):353-365

[20] Onu FM, Ogu E, Ikehi ME. Use of neem and garlic dried plant powders for controlling some stored grains pests. Egyptian Journal Of Biological Pest Control. 2015;**25**(2):507-512

[21] Vázquez-Fresno R, Rosana ARR, Sajed T, Onookome-Okome T, Wishart NA, Wishart DS. Herbs and spices- biomarkers of intake based on human intervention studies - a systematic review. Genes & Nutrition. 2019;**14**(1):1-27

[22] Pinto T, Aires A, Cosme F, Bacelar E, Morais MC, Oliveira I, et al. Bioactive (poly)phenols, volatile compounds from vegetables, medicinal and aromatic plants. Food. 2021;**10**(1):106-135

[23] Molina AK, Corr CG, Prieto MA, Pereira C. Bioactive natural pigments' extraction, isolation, and stability in food applications. Molecules. 2023;**28**(3):1200

[24] Senay TL. Systematic review on spices and herbs used in food industry. American Journal of Ethnomedicine. 2020;7(5):1-10

[25] Kwaśniewska M, Pikala M, Grygorczuk O, Waśkiewicz A, Stepaniak U, Pająk A, et al. Dietary antioxidants, quality of nutrition and cardiovascular characteristics among omnivores, flexitarians and vegetarians in Poland—The results of Multicenter National Representative Survey WOBASZ. Antioxidants. 2023;**12**(2):222

[26] Hajhashemi V, Vaseghi G, Pourfarzam M, Abdollahi A. Are antioxidants helpful for disease prevention? Research in Pharmaceutical Sciences. 2010;**5**(1):5-12

[27] Dolas AS, Gotmare SR. The health benefits and risks of antioxidants. An International Research Journal. 2015;**6**(1):25-30

Chapter 6

# From Plant to Patient: The Metabolic Benefits of Phytoestrogens

*Manish Kumar Verma, Brijesh Kumar Singh and Madhulika Tripathi*

## Abstract

In recent years, there has been a growing interest in the potential health benefits of phytoestrogens, a diverse group of naturally occurring compounds found in plants. These compounds, with structural similarities to estrogen, are found in foods such as soybeans, flaxseeds and sesame seeds, and are studied for their roles in modulating hormonal activity and antioxidant properties. Their estrogen-like functions make them particularly interesting in managing conditions exacerbated by hormonal imbalances, such as type 2 diabetes, obesity, cardiovascular diseases, and osteoporosis. Phytoestrogens enhance insulin sensitivity and glucose metabolism, while their anti-oxidant properties neutralize free radicals and inhibit lipid peroxidation. They also influence fat metabolism and reduce adipogenesis, addressing oxidative stress and inflammation. Their estrogenic effects improve bone mineral density and reduce fracture risks, particularly in postmenopausal women, highlighting their broad potential in metabolic disease management. This growing body of research suggests that phytoestrogens, due to their unique properties and diverse mechanisms of action, could be a valuable addition to dietary and therapeutic strategies in the prevention and management of various metabolic diseases. This chapter delves into the multifaceted roles of phytoestrogens in metabolic disorders, with a special focus on their antioxidant properties.

**Keywords:** antioxidant potential, autophagy, mitophagy, cytotoxicity, natural compound

## 1. Introduction

The potential health benefits of phytoestrogens have grown exponentially. These naturally occurring compounds, abundant in plants (**Table 1**) like soybeans (*Glycine max* L.) rich in isoflavones such as genistein and daidzein, and flaxseeds (*Linum usitatissimum* L.) abundant in lignans, like secoisolariciresinol diglucoside, share structural similarities with the hormone estrogen [1]. Moreover, red clover's (*Trifolium pratense* L.) isoflavones, including formononetin and biochanin A, black cohosh's (*Actaea racemosa* L.) triterpene glycosides, like cimicifugoside and actein, these phytoestrogens have garnered attention for their potential therapeutic applications [2] and have spurred extensive research into their effects on various metabolic diseases.

| **Plant's common name (scientific name)** | **Compound's nature (compound names)** |
|---|---|
| Alfalfa (*Medicago sativa* L.): | Coumestans (e.g., medicagol) |
| Apricots (*Prunus armeniaca* L.): | Coumarins (e.g., bergapten) |
| Barley (*Hordeum vulgare* L.L.): | Lignans (e.g., matairesinol) |
| Black Cohosh (*Actaea racemosa* L.): | Triterpene glycosides (e.g., cimicifugoside and actein) |
| Buckwheat (*Fagopyrum esculentum Moench*): | Lignans (e.g., secoisolariciresinol) |
| Cabbage (*Brassica oleracea* var. *acephala* var. *capitata* L.): | Isoflavones (e.g., coumestrol) |
| Chickpeas (*Cicer arietinum* L.): | Isoflavones (e.g., genistein and daidzein) |
| Coffee (*Coffea arabica* L.): | Lignans (e.g., pinoresinol) |
| Dong Quai (*Angelica sinensis (*Oliv.*) Diels.*): | Coumarins (e.g., ligustilide) |
| Fennel (*Foeniculum vulgare Mill.*): | Coumarins (e.g., bergapten) |
| Flaxseeds (*Linum usitatissimum* L.): | Lignans (e.g., secoisolariciresinol diglucoside, and matairesinol) |
| Garbanzo Beans (*Cicer arietinum* L.): | Isoflavones (e.g., kaempferol) |
| Green Beans (*Phaseolus vulgaris* L.): | Isoflavones (e.g., pratensein) |
| Green Tea (*Camellia sinensis (*L.*) Kuntze*): | Isoflavones (e.g., catechins) |
| Hops (*Humulus lupulus* L.): | Flavonoids (e.g., xanthohumol) |
| Kale (*Brassica oleracea* var. *acephala*): | Isoflavones (e.g., coumestrol) |
| Lentils (*Lens culinaris Medik.*): | Isoflavones (e.g., biochanin A) |
| Licorice Root (*Glycyrrhiza glabra* L.): | Isoflavans (e.g., glabridin) |
| Lima Beans (*Phaseolus lunatus* L.): | Isoflavones (e.g., diosmetin) |
| Lupin Beans (*Lupinus* spp.): | Isoflavones (e.g., lupinifolin) |
| Mung Beans (*Vigna radiata (*L.*) R. Wilczek*): | Isoflavones (e.g., formononetin) |
| Oats (*Avena sativa* L.): | Lignans (e.g., enterolactone) |
| Pueraria (Kudzu) (*Pueraria lobata (*Willd.*) Ohwi*): | Isoflavones (e.g., puerarin and daidzein) |
| Pumpkin (*Cucurbita pepo*): | Lignans (e.g., lariciresinol and medioresinol) |
| Red Clover (*Trifolium pratense* L.): | Isoflavones (e.g., formononetin and biochanin A) |
| Red Wine Grape (*Vitis vinifera* L.): | Lignans (e.g., pinoresinol) |
| Sage (*Salvia officinalis* L.): | Flavonoids (e.g., apigenin and luteolin) |
| Sesame Seeds (*Sesamum indicum* L.): | Lignans (e.g., sesamin and sesamolin) |
| Sesbania Flower (*Sesbania grandiflora (*L.*) Poir.*): | Isoflavones (e.g., genistein and daidzein) |
| Soybeans (*Glycine max* L.): | Isoflavones (e.g., genistein, daidzein, and glycitein) |
| Sunflower Seeds (*Helianthus annuus* L.): | Lignans (e.g., lariciresinol) |

**Table 1.**
*Common plants rich in phytoestrogens along with their corresponding compounds.*

In this chapter, we explore the complex web of interactions orchestrated by phytoestrogens at an organismal level during metabolic diseases [3, 4]. Beyond their estrogenic properties, which hold the key to their potential therapeutic applications, we place a special emphasis on their often overlooked yet equally significant role as potent antioxidants [5, 6]. These dual capabilities of phytoestrogens not only show promise in addressing metabolic disorders but also hint at a broader potential for promoting overall health and well-being. Throughout this exploration, we uncover the possible mechanisms by which phytoestrogens interact with cellular receptors, shedding light on their potential to modulate hormonal responses. Additionally, we scrutinize accumulating evidence that supports their efficacy in mitigating metabolic diseases, ranging from type 2 diabetes to obesity and cardiovascular conditions. Alongside this, we illuminate their crucial role in buffering against oxidative stress, a cornerstone of many chronic health challenges. This chapter also aims to unravel the multifaceted roles of phytoestrogens, positioning them not merely as dietary components but as potential allies in the pursuit of metabolic health and longevity. By doing so, we hope to foster a deeper understanding of the profound impact that these natural compounds can have on our well-being and inspire further research into their therapeutic applications.

These plants (**Table 1**) contain varying concentrations of phytoestrogens, and their effects on the body can differ depending on factors such as dosage, individual physiology, and overall diet. It is important to note that while these compounds can exert estrogenic effects, they are much milder than the natural hormone estrogen, and their impact may vary from person to person.

## 2. Phytoestrogens: a brief overview

Phytoestrogens, a group of naturally occurring compounds derived from plants, have garnered significant interest for their potential health benefits and improvements in overall metabolism [6, 7]. These compounds share structural similarities with the hormone estrogen (**Figure 1**) and have been the subject of extensive research regarding their impact on various aspects of human health [8, 9].

Classified into three main categories—isoflavones, lignans, and coumestans—phytoestrogens are widely distributed in numerous plant-based foods [10]. These include soybeans, flaxseeds, legumes, and an array of fruits and vegetables. The diversity of their sources highlights the potential for incorporating phytoestrogens into a balanced diet (**Table 1**).

### 2.1 Classification, sources, and bioavailability of phytoestrogens

Phytoestrogens belong to the group of polyphenolic compounds and can be categorized into three main classes: isoflavones, lignans, and coumestans [9]. Prominent dietary sources of phytoestrogens include soybeans, flaxseeds, legumes, and various fruits and vegetables [6]. Phytoestrogens have potential health benefits due to their unique properties [7]. The most prominent phytoestrogens are as follows.

#### *2.1.1 Isoflavones*

These are perhaps the most well-known phytoestrogens and are commonly found in soy-based products such as soybeans and tofu. Isoflavones are characterized by their distinct arrangement of phenolic rings. Prominent isoflavones include genistein, daidzein, and glycitein. In addition to soy, isoflavones are present in various legumes

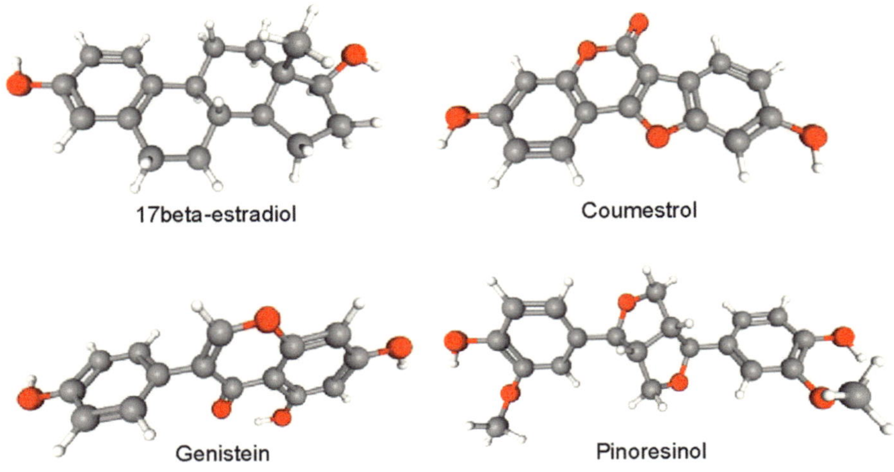


**Figure 1.**
*Structural similarity between naturally produced estrogen hormone (17beta-estradiol, E2) and phytoestrogens coumesterol (coumestans) genistein (isoflavones), and pinoresinol (lignan). The structures are from PubChem compounds (https://pubchem.ncbi.nlm.nih.gov).*

and some fruits. Isoflavones are known for their potent estrogenic effects and numerous health benefits.

While isoflavones have health benefits, they may also have some potential side effects for certain individuals. These side effects include hormonal disruptions, gastrointestinal discomfort, rare allergic reactions, possible thyroid function interference, and reproductive effects, especially in supplement form. Isoflavones may interact with medications (levothyroxine, tamoxifen, anticoagulants, etc.), affecting hormone levels and raising concerns about breast tissue effects, although conclusive evidence is lacking. Consuming isoflavones in moderation within a balanced diet is recommended, as they offer potential health advantages, like reducing the risk of specific chronic diseases.

### *2.1.2 Lignans*

Lignans, another class of phytoestrogens, are abundant in seeds and grains, with flaxseeds being one of the richest sources. Secoisolariciresinol diglucoside and matairesinol are notable lignans. Beyond flaxseeds, lignans are found in sesame seeds, whole grains, and certain vegetables. Lignans are prominent polyphenols recognized for their complex chemical structures, which often comprise multiple phenolic rings. Lignans exhibit remarkable antioxidant properties and have garnered attention for their potential impact on human health.

Lignans are generally considered safe and beneficial for health. However, while they exhibit mild estrogenic properties, the side effects of lignans are typically minimal and less common compared to other phytoestrogens. Some individuals may experience temporary gastrointestinal discomfort like gas or bloating when consuming large amounts of lignan-rich foods, and rare allergic reactions are possible. Lignans may also interact with medications (antiplatelet and anticoagulant, Synthroid, etc.), necessitating consultation with a healthcare provider if one is on such medications. Overall, lignans are known for their antioxidant properties and potential contributions to heart health and disease prevention.

*DOI: http://dx.doi.org/10.5772/intechopen.1004342*

#### *2.1.3 Coumestans*

While less studied than isoflavones and lignans, coumestans are present in various plants, including clover, alfalfa, and bean sprouts. Formononetin and coumestrol are representative coumestans. These compounds contribute to the phytoestrogen content of leguminous plants. However, less studied compared to isoflavones and lignans, coumestans like coumestrol have shown promise in research for their potential health benefits.

While there is less research on their side effects than isoflavones and lignans, excessive consumption of coumestan-rich foods may lead to mild gastrointestinal discomfort. However, this is less common than other phytoestrogens, and rare allergic reactions are possible. Coumestans may also interact with medications affecting hormone levels, especially during hormone replacement therapy, necessitating consultation with healthcare providers for those on such medications. Further, coumestans with antioxidant supplements, such as vitamin C or vitamin E, may have a combinatorial effect. Despite limited research, coumestans like coumestrol have shown promise for their antioxidant properties and potential health contributions.

#### *2.1.4 Daily recommended dosage of phytoestrogens*

It is important to note that there is no specific recommended daily allowance (RDA) for phytoestrogens and intake recommendations may vary among health organizations and countries. Additionally, individual tolerance and response to isoflavones can vary. Some people may consume higher amounts without adverse effects, while others may be more sensitive.

The recommended daily dose of isoflavones and lignans can vary depending on factors such as age, gender, health status, and individual dietary preferences. Isoflavones and lignans are typically consumed through dietary sources rather than supplements, and their intake can vary widely among different populations, such as one to two servings of soy-based foods (tofu, soy milk, edamame, or tempeh), legumes, ground flaxseeds, or sesame seeds per day can provide a reasonable amount of isoflavones. A typical serving size is around 3 to 4 ounces (85 to 113 grams). Coumestans are generally not consumed in isolation, and they are not commonly available as supplements. Clover and Alfalfa Sprouts can be added to salads and sandwiches, while beans and legumes can be consumed to fulfill daily requirements for coumestans.

#### *2.1.5 Improving digestibility and bioavailability of phytoestrogens*

Phytoestrogens, lignans in particular, are recognized for their antioxidant properties, which may help protect cells from oxidative damage caused by free radicals. While they are strong antioxidants *in vitro*, the extent of their antioxidant activity in the human body can vary depending on factors such as food preparation and digestion. The bioavailability of phytoestrogens can be limited because they are bound to dietary fiber and need to be released and absorbed in the digestive tract. In this regard, various strategies can be utilized to increase their bioavailability, such as grinding flaxseeds before consumption can break down the seed's protective outer layer, making the lignans more accessible for absorption. Sprouting certain seeds, grains, or legumes can increase the bioavailability of phytoestrogens, including lignans. Consuming lignan-rich foods with sources of healthy fats, such as nuts, seeds, or avocados, may enhance the absorption of fat-soluble lignans. Fermentation can also increase the availability of phytoestrogens.

Beyond the aforementioned examples, various fruits, vegetables, nuts, and seeds also contribute to our dietary intake of phytoestrogens. This diverse range of sources highlights the prevalence of these compounds in our diets and focuses on the need for further research to understand their potential impact on human health. Next, we will explore the distinct properties and biological activities of these phytoestrogen classes, shedding light on their role in metabolic diseases and their antioxidant capabilities.

### 2.2 Estrogenic activity

Phytoestrogens closely resemble the structure of estrogen (**Figure 1**) and are currently a major focus of research in the field of metabolic health. These natural substances, which are present in a wide range of plants, interact with the body's estrogen receptors, thereby influencing hormonal functions and physiological reactions. This characteristic forms the basis of their potential utility in managing metabolic diseases. Isoflavones, notably genistein and daidzein found in soybeans, are prime examples of phytoestrogens with significant estrogenic activity [2, 8]. Their ability to bind to estrogen receptors facilitates cellular responses similar to those triggered by natural estrogen. This interaction holds implications for various health aspects, ranging from bone density modulation to cholesterol level regulation [11].

Lignans, particularly abundant in flaxseeds, such as secoisolariciresinol and matairesinol, also exhibit estrogenic potential. Their interaction with estrogen receptors can mimic estrogen's actions, contributing to hormonal balance and potentially offering protection against conditions like osteoporosis [12]. Red clover, another key source, contains isoflavones like formononetin and biochanin A, which exhibit estrogenic effects. Additionally, the presence of coumestrol, a coumestan found in various clover species, though less studied, indicates its potential role in hormonal health due to its affinity for estrogen receptors [13]. Hops, particularly rich in the flavonoid xanthohumol, expand the diversity of phytoestrogens in the diet [14]. These compounds interact with estrogen receptors, adding to the overall estrogenic activity observed in phytoestrogen-rich foods.

The unique interaction of phytoestrogens with estrogen receptors has sparked considerable scientific interest, especially regarding their potential therapeutic applications in metabolic diseases [3, 4, 11]. Research has been exploring how these compounds, through their estrogenic activity, may enhance insulin sensitivity, regulate glucose metabolism, and affect adipose tissue function [15]. Their role in cardiovascular health, particularly in regulating lipid profiles, blood pressure, and endothelial function, is also under investigation [16]. Moreover, the estrogen-like effects of phytoestrogens on bone health offer potential benefits in preventing osteoporosis [17].

The complex interplay between phytoestrogens and estrogen receptors demonstrates a delicate balance within the body's endocrine system. This relationship not only underlines the potential therapeutic applications of phytoestrogens but also highlights the importance of dietary sources rich in these compounds. Foods like soybeans, flaxseeds, clover, and hops are not just nutritionally valuable but also offer potential health benefits due to their phytoestrogen content. As research progresses, our understanding of the multifaceted effects of phytoestrogens is expected to deepen. This knowledge will be crucial for developing innovative strategies to leverage phytoestrogens in promoting health and well-being, particularly for individuals dealing with metabolic disorders. The exploration of phytoestrogens thus represents a significant area of interest in nutritional science and therapeutic research, with the potential to impact public health positively.

## 3. Phytoestrogens and metabolic diseases

Phytoestrogens, natural compounds found in plants, have shown promise in influencing metabolic health [8]. Notably, isoflavones, prevalent in soybeans and legumes, exhibit potential in improving insulin sensitivity, holding promise for type 2 diabetes management [15]. Additionally, phytoestrogens may play a role in regulating adipose tissue metabolism, suggesting their relevance in addressing obesity [4]. Their potential impact on cardiovascular health is also noteworthy, with evidence suggesting benefits in blood pressure regulation and cholesterol levels [18]. Moreover, lignans from sources like flaxseeds and sesame seeds show promise in supporting bone health, indicating potential adjunctive treatments for osteoporosis. This section illuminates the multifaceted contributions of phytoestrogens to various facets of metabolic well-being.

### 3.1 Type 2 diabetes mellitus

Type 2 diabetes mellitus (T2DM) represents a significant global health challenge characterized by insulin resistance and impaired glucose metabolism. In this context, phytoestrogens have emerged as a focal point of research, particularly regarding their potential to improve insulin sensitivity and glucose regulation. Among the diverse phytoestrogen sources, soybeans and flaxseeds have shown considerable promise in mitigating the complexities associated with T2DM [18].

Isoflavones, primarily found in soybeans (*Glycine max* L.), such as genistein and daidzein, have been the subject of extensive studies. These compounds have been observed to enhance insulin sensitivity and modulate glucose metabolism, acting as insulin sensitizers and potentially reducing insulin resistance. Flaxseeds (*Linum usitatissimum* L.), rich in lignans like secoisolariciresinol diglucoside and matairesinol, present an alternative phytoestrogen source with demonstrated efficacy in T2DM. Studies also indicate that lignan supplementation in T2DM patients led to improved glycemic control, highlighting their potential role in managing this metabolic disorder [19]. Expanding beyond soybeans and flaxseeds, red clover (*Trifolium pratense* L.), with its rich content of isoflavones like formononetin and biochanin A, has also shown potential in improving insulin resistance. This was evidenced in research by Liu et al. in 2013 (published in the *American Journal of Clinical Nutrition*), suggesting that red clover isoflavones could offer additional benefits in T2DM treatment strategies [20].

The inclusion of phytoestrogen-rich foods like soybeans, flaxseeds, and red clover in dietary regimes may provide comprehensive benefits in the management of T2DM. Furthermore, exploring other sources such as pumpkin seeds (*Cucurbita pepo*), known for their lignan content, and chia seeds (*Salvia hispanica*), rich in alpha-linolenic acid and lignans, broadens the scope for personalized dietary approaches to metabolic health. This expanding research on phytoestrogens and their influence on metabolic disorders like T2DM highlights their potential as adjunctive therapeutic agents [19]. As our understanding deepens, integrating these phytoestrogens into dietary and therapeutic strategies presents a promising frontier in the holistic management of T2DM, potentially enhancing standard treatment protocols and improving patient outcomes.

*Potential Contraindications:* Some diabetes medications may interact with phytoestrogens in foods like soy and supplements. Sulfonylureas, such as glipizide and glyburide, stimulate insulin release and excessive phytoestrogen intake can affect blood sugar levels, potentially leading to hypoglycemia. Thiazolidinediones like pioglitazone and rosiglitazone improve insulin sensitivity, and phytoestrogens, particularly isoflavones, may have similar effects, potentially influencing medication efficacy. Alpha-glucosidase Inhibitors, such as acarbose and miglitol, slow carbohydrate digestion, but excessive dietary fiber from sources like flaxseeds or whole grains can impact medication absorption. While DPP-4 inhibitors like sitagliptin and saxagliptin enhance insulin release, monitoring blood glucose levels is advisable if significant dietary changes involving phytoestrogens are made.

### 3.2 Obesity

Phytoestrogens, naturally occurring compounds found in various plants, have become a focal point in obesity research due to their potential role in modulating adipose tissue metabolism and reducing fat accumulation. Their effects on lipid profiles and adipogenesis suggest they could be valuable in combating obesity.

Soybeans (*Glycine max* L.), known for their rich isoflavone content, including genistein and daidzein, have shown promise in regulating adipose tissue. Research has particularly highlighted genistein's ability to modulate adipogenesis, which is the process of developing mature fat cells from precursor cells. By inhibiting adipocyte differentiation, genistein may help regulate fat accumulation, offering a potential pathway for managing obesity [21]. Flaxseeds (*Linum usitatissimum* L.) are another significant source of phytoestrogens, especially lignans like secoisolariciresinol diglucoside and matairesinol. Studies indicate that these lignans may positively influence body composition by reducing fat mass, suggesting their utility in weight management strategies [22].

Red clover (*Trifolium pratense* L.), with its isoflavones formononetin and biochanin A, also emerges as a noteworthy player in the obesity context. Research suggests that these compounds may impact lipid metabolism, contributing to decreased fat accumulation [23]. Additionally, Pueraria (Kudzu) (*Pueraria lobata* (Willd.) Ohwi), rich in isoflavones like puerarin and daidzein, has attracted attention for its potential to improve lipid profiles. Studies demonstrate that puerarin might reduce total cholesterol and triglycerides, positioning it as a promising compound for obesity management [24].

The scope of phytoestrogens in tackling obesity extends further. Chia seeds (*Salvia hispanica*), known for their alpha-linolenic acid and lignan content, have been recognized for their potential in weight management [25]. These seeds may aid in improving metabolic markers and reducing fat accumulation. Alfalfa (*Medicago sativa* L.), containing coumestans such as medicagol, presents another interesting avenue in obesity research [26]. While coumestans are less studied than isoflavones and lignans, they offer a novel area of exploration in the context of metabolic health.

The growing body of evidence underscores the potential of a variety of phytoestrogen-rich plants in addressing obesity. As such, there is no specific recommended dose of phytoestrogens for managing obesity because individual responses can differ, and more research is needed to establish clear guidelines. In animal studies focused on obesity and diabetes, soy protein has demonstrated the ability to lower serum insulin levels and improve insulin sensitivity. Human studies, both with and without diabetes, have also indicated that soy protein can help regulate blood sugar levels, reduce body weight, lower elevated lipid levels, and decrease excessive insulin production. These findings suggest favorable effects of soy protein on both obesity and diabetes. However, it is important to note that most of these clinical trials were relatively short in duration and involved a limited number of participants. Furthermore, the specific components of soy protein and flaxseed responsible for these benefits remain uncertain. It is unclear whether isoflavones (such as daidzein and genistein), lignans (including matairesinol and secoisolariciresinol), or other compounds play the primary role. Isoflavones and lignans are thought to exert their effects through diverse mechanisms, including the modulation of insulin secretion in the pancreas and antioxidant actions. Each plant offers unique phytochemicals, contributing to a diverse spectrum of bioactive compounds that may aid in the management of obesity. This expanding field of research highlights the importance of phytoestrogens in developing comprehensive dietary and therapeutic strategies for obesity management, enhancing our understanding of their role in metabolic health.

*Potential contraindications:* While there is generally no direct contraindication, individuals taking obesity medications such as orlistat, phentermine-topiramate, bupropion-naltrexone, or liraglutide should be cautious of potential interactions with phytoestrogen-containing products.

### 3.3 Cardiovascular diseases

The role of phytoestrogens in cardiovascular disease prevention is gaining increasing attention due to their varied and significant effects on heart health. These naturally occurring compounds, found in a variety of plants, have been linked to multiple benefits that collectively reduce the risk of cardiovascular diseases. Isoflavones from soybeans, particularly genistein and daidzein, have been extensively studied for their cardiovascular benefits. Genistein is known to improve endothelial function, which is crucial for maintaining the health and elasticity of blood vessels. This improvement in endothelial function can lead to better blood flow and reduced risk of arterial blockages. Genistein has also been observed to lower LDL cholesterol levels, which is essential in preventing the buildup of plaques in the arteries, a major risk factor for heart disease. Lignans found in flaxseeds are another group of phytoestrogens with cardiovascular benefits. These compounds have been linked to improved blood pressure regulation and healthier lipid profiles. By influencing these factors, lignans can significantly reduce the risk of hypertension and atherosclerosis, both of which are key contributors to cardiovascular disease.

Genistein has been the subject of numerous investigations. Studies have indicated that genistein may improve endothelial function, a critical factor in maintaining healthy blood vessels [27]. This property of genistein, along with its potential to reduce LDL cholesterol levels, underscores its significance in cardiovascular health. Flaxseeds (*Linum usitatissimum* L.), a potent source of lignans, like secoisolariciresinol diglucoside and matairesinol, contribute to the multifaceted approach in mitigating cardiovascular risks [28]. Lignans from flaxseeds may contribute to blood pressure regulation. Additionally, lignans have been associated with favorable effects on lipid profiles, further solidifying their role in cardiovascular health [28]. Red clover (*Trifolium pratense* L.), known for its isoflavones formononetin and biochanin A, provides another avenue in the pursuit of cardiovascular wellness. Research demonstrated that red clover isoflavones may lead to modest reductions in LDL cholesterol levels, a pivotal factor in preventing atherosclerosis [29]. Hops (*Humulus lupulus* L.), a plant rich in the compound xanthohumol, shows promise in cardiovascular health. Studies suggest that xanthohumol may contribute to reducing LDL cholesterol oxidation, an essential step in preventing atherosclerosis [30, 31]. Pomegranate (*Punica granatum*), known for its rich content of ellagitannins, offers potential benefits for cardiovascular health [32]. Research indicated that pomegranate polyphenols may contribute to the inhibition of LDL cholesterol oxidation, a crucial step in preventing plaque formation [33].

Additionally, Black cohosh (*Actaea racemosa* L.), containing triterpene glycosides, like cimicifugoside and actein, may also play a role in cardiovascular health [34]. Though less studied in this context, emerging research suggests that compounds within black cohosh may contribute to cardiovascular wellness through their diverse biological activities [35]. Alfalfa (*Medicago sativa* L.), a plant rich in coumestans, like medicagol, may contribute to cardiovascular health [36].

The diverse range of phytoestrogens, each with unique properties, collectively contribute to heart health. They work by improving endothelial function, lowering harmful cholesterol levels, reducing blood pressure, and preventing the oxidation of LDL cholesterol. Incorporating phytoestrogen-rich foods into the diet could be a valuable strategy for reducing the risk of cardiovascular diseases.

*Potential Contraindications:* Potential interactions and contraindications exist between certain cardiovascular medications and phytoestrogens. For instance, warfarin showed interaction with phytoestrogen-rich foods. Grapefruit juice, containing phytoestrogens, can interact with statins like atorvastatin and simvastatin. Beta-blockers such as metoprolol and atenolol could be affected by the mild blood pressure-lowering effects of isoflavones in soy. While there are no direct contraindications, individuals on antihypertensive medications like angiotensin-converting enzyme (ACE) inhibitors and calcium channel blockers need to be mindful of potential interactions when consuming phytoestrogen-rich foods, monitoring blood pressure as needed.

### 3.4 Osteoporosis

Osteoporosis is a prevalent bone disease characterized by reduced bone mass and deteriorating bone tissue, leading to increased fracture risk [37]. This condition is particularly common among postmenopausal women due to the decrease in estrogen levels, a hormone vital for bone health [38]. Phytoestrogens have shown potential in mitigating the risk and progression of osteoporosis, primarily due to their estrogen-like effects. The mechanism behind the beneficial effects of phytoestrogens on bone health is primarily linked to their ability to bind to estrogen receptors in the body. This binding can help compensate for the decreased estrogen levels in postmenopausal women, a key factor in the development of osteoporosis. Phytoestrogens can modulate the activity of osteoclasts (cells that break down bone tissue) and osteoblasts (cells that build new bone), helping to maintain a balance in bone remodeling. Moreover, phytoestrogens have antioxidant properties (explained in the later section of this chapter), which may help protect bone tissue from oxidative stress and inflammation, further contributing to their osteoprotective effects.

Isoflavones are the most studied phytoestrogens in the context of bone health. They have been shown to positively influence bone mineral density (BMD). Studies suggest that isoflavones like genistein and daidzein may help in preserving BMD in postmenopausal women, potentially reducing the risk of osteoporotic fractures [39]. Lignans are abundant in flaxseeds, sesame seeds, and whole grains. They are converted into enterolignans by intestinal bacteria, which then exert estrogenic effects. Research indicates that lignan intake is associated with improved bone turnover markers and could be beneficial in maintaining bone density, particularly in postmenopausal women [40]. Coumestans, less common but present in foods like alfalfa and clover, also exhibit estrogenic properties and may play a role in bone health, although more research is needed in this area [41].

In addition to flaxseeds and sesame seeds, chia seeds (*Salvia hispanica*) represent another notable source of lignans. The presence of alpha-linolenic acid, an omega-3 fatty acid, in chia seeds further complements their potential in supporting bone health. Omega-3 fatty acids have been associated with anti-inflammatory effects, which can contribute to a favorable bone environment [42]. Hops (*Humulus lupulus* L.) has also shown promise in bone health. Studies suggest that xanthohumol may have a protective effect on bone density, making it an intriguing subject for further investigation [43]. Red clover (*Trifolium pratense* L.) may have a positive impact on bone health in postmenopausal women [44].

It is important to note that while phytoestrogens, including lignans, present promising potential in supporting bone health, they are most effective when integrated into a comprehensive approach to osteoporosis management. This approach includes a balanced diet rich in essential nutrients for bone health, weight-bearing exercises, and, when necessary, conventional medical interventions. The diverse array of plants and their corresponding phytoestrogen compounds offer intriguing possibilities for the preservation of bone integrity. Lignans from flaxseeds, sesame seeds, and chia seeds, along with compounds from hops and red clover, exemplify nature's potential contributions to the fight against osteoporosis.

*Potential Contraindications:* Potential interactions and contraindications may exist between certain osteoporosis medications and phytoestrogens. Bisphosphonates like alendronate and risedronate may be affected by excessive calcium intake from soy, potentially impacting medication absorption. Selective estrogen receptor modulators (SERMs) like raloxifene, which have estrogenic effects, may interact with phytoestrogens, possibly leading to additive estrogenic effects on bone health.

## 4. Antioxidant roles of phytoestrogens

Phytoestrogens, derived from plants, possess robust antioxidant properties. They excel in neutralizing harmful free radicals, inhibiting lipid peroxidation, and bolstering the body's enzymatic and nonenzymatic antioxidant defenses. These actions collectively mitigate oxidative stress, a critical factor in metabolic diseases [5].

### 4.1 Mechanisms of antioxidant action

Phytoestrogens possess powerful antioxidant properties, which stem from their ability to scavenge free radicals, inhibit lipid peroxidation, and modulate enzymatic antioxidant defenses. These actions collectively contribute to a reduction in oxidative stress.

The antioxidant actions of phytoestrogens are multifaceted, involving several key mechanisms:

#### 4.1.1 Free radical scavenging

Phytoestrogens are adept at scavenging free radicals, highly reactive molecules that can cause cellular damage. By neutralizing these radicals, phytoestrogens help prevent the chain reactions that lead to oxidative stress.

#### 4.1.2 Inhibition of lipid peroxidation

Lipid peroxidation is a destructive process where free radicals attack lipids in cell membranes, leading to cellular dysfunction. Phytoestrogens, through their chemical structure and reactivity, have the capacity to interrupt this chain of events, preserving cellular integrity.

#### 4.1.3 Modulation of enzymatic antioxidant defenses

Phytoestrogens can influence the activity of enzymatic antioxidants within cells. These enzymes, such as superoxide dismutase and catalase, are crucial in neutralizing harmful reactive oxygen species (ROS). By enhancing their activity, phytoestrogens bolster the cell's defense against oxidative damage.

#### 4.1.4 Enhancement of nonenzymatic antioxidants

In addition to affecting enzymatic defenses, phytoestrogens can boost the levels of nonenzymatic antioxidants, such as glutathione and vitamins C and E. These molecules play pivotal roles in neutralizing ROS and preventing cellular harm.

Notably, different phytoestrogen compounds exhibit varying degrees of antioxidant efficacy. For instance, isoflavones, prevalent in soybeans and legumes, are recognized for their potent antioxidant activity [45]. Genistein has been extensively studied for its ability to quench free radicals and protect cells from oxidative damage [46]. Lignans, found abundantly in flaxseeds, sesame seeds, and whole grains, also contribute significantly to antioxidant defenses [47]. Their structural attributes allow them to intercept free radicals and impede oxidative stress. Coumestans, present in various legumes and sprouts, possess antioxidant capabilities [48]. Compounds like coumestrol have demonstrated the capacity to mitigate oxidative damage by neutralizing free radicals [49]. Stilbenes, such as resveratrol found in red grapes, berries, and peanuts, are recognized for their potent antioxidant effects [50]. They are particularly adept at inhibiting lipid peroxidation and preserving cellular integrity.

Thus, the antioxidant roles of phytoestrogens represent a crucial aspect of their potential health benefits. Their ability to scavenge free radicals, inhibit lipid peroxidation and modulate enzymatic and nonenzymatic antioxidant defenses collectively contribute to reducing oxidative stress. Understanding these mechanisms provides valuable insights into the potential applications of phytoestrogens in combating metabolic diseases and promoting overall health.

### 4.2 Roles of phytoestrogen's antioxidant activity in metabolic diseases

The antioxidant capabilities of phytoestrogens extend their benefits beyond mere hormonal mimicry, playing a vital role in combating metabolic diseases where oxidative stress is a key perpetrator [51, 52]. This stress, characterized by an overabundance of reactive oxygen species (ROS), is a critical factor in diseases like type 2 diabetes, obesity, cardiovascular diseases, and osteoporosis.

In type 2 diabetes, phytoestrogens, especially isoflavones from soybeans, show promise in enhancing insulin sensitivity and improving glucose metabolism [53]. They achieve this by counteracting oxidative stress through the neutralization of free radicals and the inhibition of lipid peroxidation. Genistein, in particular, stands out for its effectiveness in protecting pancreatic beta cells, which are crucial for insulin production [54].

In the realm of obesity, a condition deeply intertwined with oxidative stress, phytoestrogens demonstrate a capability to modulate fat metabolism and suppress adipogenesis [55]. This effect is particularly significant in maintaining the integrity of cells in adipose tissue. The antioxidant properties of compounds like isoflavones, lignans from sources like flaxseeds, and stilbenes from red grapes contribute to these protective effects [56].

Regarding cardiovascular diseases, oxidative stress is a well-known contributor. Phytoestrogens, through their antioxidant actions, scavenge free radicals, inhibit lipid peroxidation, and boost enzymatic antioxidant defenses, offering a multipronged shield against cardiovascular complications. Specific phytoestrogens like isoflavones from soybeans, resveratrol from red grapes, and epigallocatechin gallate (EGCG) from green tea have been shown to be particularly effective. These compounds aid in lowering blood pressure, reducing levels of harmful LDL cholesterol, and enhancing endothelial function, which is pivotal for vascular health [57, 58].

For osteoporosis, characterized by reduced bone density and increased fracture risk, phytoestrogens, particularly lignans found in flaxseeds and sesame seeds, are of great interest. They safeguard bone health by protecting bone cells against oxidative damage. Secoisolariciresinol, a prominent lignan, is notably effective in this regard [59, 60].

Overall, the antioxidant activity of phytoestrogens offers a comprehensive approach to mitigating oxidative stress, a central factor in various metabolic diseases. By neutralizing harmful free radicals, preventing lipid peroxidation, and enhancing the body's own antioxidant defenses, phytoestrogens present a natural and multifaceted strategy to address oxidative damage. This highlights the potential of incorporating phytoestrogen-rich foods and supplements as valuable components in the prevention and management of these metabolic diseases.

## 5. Conclusion

Phytoestrogens, in the context of metabolic diseases, reveals a fascinating spectrum of natural compounds with significant health benefits. These substances, structurally similar to estrogen, exhibit both estrogenic and antioxidant activities, presenting themselves as potential therapeutic agents in various metabolic disorders. This chapter highlighted the roles of different phytoestrogens-isoflavones, lignans, and coumestans, in combating metabolic diseases. Their ability to mimic estrogenic activity and interact with estrogen receptors offers promising therapeutic avenues in conditions like type 2 diabetes, obesity, cardiovascular diseases, and osteoporosis. For instance, isoflavones in type 2 diabetes improve insulin sensitivity, while lignans in osteoporosis enhance bone mineral density.

The antioxidant capabilities of phytoestrogens further broaden their therapeutic scope. By scavenging free radicals, inhibiting lipid peroxidation, and bolstering enzymatic antioxidant defenses, they offer protection against oxidative stress, a common factor in many metabolic diseases. Beyond these, the exploration of a variety of phytoestrogen-rich plants like grapes, berries, green tea, and alfalfa showcases the vast potential of these compounds in metabolic health. Thus, phytoestrogens represent a burgeoning field in the management of metabolic diseases. Their unique properties and diverse mechanisms of action open up new possibilities for integrating natural approaches with conventional medical treatments. As research advances, the role of phytoestrogens in metabolic health is set to become increasingly significant, offering a harmonious blend of nature and science in combating these conditions.

## Acknowledgements

The authors are thankful to the Ministry of Health (MOH) and the National Medical Research Council (NMRC), Singapore, for financial assistance grant number NMRC/OFYIRG/0002/2016 and MOH-000319 (MOH-OFIRG19may-0002), Khoo Bridge Funding Award (DUKE-NUS-KBRFA/2023/0075), and Duke/Duke-NUS Collaboration Pilot Project award (RECA(Pilot)/2022/0060) to BKS; NMRC/OFYIRG/077/2018 to MT.

## Conflict of interest

The authors declare no conflict of interest.

## Author details

Manish Kumar Verma[1], Brijesh Kumar Singh[2*] and Madhulika Tripathi[2*]

1 Raghuveer Singh Government (P.G.) College, Lalitpur, Uttar Pradesh, India

2 Cardiovascular and Metabolic Disorders Program, Duke-NUS Medical School, Singapore

*Address all correspondence to: singhbrijeshk@duke-nus.edu.sg
and madhulika.triptahi@duke-nus.edu.sg

## References

[1] Chen LR, Chen KH. Utilization of isoflavones in soybeans for women with menopausal syndrome: An overview. International Journal of Molecular Sciences. 2021;**22**(6):3212

[2] Alshehri MM et al. Therapeutic potential of isoflavones with an emphasis on Daidzein. Oxidative Medicine and Cellular Longevity. 2021;**2021**:6331630

[3] Duncan AM et al. Soy isoflavones exert modest hormonal effects in premenopausal women. The Journal of Clinical Endocrinology and Metabolism. 1999;**84**(1):192-197

[4] Taku K et al. Soy isoflavones lower serum total and LDL cholesterol in humans: A meta-analysis of 11 randomized controlled trials. The American Journal of Clinical Nutrition. 2007;**85**(4):1148-1156

[5] Halliwell B. Free radicals, antioxidants, and human disease: Curiosity, cause, or consequence? Lancet. 1994;**344**(8924):721-724

[6] Patisaul HB, Jefferson W. The pros and cons of phytoestrogens. Frontiers in Neuroendocrinology. 2010;**31**(4):400-419

[7] Sirotkin AV, Harrath AH. Phytoestrogens and their effects. European Journal of Pharmacology. 2014;**741**:230-236

[8] Adlercreutz H et al. Effect of dietary components, including lignans and phytoestrogens, on enterohepatic circulation and liver metabolism of estrogens and on sex hormone binding globulin (SHBG). Journal of Steroid Biochemistry. 1987;**27**(4-6):1135-1144

[9] Setchell KD et al. Nonsteroidal estrogens of dietary origin: Possible roles in hormone-dependent disease. The American Journal of Clinical Nutrition. 1984;**40**(3):569-578

[10] Patra S et al. A review on phytoestrogens: Current status and future direction. Phytotherapy Research. 2023;**37**(7):3097-3120

[11] Teixeira SR et al. Isolated soy protein consumption reduces urinary albumin excretion and improves the serum lipid profile in men with type 2 diabetes mellitus and nephropathy. The Journal of Nutrition. 2004;**134**(8):1874-1880

[12] Zhang Y et al. Soy isoflavones and their bone protective effects. Inflammopharmacology. 2008;**16**(5):213-215

[13] Overk CR et al. Comparison of the in vitro estrogenic activities of compounds from hops (*Humulus lupulus* L.) and red clover (*Trifolium pratense* L.). Journal of Agricultural and Food Chemistry. 2005;**53**(16):6246-6253

[14] Milligan SR et al. Identification of a potent phytoestrogen in hops *(Humulus lupulus* L. L.) and beer. The Journal of Clinical Endocrinology and Metabolism. 1999;**84**(6):2249-2252

[15] Villegas R et al. Legume and soy food intake and the incidence of type 2 diabetes in the Shanghai women's health study. The American Journal of Clinical Nutrition. 2008;**87**(1):162-167

[16] Teede HJ et al. Dietary soy has both beneficial and potentially adverse cardiovascular effects: A placebo-controlled study in men and postmenopausal women. The Journal of

Clinical Endocrinology and Metabolism. 2001;**86**(7):3053-3060

[17] Marini H et al. Effects of the phytoestrogen genistein on bone metabolism in osteopenic postmenopausal women: A randomized trial. Annals of Internal Medicine. 2007;**146**(12):839-847

[18] Liu ZM et al. Whole soy, but not purified daidzein, had a favorable effect on improvement of cardiovascular risks: A 6-month randomized, double-blind, and placebo-controlled trial in equol-producing postmenopausal women. Molecular Nutrition & Food Research. 2014;**58**(4):709-717

[19] Llaneza P et al. Soy isoflavones improve insulin sensitivity without changing serum leptin among postmenopausal women. Climacteric. 2012;**15**(6):611-620

[20] Liu K et al. Effect of green tea on glucose control and insulin sensitivity: A meta-analysis of 17 randomized controlled trials. The American Journal of Clinical Nutrition. 2013;**98**(2):340-348

[21] Choi YR, Shim J, Kim MJ. Genistin: A novel potent anti-adipogenic and anti-lipogenic agent. Molecules. 2020;**25**(9):2042

[22] Nakashima KI et al. Regulation of adipogenesis through retinoid X receptor and/or peroxisome proliferator-activated receptor by designed lignans based on natural products in 3T3-L1 cells. Journal of Natural Medicines. 2023;**77**(2):315-326

[23] Jungbauer A, Medjakovic S. Phytoestrogens and the metabolic syndrome. The Journal of Steroid Biochemistry and Molecular Biology. 2014;**139**:277-289

[24] Yang M et al. Puerarin ameliorates metabolic dysfunction-associated fatty liver disease by inhibiting ferroptosis and inflammation. Lipids in Health and Disease. 2023;**22**(1):202

[25] Vuksan V et al. Salba-chia (*Salvia hispanica* L.) in the treatment of overweight and obese patients with type 2 diabetes: A double-blind randomized controlled trial. Nutrition, Metabolism, and Cardiovascular Diseases. 2017;**27**(2):138-146

[26] Gomez-Zorita S et al. Scientific evidence supporting the beneficial effects of isoflavones on human health. Nutrients. 2020;**12**(12):3853

[27] Wei TT et al. Cannabinoid receptor 1 antagonist genistein attenuates marijuana-induced vascular inflammation. Cell. 2022;**185**(13):2387-2389

[28] Shayan M et al. Flaxseed for health and disease: Review of clinical trials. Combinatorial Chemistry & High Throughput Screening. 2020;**23**(8):699-722

[29] Nestel PJ et al. Isoflavones from red clover improve systemic arterial compliance but not plasma lipids in menopausal women. The Journal of Clinical Endocrinology and Metabolism. 1999;**84**(3):895-898

[30] Hirata H et al. Xanthohumol, a hop-derived prenylated flavonoid, promotes macrophage reverse cholesterol transport. The Journal of Nutritional Biochemistry. 2017;**47**:29-34

[31] Liou SF et al. The preventive effects of xanthohumol on vascular calcification induced by vitamin D(3) plus nicotine. Antioxidants (Basel). 2020;**9**(10):956

[32] Alami M et al. The modulatory bioeffects of pomegranate

(*Punica granatum* L.) polyphenols on metabolic disorders: Understanding their preventive role against metabolic syndrome. Nutrients. 2023;**15**(23):4879

[33] Franczyk-Zarow M et al. Pomegranate seed oil as a source of conjugated linolenic acid (CLnA) has no effect on atherosclerosis development but improves lipid profile and affects the expression of lipid metabolism genes in apoE/LDLR(−/−) mice. International Journal of Molecular Sciences. 2023;**24**(2):1737

[34] Lu NH et al. Yunnanterpene G, a spiro-triterpene from the roots of *Cimicifuga foetida*, downregulates the expression of CD147 and MMPs in PMA differentiated THP-1 cells. RSC Advances. 2018;**8**(27):15036-15043

[35] Wang BH, Ou-Yang JP. Pharmacological actions of sodium ferulate in cardiovascular system. Cardiovascular Drug Reviews. 2005;**23**(2):161-172

[36] Molgaard J, von Schenck H, Olsson AG. Alfalfa seeds lower low density lipoprotein cholesterol and apolipoprotein B concentrations in patients with type II hyperlipoproteinemia. Atherosclerosis. 1987;**65**(1-2):173-179

[37] Zhao Z et al. Mechanism and physical activities in bone-skeletal muscle crosstalk. Frontiers in Endocrinology (Lausanne). 2023;**14**:1287972

[38] Huang C et al. Identifying potential ferroptosis key genes for diagnosis and treatment of postmenopausal osteoporosis through competitive endogenous RNA network analysis. Heliyon. 2024;**10**(1):e23672

[39] Kim SB et al. Combined effects of soy isoflavone and lecithin on bone loss in ovariectomized mice. Nutrition Research and Practice. 2021;**15**(5):541-554

[40] Yang XY et al. Anti-osteoporosis effects of mammalian lignans and their precursors from flaxseed and safflower seed using zebrafish model. Journal of Food Science. 2023;**88**(12):5278-5290

[41] Zhai Y et al. Psoralidin, a prenylated coumestan, as a novel anti-osteoporosis candidate to enhance bone formation of osteoblasts and decrease bone resorption of osteoclasts. European Journal of Pharmacology. 2017;**801**:62-71

[42] Montes Chani EM et al. Long-term dietary intake of chia seed is associated with increased bone mineral content and improved hepatic and intestinal morphology in Sprague-Dawley rats. Nutrients. 2018;**10**(7):922

[43] Jackson MK et al. Dietary inflammatory potential and bone outcomes in Midwestern post-menopausal women. Nutrients. 2023;**15**(19):4277

[44] Dietz BM et al. Botanicals and their bioactive phytochemicals for women's health. Pharmacological Reviews. 2016;**68**(4):1026-1073

[45] Shao Y et al. Soybean isoflavones ameliorates lactation performance in postpartum mice by alleviating oxidative stress and regulating gut microflora. Molecular Nutrition & Food Research. 2024:e2300184

[46] Xu K et al. Anti-oxidation effect of genistein in vascular endothelial cell after $H_2O_2$ stress. Molecular Medicine Reports. 2024;**29**(1):2

[47] Jarosova M et al. Comparison of yield characteristics, chemical composition, lignans content and antioxidant potential of experimentally grown six linseed (*Linum usitatissimum* L. L.) cultivars. Plant Foods for Human Nutrition. 2024. DOI: 10.1007/s11130-023-01136-9 [Ahead of print]

[48] Chen M et al. Theoretical study on the free radical scavenging potency and mechanism of natural coumestans: Roles of substituent, noncovalent interaction and solvent. Phytochemistry. 2023;**207**:113580

[49] Zywno H et al. The influence of coumestrol on sphingolipid signaling pathway and insulin resistance development in primary rat hepatocytes. Biomolecules. 2021;**11**(2):268

[50] Rashid H et al. Molecular mechanisms of oxidative stress in acute kidney injury: Targeting the loci by resveratrol. International Journal of Molecular Sciences. 2023;**25**(1):3

[51] Manzoor MF et al. Oxidative stress and metabolic diseases: Relevance and therapeutic strategies. Frontiers in Nutrition. 2022;**9**:994309

[52] Rani V et al. Oxidative stress and metabolic disorders: Pathogenesis and therapeutic strategies. Life Sciences. 2016;**148**:183-193

[53] Talaei M, Pan A. Role of phytoestrogens in prevention and management of type 2 diabetes. World Journal of Diabetes. 2015;**6**(2):271-283

[54] Jiang T et al. Underlying mechanisms and molecular targets of genistein in the management of type 2 diabetes mellitus and related complications. Critical Reviews in Food Science and Nutrition. 2023:1-13. DOI: 10.1080/10408398.2023.2240886 [Ahead of print]

[55] Bhathena SJ, Velasquez MT. Beneficial role of dietary phytoestrogens in obesity and diabetes. The American Journal of Clinical Nutrition. 2002;**76**(6):1191-1201

[56] Hamden K et al. Phytoestrogens inhibit key-enzymes linked to obesity, type 2 diabetes and liver-kidney toxicity in high fructose-fat diet in mice. Archives of Physiology and Biochemistry. 2019;**125**(5):423-429

[57] Anthony MS. Phytoestrogens and cardiovascular disease: Where's the meat? Arteriosclerosis, Thrombosis, and Vascular Biology. 2002;**22**(8):1245-1247

[58] Clarkson TB, Anthony MS. Phytoestrogens and coronary heart disease. Baillière's Clinical Endocrinology and Metabolism. 1998;**12**(4):589-604

[59] Inpan R et al. Isoflavone intervention and its impact on bone mineral density in postmenopausal women: A systematic review and meta-analysis of randomized controlled trials. Osteoporosis International. 2023. DOI: 10.1007/s00198-023-06944-y [Ahead of print]

[60] Su Z et al. Polyphenols as potential preventers of osteoporosis: A comprehensive review on antioxidant and anti-inflammatory effects, molecular mechanisms, and signal pathways in bone metabolism. The Journal of Nutritional Biochemistry. 2024;**123**:109488

Chapter 7

# Immune-Enhancing Medicinal Plants: Are They a One Health, One Welfare Actor?

*Aurel Vasiu, Vasile Cozma, Anamaria Cozma-Petruț, Mihai Băieș, Marina Spînu, Emöke Páll, Diana Olah, Carmen Dana Șandru, Gheorghiță Duca, Köbölkuti Lorand and Gabriel Gati*

## Abstract

Medicinal plants have represented accessible and highly bioavailable remedies in traditional therapeutic and preventive practices of numerous populations worldwide. Veterinary treatments based on medicinal plants are also widespread, mainly targeting the control or prevention of parasitic diseases. Scientific support of the immune-stimulating efficacy of plants or their extracts in animals is less documented. The immunological activity of alcoholic plant extracts was investigated in numerous animal classes, starting from *Pisces*, through *Reptilia* and *Aves* and reaching *Mammalia*, envisaging their effects on innate and adaptive cell-mediated immunity, which the authors mean to share in this chapter, also providing a comparison of variable reactivity within and between the classes.

**Keywords:** plant active principles, cell-mediated immunity, *Pisces*, *Reptilia*, *Aves*, *Mammalia*

## 1. Introduction

Medicinal plants have long served as readily available and highly bioavailable remedies in the traditional therapeutic and preventive approaches of various global populations. For centuries, a plethora of natural phytochemical compounds derived from plants have demonstrated medicinal properties, aiding in healing and disease prevention [1]. The application of phytotherapy in veterinary treatments is also widespread, primarily focused on managing or preventing parasitic and infectious diseases. There are multiple factors contributing to the growing interest in herbal medications within the field of veterinary medicine. One significant factor is the widely held belief in the population that medicinal plants are not only effective but also safer when compared to synthetic compounds. Additionally, a key driver is the economic aspect, given that herbal treatments often prove to be more cost-effective than traditional interventions [2, 3]. Phytotherapeutic remedies offer a viable alternative for treating organic livestock, eliminating the need for synthetic drugs. However, there is comparatively limited scientific documentation supporting the immune-stimulating effectiveness of plants or their extracts in animals.

Medicinal plants, rich in various bioactive components, have found widespread application in the development and synthesis of drugs, such as aspirin. The field of natural product chemistry has expanded globally, aiming to explore increasingly potent and cost-effective drugs with minimized side effects [4–6]. Medicinal plants have the capability to regulate the innate immune system by boosting the activity of protease inhibitors and lytic enzymes within immune cells and molecules, enabling them to respond effectively against invading pathogens [4, 7, 8]. The bioactive compounds commonly found in aromatic plants typically exist in mixtures, predominantly composed of phenolics and terpenes, which are chemically identified by their aromatic rings [9].

Phytogenics sourced from the Lamiaceae family, formerly called Labiatae [10] are widely used plant-based additives in livestock and aquaculture [11–14]. The Lamiaceae family, consisting of over 7000 species grouped into approximately 240 genera, is widely distributed around the globe, exhibiting diverse heights and habitats [10, 15]. Members of this botanical family display diverse morphological characteristics, appearing as herbs, herbaceous plants, shrubs, or tree species [16]. Within them, numerous aromatic and medicinal plants are found, widely utilized across traditional and modern medicine, the food industry, and cosmetics [17, 18]. Extracts prepared from these plants yield bioactive compounds utilized for their growth promotion, antimicrobial, antioxidant, antitumoral, immunostimulant, anti-inflammatory, antiviral, antifungal, insecticidal, sedative, antiangiogenic, neuroprotective, and carminative potential [10, 12, 19, 20]. Phenolics and terpenes represent a category of volatile bioactive compounds derived from plants, possessing medicinal and biotechnological significance. Phenolic compounds, including monoterpenes such as carvacrol and its isomer thymol, have been subject to extensive study [12, 21, 22]. These compounds serve as the primary constituents in essential oils extracted from a variety of aromatic plants belonging to the Lamiaceae family. The most well-known members of this family include a variety of aromatic spices like thyme, mint, oregano, basil, sage, savory, rosemary, self-heal, hyssop, lemon balm, and some others with more limited use [23]. Animal feeds can be enhanced by incorporating natural additives such as oregano (*Origanum vulgare* L) and thyme (*Thymus vulgaris*-L). These aromatic herbs, rich in bioactive compounds, not only contribute to the flavor profile of animal products but also offer potential health benefits. Research suggests that certain components in oregano and thyme, such as phenolics and terpenes, may have antimicrobial, antioxidant, and immunomodulatory properties, making them valuable supplements in animal nutrition. Including these herbs in animal diets can be a holistic approach to promoting both palatability and potential health advantages in livestock and poultry.

Immunosuppressive diseases present a significant risk to the well-being and efficiency of livestock, including pigs and poultry, leading to financial losses for producers. These diseases can be initiated by diverse factors, including viral, bacterial, and parasitic infections, as well as stressors related to the environment and management, such as insufficient nutrition or overcrowding [1]. These illnesses can weaken the animal's immune system, making it more vulnerable to secondary infections and reducing its ability to fight against pathogens [24].

Understanding the mechanisms of immune processes and improving the possibilities for their control are particularly important [25]. The immune system comprises an intricate network of both innate and adaptive components, possessing an exceptional ability to adapt and respond to a wide array of challenges [26].

The inherent physiological reactions of all living organisms against invading pathogens represent fundamental host responses. Prokaryotes, in particular, protect themselves through the utilization of restriction enzymes and clustered regularly interspaced palindromic repeats (CRISPRs), enabling the degradation of foreign pathogens attempting to invade [27].

The training and development of key components in the host's innate and adaptive immune system are significantly influenced by the microbiome [26, 28–30]. The emergence of specific branches within the immune system, especially those linked to adaptive immunity [30], has occurred alongside the evolution of a complex microbiota [26]. The close interplay between the microbiome and the immune system implies that enhancing the resident microbiome, whether through pre/probiotic supplements or appropriate interventions to prevent the disruption of microbial communities, can strengthen immune defenses [28, 31].

Some of the best known immunologically active plant extracts are the lectins, such as phytohemagglutinin M or P (American pokeweed, Fam. Phytolaccaceae, Genus *Phytolacca*) or concanavalin A obtained from *Canavalia ensiformis* (L.) DC. (jackbean, Fam. Fabaceae, Genus *Canavalia*), as a mitogen to prompt T-lymphocyte cell division, stimulating proliferation by causing an increase in cAMP [32].

Similarly, the nettle family, Urticaceae, include a numerous number of well-known and useful plants, including the genus *Urtica*, *Boehmeria*, *Pipturus,* and *Debregeasia* (about 2625 species, 53 genera). *Urtica dioica* L., as one of the best-known representatives, contains numerous compounds, including phenols, sterols, fatty acids, alkaloids, flavonoids, lignans, and terpenoids. Ther plant is used for its antioxidant, anti-inflammatory, antimicrobial, antiparasitic, protective for cardiovascular, nervous and digestive systems, antidiabetic and antiaging properties [33].

The immunological activity of alcoholic plant extracts was investigated in numerous animal classes, starting from Pisces, through Reptilia and Aves and reaching Mammalia, envisaging their effects on both innate and adaptive immune system branches, which the authors mean to share in this chapter, also providing a comparison of variable reactivity within and between the classes.

## 2. Plant active principles and innate cell-mediated immune response in fish

The immune system in fish, similar to higher vertebrates, includes both the innate and adaptive immune systems [5, 34]. The defense system of fish bears numerous similarities to the immune system of higher vertebrates, being well developed and integrated. In fish, the season and temperature changes are two factors that strongly influence both the immune response and antibody synthesis [35]. Although significant progress has been made in recent years regarding the immune system in fish, it is necessary to expand research on characterizing defense cell populations and their mechanisms of action. The thymus, spleen, kidneys, and the ontogenetic lymphoid process are the main lymphoid organs in fish [36]. Immunity in fish, like in other animal categories, represents their ability to cope with the attack of specific pathogens, whether they are viruses, bacteria, or parasites [37]. The environmental temperature significantly affects the immune response of fish, as they are among the most primitive organisms that possess an adaptive immune system consisting of lymphocytes, immunoglobulins, major histocompatibility complex (MHC), and, last but not least, T-cell receptors (TCR) [36]. According to studies, pathogens on the body surface or those that enter the immune organisms can be destroyed or only attenuated. On the other hand, it seems that the state of immunity is what limits the possibility of attachment of ectoparasites or intestinal parasites [38]. Researchers believe that humoral factors are present in tissues, internal fluids, and the mucus layer, including lysozyme, alexin, interferon, and properdin. Fish produce lysozyme, lectins, complement, and acute-phase proteins (C-reactive proteins, amyloid A and P). The complement system intervenes against pathogens and inflammatory processes, playing a crucial role in both humoral and cellular immunity [39]. The presence of complement in the tegument mucus has been highlighted, providing it the ability to act as a primary defense barrier for the organism [40].

In several fish species, the use of monoclonal antibodies has allowed the identification of the two types of lymphocytes, namely B lymphocytes and T lymphocytes [34]. According to studies by Scapigliati et al. [41], few recognize peripheral T cells, and a majority of these antibodies are directed against immunoglobulins and B cells that carry Ig on their surface.

Due to limited data on the ontogeny and differentiation of lymphoid cells in fish, researchers have employed monoclonal antibodies for the superior phenotypic characterization of these lymphocytes and various cellular subpopulations [42].

Knowledge of the degree of development of both the immune system and its functions play a crucial role. The organization and functioning of the immune system in all its complexity are directly influenced by the phylogenetic development of the organism [26]. The presence of suitable receptors and effectors is what determines both the functioning and accuracy of the immune response [43, 44].

Phagocytosis in fish represents one of the main innate immune mechanisms targeting immediate defense. The *in vitro* testing of phagocytosis can not only reveal the antimicrobial reactivity of *Pisces* but also help in identifying cellular responses to environmental compound, thus testing the effects of plant chemical compounds on fish phagocytes. Farmed fish (*Onchorhyncus mykis*) are exposed to farming technology (housing, feeding, nests, grouping, anti-parasitic and other treatments, etc.) and temperature changes in the aquatic habitat due to microclimate changes, which will presumably induce stress and impaired leukocyte function. Under these circumstances, we hypothesized that an immune-modulating treatment should stimulate the innate cell-mediated immune response. Mint, present on the banks of the rivers where trout species naturally dwell, containing as biologically active ingredients volatile oils (comprising menthol, menthone, menthyl acetate, pinene, limonene, cineole, and flavonoids such as rutin and rosmarinic acid) could alleviate the immune suppression induced by thermal stress in farmed rainbow trout.

In an *in vitro* experiment using *Mentha piperita* L. (Order Lamiales, Fam. Lamiaceae, Genus *Mentha*) alcoholic extract prepared in concordance with the provision of the German Pharmacopeia in a carbon particle inclusion test [45], we demonstrated (**Figure 1**) that temperature of the season significantly influenced the leukocyte numbers and subpopulations; therefore, the immune cell-stimulating activity of the alcoholic mint due to its active principles proved to be a temperature-boosted process, which probably stimulated both the monocytes and the heterophiles, more intensely during summer than during winter ($p < 0.05$) [46].

The mint alcoholic extract was used relying on no harmful effects reported in the literature and in dosages defined in previous *in vitro* studies, (unpublished data), in an amount of 20 μL/2 mL (v/v) [46].

Similarly, the *in vitro* lymphocyte blast transformation test offers valuable information on how the adaptive immune system is ready to fight infections, by contact between the lymphoid cells and lectins, PHA and ConA. Testing the capacity of fish leukocytes to respond in an *in vitro* whole blood system (**Figure 2**) [45] to stimulation with lectins

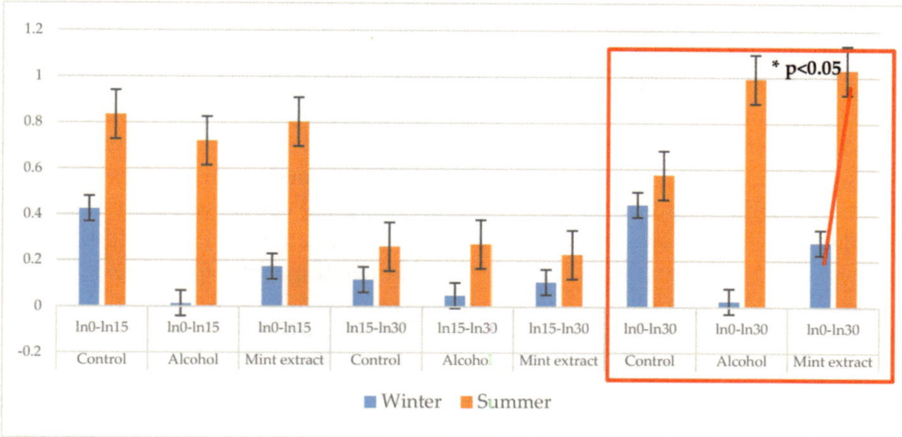


**Figure 1.**
*The phagocytic activity induced by the alcoholic mint extract in fish: There was a statistically significant increase ($p < 0.05$) in the carbon particle engulfment activity over the 0 to 30-minute in vitro testing period in summer (columns in orange) versus winter (blue columns), which indicated a potential stress-alleviating activity of the plant. Phagocytosis was measured during three periods of incubation 0 to 15 min and 15 to 30 min and overall activity from 0 to 30 min.*

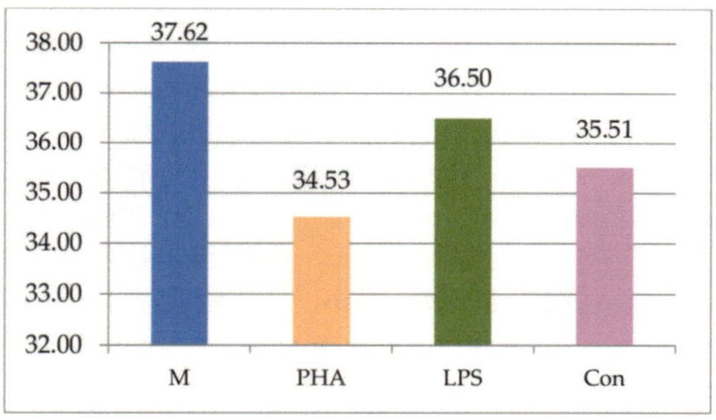


**Figure 2.**
*The in vitro cell-mediated response to plant and bacterial lectins in fish: PHA and ConA induce non-significantly lower activation of fish lymphocytes when compared to bacterial lectin LPS (when compared to control culture M, there was a 8.2% decrease in PHA-treated variant, while the lectins induced only a 2.9% - LPS and 5.6% - PHA decrease, respectively).*

indicated that there were no significant differences between the mitogens used (PHA, ConA, and bacterial lipopolysaccharide), but their efficacy in fish was poor.

## 3. Reptilia and their immune reactivity to plant extracts

The immune system of reptiles has been studied to a much-reduced extent compared to that of mammals, birds, or even fish and amphibians, probably due to their reduced economic impact. Further, the possibility of working with a high number of these individuals is minimal, for reasons such as their conservation and protection, the difficulty of keeping them in captivity, and so on [47]. Still, several research have investigated the ability of the immune system of reptiles to respond to antigenic challenges indicating the presence of agglutinating and neutralizing antibodies to various bacteria and viruses; all responses appear to be T-lymphocyte dependent [48]. In snakes, the immune response can vary depending on a number of extrinsic factors, such as nutritional status, environmental temperature, population density, season, and other ecological factors, while stress caused by capture or transport, various infections, and metabolic imbalances can also diminish immune processes [49]. It appears that the kinetics and amplitude of the humoral immune response correlated with the development of lymphoid tissues are affected by seasonal variations [50]. Secondary immune response has also been highlighted in some reptiles. Antibody-secreting cells have been demonstrated in the spleen, bone marrow, and peripheral blood but not in other lymphoid tissues such as the thymus or the cloacal complex [48]. Two classes of immunoglobulins IgM and a non-IgM (IgY)—the homolog of IgA found in mammals—have been described [51], the syntheses of which are temperature dependent [52–56]. Immunoglobulin classes of IgA, IgD, and IgE have not been identified in reptiles [57].

Cell-mediated immunity has been investigated by means of spleen cell transplantation and skin allografts and xenografts [53, 54, 58]. It must be taken into account that the conditions encountered by snakes in laboratories are not similar to those in nature (in terms of temperature conditions, day-night cycles, microhabitat, food, etc.); these "immuno-ecological" considerations can profoundly affect the "performance" of the immune system [47].

Testing the influence of lectins on two different species of turtles, *Trachemys scripta elegans* and *Emys orbicularis* (**Figure 3**), revealed a temperature-dependent response, with similar patterns in both species for the same season. Both lectins similarly increased the lymphocyte responses to the lectins when comparing the same season, while significant differences were observed due to the presence or absence of hibernation and between winter versus spring/summer ($p < 0.001$)(Köbölkuti, unpublished data).

In snakes (*Vipera berus, Vipera amodytes*), the PHA exerted a non-significantly different stimulation, still being useful to evaluate the response potential of the cell-mediated immunity [59]. No statistically significant differences were recorded between cultivation variants (**Figure 4**) or species, while the animals were hibernating, at an incubation temperature of 20°C, but there were differences between species ($p = 0.002$ in *V. berus* and $p = 0.007$ in *V. ammodytes*).

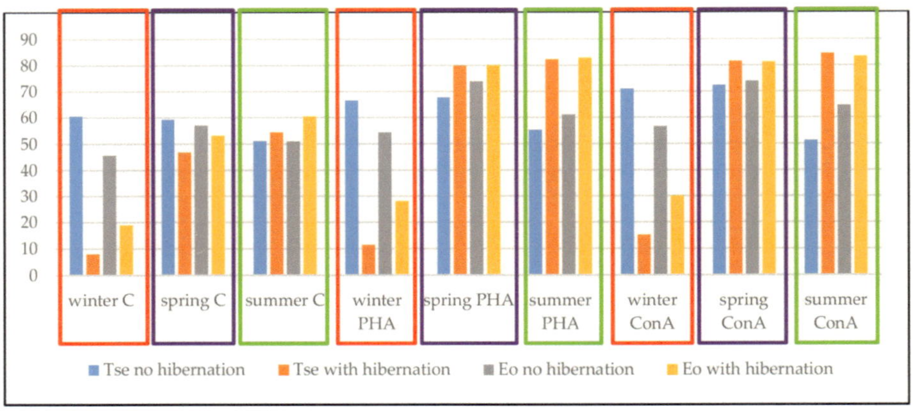


**Figure 3.**
*In vitro cell-mediated seasonal response to plant lectins in turtles: PHA and ConA induced comparable activation of turtle lymphocytes during spring and summer but significantly ($p < 0.01$) decreased in winter. The activation (%) values were compared to untreated controls © in all seasons (red frame = winter, purple frame = spring, green frame = summer). Tse = Trachemys scripta elegans, Eo = Emys orbicularis; the blue and orange squares next to Tse indicate the presence and absence of hibernation, respectively; the same applies for gray and yellow squares next to Eo.*

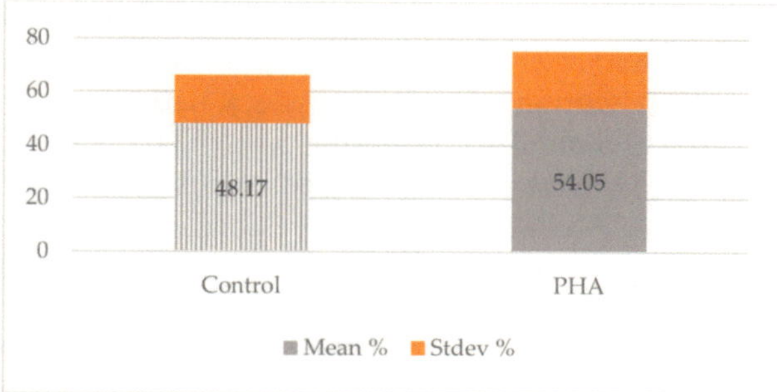


**Figure 4.**
*Stimulation indices recorded by the in vitro blast transformation test for Viperinae lymphocyte growth: no significant differences between the variants during hibernation (there was a slight increase of 12.12% in the PHA co-cultivated variant versus untreated control).*

Thus, it was concluded that the lifecycle rather than the species/variety influences the immune response to plant lectins in *Viperinae*.

## 4. Aves: a new evolutionary approach to improved immune responsiveness

In birds, in contrast to mammals, one of the primary lymphoid organs is the bursa of Fabricius, a gut-associated lymphoid tissue [60], characterized by 12–15 folds containing follicles, an organ designed for the maturation of B lymphocytes, with lymph nodes being absent [61]. The medullary portion of the follicles develops between embryonic days 11 and 13, while the cortical region's development begins around the time of hatching and concludes by the end of the second week of life. However, regarding the immune system, the differences between birds and mammals are not limited to this aspect [62]. Due to the lack of cross-reactivity for the three classes of immunoglobulins, namely IgA, IgM, and IgG, distinct names have been recommended, despite the immunoglobulin structures of birds being similar to those of mammals. Additionally, in birds, some immune-competent cells, such as T lymphocytes that lack receptors for erythrocytes, exhibit differences compared to those in mammals. Since the immune response in birds is conditioned by a set of factors different from those encountered in mammals, there is a need for organs, effectors, and regulatory mechanisms to ensure the appropriate amplitude of the immune response. Within the bursa of Fabricius, where B lymphocytes undergo differentiation, researchers have established that the fundamental mechanism of the immune response in birds closely resembles that in mammals [63]. Bursa-dependent B cells exclusively handle antibody production in birds. The chicken bursa serves as a valuable animal model for exploring B cell differentiation and function due to its distinct role in B cell development and its convenient anatomical accessibility on specific days [61, 64]. Although there is currently a wealth of data regarding the complexity of effector immune mechanisms in birds, many aspects related to the mode of action, especially lymphocytic interaction, in assessing the exact role of bursa of Fabricius, and elucidating the evolution of the B lymphocyte, remain not fully understood [65]. It is possible that the unresolved aspects concerning the biology of avian lymphocytes may be due to the lack of suitable markers, such as the formation of E rosettes, which could serve as a basis for preparative procedures [66]. The antibodies found in birds, especially in chickens, do not activate the human complement system and do not react with rheumatoid factor, nor with murine anti-IgG antibodies or human or bacterial Fc receptors, unlike those in mammals [66, 67]. Among the immune effectors found in birds are T lymphocytes, B lymphocytes, macrophages, natural killer cells (NK cells), cytotoxic T lymphocytes, and antigen-presenting cells (APC). Research has highlighted the presence of two functionally distinct cell types in birds, analogous to those in mice, namely TH (helper T cells) and TS (suppressor T cells). These cells can interact with both T and B lymphocytes [61, 68]. In contrast to other organisms, in birds, activated B lymphocytes at the plasmocyte stage produce three types of antibodies: IgG/IgY (serum/yolk), IgM, and IgA [69]. In birds, antibody-dependent cellular cytotoxicity (ADCC) is the ability of an adherent cell with phagocytic capacity, non-T, non-B, equipped with Fc receptors, to exhibit activity against syngeneic or allogeneic erythrocytes. Antigen-presenting cells are activated under the influence of molecular factors and cytokines [70]. In the absence of antigenic stimulation, the transformation of the immune system appears to follow a well-established pattern [71].

*Urtica dioica* (stinging nettle), a plant well known for the biological properties of its phenolic compounds, could be considered as a natural alternative source for possible applications in industrial areas, such as food/feed, cosmetics, and also phytomedicine. Its anti-oxidative capacity was considered the foundation of its biological activities, which still depends on the habitat and external factors that influence its composition.

In a study aiming to investigate the effect of pollution on the efficacy of the active compound of the stinging nettle [72], changes in body weight, total leukocyte numbers (TL), and delayed type hypersensitivity (DTH) were monitored in 28 day-old chickens (n = 12/group) injected subcutaneously twice (days 0 and 7), with 0.5 ml of alcoholic nettle extracts harvested from both unpolluted and polluted areas, against untreated and $70^0$ alcohol-treated controls. No significant changes were recorded for the weight gain/period between groups from polluted and unpolluted areas. Nevertheless, the total leukocyte numbers decreased in the group from polluted areas, this change being reflected by the local hypersensitivity test (wattle test), the size of the infiltration being diminished by the extract from the unpolluted area.

Investigations of the *in vitro* effects of lectins in adult hens vaccinated against Newcastle disease (day 0, day 14) during a three-week experiment indicated that the reactivity of the birds increased gradually versus PHA and ConA toward the end of the experiment, with a more pronounced stimulation capacity of the *Canavalia ensiformis* (L.) DC. than that of *Phytolacca americana* L. (unpublished data, **Table 1**).

A comparison of reactivity in Pisces, Reptilia, and Aves to the lectins of *Canavalia ensiformis (L.) DC.* and *Phytolacca americana* L. indicated a gradual increase paralleling the phylogenetic progress of the species.

| Rec. | *Control* | PHA | LPS | Con |
|---|---|---|---|---|
| Day 0 | 69.27 ± 13.31 | 64.20 ± 81.90 | 68.66 ± 10.67 | 67.12 ± 12.79 |
| Day 9 | 59.57 ± 11.20 | 52.99 ± 9.09 | 54.87 ± 11.1 | 56.21 ± 7.28 |
| Day 21 | 67.37 ± 6.11 | 70.93 ± 12.73 | 77.23 ± 21.2 | 76.22 ± 12.23 |

*Rec = sampling on days 0, 9, and 21; Control = untreated variant; PHA = phytohaemagglutinin M of* Phytolacca americana *L.; LPS = lipopolysaccharide of bacterial cell wall; Con = concanavaline A of* Canavalia ensiformis *(L.) DC. Values are expressed as arithmetic means ± standard error.*

**Table 1.**
*Stimulation indices in hens subjected in vitro to antigenic stimulation: The increase in the LPS activity when compared to that of PHA and ConA indicated an increased activity of B lymphocytes.*

## 5. Mammalia: are they top responders?

In mammals, the initial defense mechanism consists of physical barriers such as the skin and mucosal membranes [73, 74]. The epithelial cells from respiratory, gastrointestinal, and urogenital tracts not only create mechanical barriers but also release various antimicrobial factors, such as antimicrobial peptides and defensins, with a crucial role in the innate immune response [74, 75]. Other vital cellular components of the innate immune system in cattle include neutrophils, NK cells, dendritic cells (DC), gamma delta T cells, mucosa-associated invariant, macrophages, and granulocytes [74, 76].

The development of the immune system commences early in gestation across all mammalian species. As the fetus grows, the immune system undergoes a multitude of changes, with cells emerging and adopting specialized functions [77, 78]. The maturation of the immune system in cattle coincides with embryonic development and subsequently progresses during fetal development, leading to an increasingly intricate immune response to antigenic stimulation. The development of the immune system in calves takes place early during the fetal period. Both sides of the immune response, cellular and humoral, develop in parallel [79].

Since 1973, Schultz and colleagues [80] demonstrated that fetal peripheral blood lymphocytes are present starting from day 45, while cells carrying IgM are detected from day 59, and those carrying IgG emerge from day 135. The stage at which the immune system matures is highly significant, allowing the developing organism to distinguish between self and non-self antigens. This differentiation helps prevent undesired immune responses against its own components in the future. Therefore, from the age of 3 to 4 months of gestation onward, the fetus does not mount an immune response to either self or non-self antigens. This immunological tolerance is achieved through the T-mediated mechanism, avoiding immune reactions against self and non-self antigens that may mistakenly appear outside the first 3 months of gestation [80]. Through *in vitro* stimulation tests, we can assess cellular reactivity. By cultivating cells in the presence of polyclonal mitogens that specifically stimulate T lymphocytes, such as phytohemagglutinin (PHA) and concanavalin A, we can demonstrate both T lymphocyte reactivity and the integrity of intercellular cooperation pathways among antigen-presenting cells (APC), T lymphocytes, and B lymphocytes (APC-T-B). The cellular immune system of calves can respond to antigenic stimulation from birth. The immune system in calves matures progressively, evolving from conception to achieving full maturity around 6 months after birth; the response varies with age and the structure of the antigen [81]. Cellular effectors, both specific and nonspecific, play a significant role in defending the organism against infections [25]. They participate in defense mechanisms through processes like phagocytosis and pinocytosis, releasing lytic enzymes that contribute to the destruction of microorganisms. Cells with the ability to capture, ingest, and digest pathogens are actively involved in the phagocytic process [82]. Studies have demonstrated a notable increase in the blood concentration of $\gamma\delta$ T lymphocytes, particularly in newborn calves, accounting for 25–30% of peripheral blood leukocytes. In contrast, adult cattle typically exhibit values ranging from 3 to 10% [83]. The epitheliochorial type of placenta found in ruminants does not allow the transplacental passage of immunoglobulins, resulting in calves being born agammaglobulinemic [84, 85]. Studies have indicated that starting from the fourth month of gestation, the fetal response to antigenic stimulation in cattle is possible, although the antibodies are IgM, specific to a primary response, and they disappear rapidly. Newborns exhibit a lower level of immune competence compared to adults, as a consequence of the transient presence of a high titer of endogenous corticosteroids during the perinatal period [86].

Similar to calves, piglets are agammaglobulinemic at birth. Their ability to survive is directly reliant on obtaining maternal immunity through the intake of colostrum and milk [87]. The protective function of maternal immunity involves various factors, particularly specific systemic humoral immunity.

The farmed subjects, representatives of the class as well as the complex structure and functionality of the mammalian immune system, trigger numerous studies targeting the value of alternative immune-stimulating therapies in those species.

In a study researching the adjuvant and immune cell-stimulating potential of certain plants from the perspective of farmed herbivores differing by their digestive physiology, alcoholic vegetal extracts were tested *in vitro* [88]. Romanian Spotted dairy cows, Angora goats, and Romanian draft horses were sampled for blood, and the samples were treated *in vitro* in the blast transformation test [45] with alcoholic extracts of *Calendula officinalis* L., *Echinacea angustifolia* L., and *E. purpurea* L. The growth indices were the lowest in goats (*C. officinalis* L.- 58.52 ± 10.02%, *E. angustifolia* L. - 50.06 ± 11.67%, *E.purpurea* L.-50.79 ± 10.98%) and higher in bovine (*C. officinalis* L.– 69.9 ± 2.65%, *E.angustifolia* L.- 74.9 ± 10.1%, $p < 0.05$) and increased toward *E. angustifolia* L. versus *C. officinalis* L.; further, the responses to *C. officinalis* L., *E. angustifolia* L. *and E.purpurea* L. were the most pronounced in horses. Thus, there were differences between the tested species, probably due to the differences in the anatomy as well as the physiology of their immunity.

It is well-known that the stress of various origins represents an important factor impeding on vaccination results in bovine. In a study aimed at investigating the anti-stress and immunological influence of the oral administration of a marigold tea (antioxidant activity 85.10 ± 5.21%) to adult bovine, vaccinated against multiple antigens with Cattle Master, it was shown that the stress levels constantly decreased in marigold tea-treated animals (1.09 ± 0.047 to 0.064 ± 0.02) and did not significantly vary in controls. The blast transformation test indicated lower SI% and a prolonged response to vaccination in the treated group (40.65 ± 14.91 to 79.82 ± 3.34 versus 50.62 ± 5.93 and 81.31 ± 2.25%) by the end of the experiment. It could be concluded that the marigold tea exerted a stress-lowering effect but prolonged the response to vaccination when orally administered as an infusion to adult bovine.

As mentioned before, the outcome of the host-pathogen interaction microbial aggression highly depends on immunity, and medicinal plants, available on the pastures, could enhance and strengthen the resistance against diseases.

Research was carried out on Mangalitza pigs (suckling, weaned piglets and sows) raised for tasty lean meat mainly on low-input farms to monitor the effects of plants readily available from local resources such as *Calendula officinalis* L., *Satureja hortensis* L., *Allium sativum* L., *Coriandrum sativum* L., and *Cucurbita maxima* L. An *in vitro* leukocyte blast transformation test was carried out by treating each 0.2 ml of the whole blood mixed with RPMI1640 culture medium (1:4 v/v) with 1.5 μL/0.2 mL of the selected alcoholic extracts (70°, v/v). The results indicated that statistically significant differences were present between the young age groups, suckling and weaned piglets ($p = 0.017$ to 0.000016) for all plants except marigold, for weaned piglets and sows ($p = 0.0001–0.0359$) for all plant extracts, while for suckling piglets/sows $p = 0.0035$ and $p = 0.0461$ were recorded for thyme and garlic, respectively. No toxic effects were present for the tested alcoholic extracts of *Calendula officinalis* L., *Satureja hortensis* L., *Allium sativum* L., *Coriandrum sativum* L., and *Cucurbita maxima* L. (unpublished data). This experiment underlined that the plant extracts used, well-known for their biological effects, impacted animals based on the age of the pigs and plant family, proving their immune-stimulating capacity [89].

## 6. Conclusions

Multiple beneficial biological effects of medicinal plants, underlined by millennia of ethno-pharmaceutical practices, prove to still hide part of their constructive potential. In animal farming and veterinary medicine, a support for One Health, One Welfare for consumers, medicinal plants could provide help, strengthening the fight in preserving both.

Further, the differentiated effect of plant extracts on different classes of animals, based on their phylogeny, impacted on individuals grounded on their level of immune development as well as age and raising technology, proving their immune-stimulating capacity allocated to their own taxonomy.

Use of plants in preserving animal health and welfare by stimulating immune functions as shown in the case of mint, marigold, pumpkin, summer savory, and garlic seems encouraging within the One Health, One Welfare scheme.

Numerous investigations, such as those described at the level of laboratory tests in the present work, provided proof that the biological activity of plant extracts was both animal and plant species dependent. Tailored treatment and feed supplementation schemes founded on the laboratory results could be beneficial in farmed animals as well, providing basis for sustainability of the enterprises at lower costs.

## Acknowledgements

We acknowledge the financial support provided by the European Union's Horizon 2020 research and innovation program, under grant agreement no. 816172, PPILOW and the Ministry of Research, Innovation and Digitization, CNCS -UEFISCDI, within PNCD III and ERANET Core Organic Co-fund ROAM Free #249/2021.

## Conflict of interest

None of the authors have any existing or possible conflict of interest, including financial, personal or any other relationship which could influence their scientific work.

**Author details**

Aurel Vasiu[1], Vasile Cozma[1], Anamaria Cozma-Petruţ[2], Mihai Băieş[1], Marina Spînu[1,3]*, Emöke Páll[1,3], Diana Olah[1], Carmen Dana Şandru[1,3], Gheorghiţă Duca[3], Köbölkuti Lorand[1] and Gabriel Gati[4,5]

1 Faculty of Veterinary Medicine, University of Agricultural Sciences and Veterinary Medicine, Cluj-Napoca, Romania

2 Faculty of Pharmacy "Iuliu Haţieganu", Department of Bromatology, Hygiene, Nutrition, University of Medicine and Pharmacy, Cluj-Napoca, Romania

3 Institute of Research and Development for Montanology, Cristian, Sibiu, Romania

4 Environmental Health Center, Cluj-Napoca, Romania

5 School of Public Health, College of Political, Administrative and Communication Sciences, Babeş-Bolyai University, Cluj-Napoca, Romania

*Address all correspondence to: marina.spinu@gmail.com

## References

[1] Nabi F, Ahmed J, Tao W, Lu Q, Bhutto ZA, Qadir A, et al. An updated review on efficiency of penthorum chinense pursh in traditional uses, toxicology, and clinical trials. BioMed Research International. 2023:4254051. DOI: 10.1155/2023/4254051

[2] Kubkomawa HI, Nafarnda DW, Tizhe MA, Daniel TK, Shua NJ, Ugwu CC, et al. Ethno-veterinary health management practices amongst livestock producers in Africa: A review. Advances in Agricultural Sciences. 2020;**6**:1-006

[3] Romero B, Susperregui J, Sahagún AM, Diez MJ, Fernández N, García JJ, et al. Use of medicinal plants by veterinary practitioners in Spain: A cross-sectional survey. Frontiers in Veterinary Science. 2022;**9**:1060738. DOI: 10.3389/fvets.2022.10607384

[4] Awaad AS, Al-Jaber NA. Antioxidant natural plant. Recent Progress in Medicinal Plants: Ethnomedicine: Source & Mechanism I. 2010;**27**:1e35

[5] Awaad E, Awaad A. Role of medicinal plants on growth performance and immune status in fish. Fish & Shellfish Immunology. 2017;**67**:40-54. DOI: 10.1016/j.fsi.2017.05.034

[6] Hassan BAR. Medicinal plants (importance and uses). Pharmaceutica Analytica Acta. 2013. DOI: 10.4172/2153-2435.1000e139

[7] Sakai M. Current research status of fish immunostimulants. Aquaculture. 1999;**172**:63e92

[8] Van Hai N. The use of medicinal plants as immunostimulants in aquaculture: A review. Aquaculture. 2015;**446**:88e96

[9] Christaki E, Giannenas I, Bonos E, Florou-Paneri P. Chapter 2 - Innovative uses of aromatic plants as natural supplements in nutrition. In: Florou-Paneri P, Christaki E, Giannenas I, editors. Feed Additives. Cambridge, Massachusetts: Academic Press; 2020. pp. 19-34. DOI: 10.1016/B978-0-12-814700-9.00002-9

[10] Ramos da Silva LR, Ferreira OO, Cruz JN, de Jesus Pereira Franco C, Oliveira Dos Anjos T, Cascaes MM, et al. Lamiaceae essential oils, phytochemical profile, antioxidant, and biological activities. Evidence-based Complementary and Alternative Medicine. 2021;**2021**:6748052. DOI: 10.1155/2021/6748052

[11] Elumalai P, Kurian A, Lakshmi S, Faggio C, Esteban MA, Ringø E. Herbal immunomodulators in aquaculture. Reviews in Fisheries Science & Aquaculture. 2020;**29**(1):33-57. DOI: 10.1080/23308249.2020.1779651

[12] Firmino JP, Galindo-Villegas J, Reyes-Lopez FE, Gisbert E. Phytogenic bioactive compounds shape fish mucosal immunity. Frontiers in Immunology. 2021;**12**:695973

[13] Franz CM, Baser KHC, Hahn-Ramssl I. Chapter 3 - herbs and aromatic plants as feed additives: Aspects of composition, safety, and registration rules. In: Florou-Paneri P, Christaki E, Giannenas I, editors. Feed Additives. Cambridge, Massachusetts: Academic Press; 2020. pp. 35-56. DOI: 10.1016/B978-0-12-814700-9.00003-0

[14] Napoli E, Siracusa L, Ruberto G. New tricks for old guys: Recent developments in the chemistry, biochemistry, applications and exploitation of

selected species from the Lamiaceae Family. Chemistry & Biodiversity. 2020;**17**(3):e1900677. DOI: 10.1002/cbdv.201900677

[15] Kallunki JA, Heywood VH. Flowering plants of the world. Brittonia. 1994;**46**:4. DOI: 10.2307/2806914

[16] Harley RM. Checklist and key of genera and species of the Lamiaceae of the Brazilian Amazon. Rodriguésia. 2012;**63**(1):129-144

[17] Nieto G. Biological activities of three essential oils of the Lamiaceae family. Medicines (Basel). 2017;**4**:63. DOI: 10.3390/medicines4030063

[18] Sun J, Sun P, Kang C, Zhang L, Guo L, Kou Y. Chemical composition and biological activities of essential oils from six lamiaceae folk medicinal plants. Frontiers in Plant Science. 2022;**13**:919294. DOI: 10.3389/fpls.2022.919294

[19] Luz TRSA, Leite JAC, de Mesquita LSS. Seasonal variation in the chemical composition and biological activity of the essential oil of Mesosphaerum suaveolens (L.) Kuntze. Industrial Crops and Products. 2020;**153**:112600

[20] Ouakouak H, Benchikha N, Hassani A, Ashour ML. Chemical composition and biological activity of Mentha citrata Ehrh., essential oils growing in southern Algeria. Journal of Food Science and Technology. 2019;**56**(12):5346-5353

[21] Kosakowska O, Węglarz Z, Pióro-Jabrucka E, Przybył JL, Kraśniewska K, Gniewosz M, et al. Antioxidant and antibacterial activity of essential oils and Hydroethanolic extracts of Greek oregano (O. Vulgare L. subsp. hirtum (link) Ietswaart) and common oregano (O. Vulgare L. subsp. vulgare). Molecules. 2021;**26**(4):988. DOI: 10.3390/molecules26040988

[22] Mączka W, Twardawska M, Grabarczyk M, Wińska K. Carvacrol-a natural phenolic compound with antimicrobial properties. Antibiotics (Basel). 2023;**12**(5):824. DOI: 10.3390/antibiotics12050824

[23] Uritu CM, Mihai CT, Stanciu GD, Dodi G, Alexa-Stratulat T, Luca A, et al. Medicinal plants of the family Lamiaceae in pain therapy: A review. Pain Research & Management. 2018;**8**:7801543. DOI: 10.1155/2018/7801543

[24] Düpjan S, Dawkins MS. Animal welfare and resistance to disease: Interaction of affective states and the immune system. Frontiers in Veterinary Science. 2022;**9**:929805. DOI: 10.3389/fvets.2022.929805

[25] Chaplin DD. Overview of the immune response. The Journal of Allergy and Clinical Immunology. 2010;**125** (2 Suppl. 2):S3-S23. DOI: 10.1016/j.jaci.2009.12.980

[26] Belkaid Y, Hand TW. Role of the microbiota in immunity and inflammation. Cell. 2014;**157**(1):121-141. DOI: 10.1016/j.cell.2014.03.011

[27] Dunin-Horkawicz S, Klaus KO, Lupas AN. Prokaryotic protein networks mediating innate immunity and apoptosis. Journal of Molecular Biology. 2014;**426**:1568-1582. DOI: 10.1016/j.jmb.2013.11.030

[28] Di Sotto A, Vitalone A, Di Giacomo S. Plant-derived nutraceuticals and immune system modulation: An evidence-based overview. Vaccines (Basel). 2020;**8**(3):468. DOI: 10.3390/vaccines8030468

[29] Kogut MH, Lee A, Santin E. Microbiome and pathogen interaction with the immune system. Poultry Science. 2020;**99**(4):1906-1913. DOI: 10.1016/j.psj.2019.12.011

[30] Zheng D, Liwinski T, Elinav E. Interaction between microbiota and immunity in health and disease. Cell Research. 2020;**30**:492-506. DOI: 10.1038/s41422-020-0332-7

[31] Mezouar S, Chantran Y, Michel J, Fabre A, Dubus J-C, Leone M, et al. Microbiome and the immune system: From a healthy steady-state to allergy associated disruption. Human Microbiome Journal. 2018;**10**:11-20. DOI: 10.1016/j.humic.2018.10.001. ISSN 2452-2317

[32] Parker CW. The role of intracellular mediators in the immune response. In: Cohen S, Pick E, Joost J, editors. Oppenheim, Biology of the Lymphokines. Amsterdam: Academic Press, Elsevier; 1979. pp. 541-583. DOI: 10.1016/B978-0-12-178250-4.50024-6

[33] Taheri Y, Quispe C, Herrera-Bravo J, Sharifi-Rad J, Ezzat SM, Merghany RM, et al. Urtica dioica-derived phytochemicals for pharmacological and therapeutic applications. Evidence-based Complementary and Alternative Medicine: eCA. 2022:4024331. DOI: 10.1155/2022/4024331

[34] Secombes C, Wang T. The innate and adaptive immune system of fish. Infectious Disease in Aquaculture Prevention and Control. 2001;**231**:3e68

[35] Cerbu CG, Spînu M, Şandru CD, Vasiu A, Olah D, Brudaşcă GF. Influence of the raising system on phagocytosis in the endangered Danube salmon (Hucho hucho). Global Ecology and Conservation. 2018;**16**:e00514

[36] Viswanath K. Fish immune system and its nutritional modulation for preventive health care. Animal Feed Science and Technology. 2012;**173**(1-2):111-133

[37] Buchmann K. Evolution of innate immunity: Clues from invertebrates via fish to mammals. Frontiers in Immunology. 2014;**5**:459. DOI: 10.3389/fimmu.2014.00459

[38] Zapata A, Diez B, Cejalvo T, Gutiérrez-de Frías C, Cortés A. Ontogeny of the immune system of fish. Fish & Shellfish Immunology. 2006;**20**(2):126-136. DOI: 10.1016/j.fsi.2004.09.005

[39] Yano T. Assays of hemolytic complement activity. In: Stolen JS, Fletcher TC, Anderson DP, Kaatari SL, Rowley AF, editors. Techniques in Fish Immunol. Vol. 2. USA: SOS Publications; 1992. pp. 131-141

[40] Lambris JD. The chemistry, biology and phylogeny of C3 complement today. Complement profiles. Karger Basel. 1993;**I**:16-45

[41] Scapigliati G, Romano N, Abelli L. Monoclonal antibodies in fish immunology: Identification, ontogeny and activity of T- and B-lymphocytes. Aquaculture. 1999;**172**:3-28

[42] Castillo A, Sánchez C, Domínguez J, Kaatari SL, Villena AJ. Ontogeny of IgM and IgM-bearing cells in rainbow trout. Developmental and Comparative Immunology. 1993;**17**:419-424

[43] Dozmorov IM, Dresser D. Immune system as a sensory system. International Journal of Biomedical Sciences. 2010;**6**(3):167-175

[44] Wu L, Qin Z, Liu H, Lin L, Ye J, Li J. Recent advances on phagocytic B cells in teleost fish. Frontiers in Immunology. 2020;**27**(11):824. DOI: 10.3389/fimmu.2020.00824

[45] Khokhlova IS, Spinu M, Krasnov BR, Degen AA. Immune response to fleas in a wild desert rodent: Effect of parasite

species, parasite burden, sex of host and host parasitological experience. The Journal of Experimental Biology. 2004;**207**(Pt. 16):2725-2733. DOI: 10.1242/jeb.01090

[46] Spînu M, Pall E, Niculae M, Vasiu A, Brudaşcă GF, Cerbu C, et al. Seasonal changes of phagocytic activity enhanced by an acoholic mint extract in cultured rainbow trout. IJPER. 2018;**52**(4s):s86-s90. DOI: 10.5530/ijper.52.4s.80

[47] Jurd RD. In: Turner RJ, editor. Immunology: A Comparative Approach. John Wiley & Sons Ltd: England; 1994. pp. 137-147

[48] Good RA, Papermaster BW. Ontogeny and phylogeny of adaptive immunity. Advances in Immunology. 1944;**4**:1-115

[49] Cooper EL, Klempau AE, Zapata AG. Reptilian immunology. In: Gans C, Billett F, Maderson PFA, editors. Biology of the Reptilia. Vol. 14. New York: Wiley-Interscience; 1985. pp. 599-678

[50] Frye FL. Biomedical and Surgical Aspects of Captive Reptile Husbandry. Florida: Krieger Publishing Company; 1991. pp. 101-149, 209-281

[51] Salantino SK, Minton SA. Immune response of snakes. Copeia. 1973;**3**:504-515. Available from: https://www.jstor.org/stable/1443116

[52] Ambrosius H. Immunoglobulins and antibody production in reptiles. In: Marchalonis JJ, editor. Comparative Immunology. Oxford, England: Blackwell; 1976. pp. 298-334

[53] Borysenko M. Changes in spleen histology in response to antigenic stimulation in the snapping turtle, Chelydra sepentina. Journal of Morphology. 1976;**149**:243-264

[54] Cooper EL. Skin transplant rejection in Apodan amphibians and Lacertan reptiles. American Zoologist. 1969;**9**:333

[55] Karstad L. Influence of low body temperature on establishment of prolonged infections in animal reservoirs. In: Sci. Proc. 100th Ann. Meeting Amer. Vet. Med- Assoc. Chicago: American Veterinary Medical Association; 1963. pp. 186-193

[56] Manning MJ, Turner RJ. Comparative Immunology. Scotland, Blackie and Son: Glasgow; 1976

[57] Kollias GV. Immunological aspects of infectious disease in reptiles and amphibians. In: Hoff GL, Frye FL, Jacobson ER, editors. Diseases of Amphibians and Reptiles. New York NY: Plenum Press; 1984

[58] Borysenko M, Cooper EL. Lymphoid tissue in the snapping turtle, Chelydra sepentina. Journal of Morphology. 1972;**138**:487-481

[59] Kobolkuti L, Cadar D, Czirjak G, Niculae M, Kiss T, Sandru C, et al. The effects of environment and physiological cyclicity on the immune system of Viperinae. The Scientific World Journal. 2012:574867. DOI: 10.1100/2012/574867

[60] Ekino S, Sonoda K. New insight into the origin of IgG-bearing cells in the bursa of Fabricius. International Review of Cell and Molecular Biology. 2014;**312**:101-137

[61] Cheng J, Lei H, Xie C, Chen J, Yi X, Zhao F, et al. B lymphocyte development in the Bursa of Fabricius of young broilers is influenced by the gut microbiota. Microbiology Spectrum. 2023;**11**(2):e0479922. DOI: 10.1128/spectrum.04799-22

[62] Davison F. The importance of the avian immune system and its unique

features. In: Avian Immunology. Amsterdam: Elsevier, Academic Press; 2014. pp. 1-9

[63] Salvante KG. Techniques for studying integrated immune function in birds. The Auk. 2006;**123**(2):575-586

[64] Funk PE, Palmer JL. Dynamic control of B lymphocyte development in the bursa of Fabricius. Archivum Immunologiae et Therapiae Experimentalis (Warsz). 2003;**51**:389-398

[65] Oláh I, Felföldi B, Benyeda Z, Nagy N, Kovács T. The morphology and differentiation of stromal cells in the cortex of follicles in the bursa of Fabricius of the chicken. Anatomical Record (Hoboken, NJ. 2007). 2022;**305**(11):3297-3306. DOI: 10.1002/ar.24893

[66] Sharma JM. The structure and function of the avian immune system. Acta Veterinaria Hungarica. 1997;**45**(3):229-238

[67] Romito M, Viljoen GJ, Du Plessis DH. Eliciting antigen-specific egg-yolk IgY with naked DNA. BioTechniques. 2001;**31**(3) 670, 672, 674-5

[68] Yamamoto Y, Komatsu T, Suzuki Y. Localization of neuropeptides in endocrine cells of the chicken thymus. The Journal of Veterinary Medical Science 1997;**59**(7):601-603

[69] Warr GW, Magor KE, Higgins DA. IgY: Clues to the origins of modern antibodies. Immunology Today. 1995;**16**(8):392-398

[70] Murtaugh MP, Foss DL. Inflammatory cytokines and antigen presenting cell activation. Veterinary Immunology and Immunopathology. 2002;**87**(3-4):109-121

[71] Carter JH. The immune system as a model for pattern recognition and classification. Journal of the American Medical Informatics Association. 2000;**7**(1):28-41. DOI: 10.1136/jamia.2000.0070028

[72] Vasiu A, Sandru C, Spinu M, Ungureanu E, Marian F, Popescu S, et al. Environmental pollution degree changes the biological activity of nettle plant extracts in chickens. Current Perspectives on Medicinal and Aromatic Plants. 2023;**6**(2):84-89. DOI: 10.38093/cupmap.1337889

[73] Rodríguez-Habibe I, Celis-GiraldoC,PatarroyoME,AvendañoC, Patarroyo MA. A comprehensive review of the immunological response against foot-and-mouth disease virus infection and its evasion mechanisms. Vaccines (Basel). 2020;**8**(4):764. DOI: 10.3390/vaccines8040764

[74] Vlasova AN, Saif LJ. Bovine immunology: Implications for dairy cattle. Frontiers in Immunology. 2021;**29**(12):643206. DOI: 10.3389/fimmu.2021.643206

[75] Hans M, Madaan HV. Epithelial antimicrobial peptides: Guardian of the oral cavity. International Journal of Peptide. 2014;**2014**:370297. DOI: 10.1155/2014/370297

[76] Burrell CJ, Howard CR, Murphy FA. Innate immunity. In: Fenner and White's Medical Virology. Amsterdam: Elsevier; 2017. pp. 57-64. DOI: 10.1016/B978-0-12-375156-0.00005-9

[77] Cortese VS. Neonatal immunology. The Veterinary Clinics of North America. Food Animal Practice. 2009;**25**(1):221-227. DOI: 10.1016/j.cvfa.2008.10.003

[78] Halliwell REW, Gorman NT. Veterinary clinical immunology. In:

Neonatal Immunology. Philadelphia: WB. Saunders Co.; 1989. pp. 193-205

[79] Barrington GM, Parish SM. Bovine neonatal immunology. The Veterinary Clinics of North America. Food Animal Practice. 2001;**17**(3):463-476. DOI: 10.1016/s0749-0720(15)30001-3

[80] Sordillo LM, Aitken SL. Impact of oxidative stress on the health and immune function of dairy cattle. Veterinary Immunology and Immunopathology. 2009;**128**(1-3):104-109. DOI: 10.1016/j.vetimm.2008.10.305

[81] Chase CC, Hurley DJ, Reber AJ. Neonatal immune development in the calf and its impact on vaccine response. The Veterinary Clinics of North America. Food Animal Practice. 2008;**24**(1):87-104. DOI: 10.1016/j.cvfa.2007.11.001

[82] Rosales C, Uribe-Querol E. Phagocytosis: A fundamental process in immunity. BioMed Research International. 2017;**2017**:9042851. DOI: 10.1155/2017/9042851

[83] Guzman E, Hope J, Taylor G, Smith AL, Cubillos-Zapata C, Charleston B. Bovine γδ T cells are a major regulatory T cell subset. Journal of Immunology. 2014;**193**(1):208-222. DOI: 10.4049/jimmunol.1303398

[84] Bragg R, Macrae A, Lycett S, Burrough E, Russell G, Corbishley A. Prevalence and risk factors associated with failure of transfer of passive immunity in spring born beef suckler calves in Great Britain. Preventive Veterinary Medicine. 2020;**181**:105059. ISSN 0167-5877

[85] Gabriël S, Geldhof P, Phiri IK, Cornillie P, Goddeeris BM, Vercruysse J. Placental transfer of immunoglobulins in cattle infected with Schistosoma mattheei. Veterinary Immunology and Immunopathology. 2005;**104**(3-4):265-272. DOI: 10.1016/j.vetimm.2004.12.010

[86] Fair T. The contribution of the maternal immune system to the establishment of pregnancy in cattle. Frontiers in Immunology. 2015;**6**:7. DOI: 10.3389/fimmu.2015.00007

[87] Salmon H, Berri M, Gerdts V, Meurens F. Humoral and cellular factors of maternal immunity in swine. Developmental & Comparative Immunology. 2009;**33**(3):384-393. ISSN 0145-305X. DOI: 10.1016/j.dci.2008.07.007

[88] Vasiu A, Sandru C, Pall E, Ungureanu E, Marian F, Popescu S, et al. Host and plant taxonomy as influential factors in controling in vitro specific cell-mediated response in herbivores. Current Perspectives on Medicinal and Aromatic Plants. 2023a;**6**(2):78-83. DOI: 10.38093/cupmap.1337829

[89] Spînu M, Rusu LA, Niculae M, Pall E, Vasiu A, Olah DI, et al. In vitro age dependent cell-mediated responses to antimicrobial plant extracts in free range Mangalitza pigs. In: The 6th International Mediterranean Symposium on Medicinal and Aromatic Plants (MESMAP-6), 16th to 19th April 2020, Izmir, Selcuk (Ephesus), Turkey

Chapter 8

# Innovative Perspectives on the Use of Herbs and Spices as Illuminators of Health in the Context of Diseases

*Jose Prakash Dharmian and Prakash Ramakrishnan*


## Abstract

This comprehensive article embarks on a global journey through the multifaceted landscape of herbs and spices in disease management. We explore the traditional wisdom and emerging scientific evidence behind the utilization of these botanical wonders. Incorporating detailed case studies from diverse regions, we unravel the nuanced interplay between cultural practices, natural remedies, and the pursuit of holistic well-being. The synthesis of traditional wisdom and modern research offers insights into the potential integration of herbs into global healthcare practices. This delves into the molecular-level chemical constituents of herbs and their potential applications in disease management across the globe. By combining insights from traditional knowledge and cutting-edge scientific research, we aim to unravel the intricate chemistry of herbs and understand how these bioactive compounds contribute to therapeutic effects.


**Keywords:** herbal medicine, curcumin, ethnobotanical studies, flavonoids, terpenoids

## 1. Introduction

Herbal medicine, rooted in ancient traditions, weaves a narrative of healing through the ages. The profound connection between humans and plants has led to the exploration of the chemical constituents within herbs, unveiling a treasure trove of therapeutic compounds [1]. It has played a pivotal role in healthcare practices across cultures for centuries. Spices and herbs have long been considered medicinal foods. The study of spices' biological activity-transmission potential is making a slow comeback in the realm of human wellbeing. Seed spices are a vital category of agricultural commodities that make significant contributions to our nation's economic success [2]. India has a long history of manufacturing spices. Rajasthan and Gujarat have contributed over 80% of the country's spice output. Spices create a diverse set of secondary metabolites known as phytonutrients. Plant-based chemicals are naturally occurring chemical substances with biological functions that occur in plants that serve as a natural defensive system. They have historically been employed as medicine, fragrances, scent compounds, dyes, and agrochemicals [3]. Chemoinformatics will help in the development of novel medicines by combining modern biology and computer science technology. These metabolites remain a major source of new drugs today. Herbs and spices are now permitted to be utilized to enhance both food appeal and health. According to World Health Organization data, 70–80% of the global population seeks primary healthcare from contemporary medicine, largely herbal and natural ones. Furthermore, as much as 60% of the world population and 80% of the population in developing nations utilize herbs and plants for medicinal purposes [4]. The antibacterial properties of cinnamon oil contrary to Bacillus anthracisspores were established in the first scientific study on the role of spices as preservatives. Population demography, a focus on health rather than illness, the trend toward self-care and self-diagnosis, and growing consumer awareness of traditional remedies are the main factors propelling the worldwide phytochemical company's growth. Throughout the world, people are using herbs and spices as food additives to enhance the organoleptic properties of food and prolong its shelf life by lowering or getting rid of foodborne bacteria. Several studies have suggested the use of dietary herbs and spices for their anti-inflammatory, immunological modulatory qualities, antimutagenic and antioxidative which have been shown to improve human health [5]. Herbs can be excellent providers of antioxidants and salt replacements, according to dietary recommendations. In today's nutrition food chain, dairy items are a distinctive carrier that has been successfully used to carry phytochemicals and other nutrients for health benefits. Furthermore, introducing spices, herbs, or extracts to specific dairy products makes them nutraceutical transmitters. As a result, the dairy business should develop innovative techniques for improving the capabilities of traditional dairy products, which might offer great value and have a good 188 impact on users. These byproducts are currently a major source of new medicines.

| S. no | Activity | Important chemical constituents | Dose in rat model | References |
|---|---|---|---|---|
| 1 | Antidiabetic action | Proanthocyanidins (type A epicatechin polymers) | 50 mg/kg | [7, 8] |
| | | Cinnamaldehyde | 20 mg/kg | [9] |
| | | Cinnamic acid | 10 mg/kg | [10] |
| | | 1,8-cineole | 500 mg/kg | [11] |
| | | Alpha-terpinyl acetate | 100 mg/kg | [12] |
| | | Diosgenin | 100 mg/kg | [13] |
| | | Galactomannan | 100 mg/kg | [14] |
| | | Trigoneosides | 50 mg/kg | [15] |
| | | Hydroxyisoleucine | 50 mg/kg | [16] |
| | | Gallic acid | 25 mg/kg | [17] |
| 2 | Antiulcer action | Gingerol | 200 mg/kg | [18, 19] |
| | | Shogaol | 200 mg/kg | [20] |
| 3 | Hypertension | 1,8-cineole | 10 mg/kg | [21, 22] |
| | | Terpinyl acetate | 30 mg/kg | [23] |
| 4 | Immunomodulatory action | Curcumin | 10 mg/kg | [24, 25] |
| 5 | Cardiovascular diseases | Allicin, | 7 mg/kg | [26, 27] |
| | | 6-shogaol | 20 mg/kg | [28] |

**Table 1.**
*Chemical composition of herbs and spices according to diseases.*

Plant-derived chemicals have recently received a lot of interest because of their vast variety of applications. This chapter aims to delve into the multifaceted world of herbal interventions in disease management, taking a global perspective. As the world grapples with an increasing burden of chronic diseases, attention turns to traditional remedies embedded in diverse cultures. Across continents, traditional healing systems have long incorporated herbs and spices for their perceived therapeutic properties. From Ayurveda in India to Traditional Chinese Medicine, the global tapestry of healing reveals a shared recognition of nature's pharmacy. At the heart of many traditional healing practices stands turmeric, renowned for its anti-inflammatory and antioxidant properties. The case study explores the extensive use of turmeric in South Asian cuisines and traditional medicine, with a focus on its role in managing conditions like arthritis and digestive issues [6]. Recent scientific studies, including clinical trials and epidemiological observations, shed light on the mechanisms and efficacy of curcumin, the active compound in turmeric as shown in **Table 1**. As we traverse the world, we witness the integration of this golden spice into culinary practices and healing traditions. Venturing to the Mediterranean, we explore the enchanting world of oregano. Beyond its culinary allure, oregano boasts antimicrobial and anti-inflammatory properties [29]. This aromatic herb, with its rich history in traditional medicine, becomes a focal point for cognitive support and anti-inflammatory benefits. Exploring the global diaspora of Middle Eastern communities, we witness how sage transcends borders, adapting to new culinary landscapes while preserving its medicinal essence [30, 31]. Understanding the molecular constituents of herbs is crucial for elucidating their mechanisms of action and optimizing their therapeutic applications. This section emphasizes the importance of molecular-level analysis in uncovering the full potential of herbal medicine.

## 2. The molecular tapestry of herbal medicine

As the prevalence of chronic diseases continues to rise globally, the importance of exploring alternative and complementary approaches becomes increasingly evident. Herbs, with their diverse bioactive compounds, present a reservoir of potential therapeutic interventions that can address the complexities of modern health challenges. Ethnobotanical studies provide insights into the intricate relationships between indigenous communities and their herbal resources [32]. The challenges associated with standardization, regulation, and cross-cultural integration is explored, along with potential opportunities for a more holistic approach to healthcare. The chemical constituents within herbs are the building blocks of their therapeutic effects. These include alkaloids, flavonoids, terpenoids, essential oils, and phenolic compounds, among others. Each constituent contributes to the herb's unique properties, offering a diverse array of compounds that interact with the human body in intricate ways. Turmeric, with its vibrant yellow pigment, owes its therapeutic prowess to curcumin. This polyphenol exhibits anti-inflammatory and antioxidant properties, making it a staple in traditional medicine for ailments ranging from joint pain to digestive issues. *Ginkgo biloba*, a revered herb in Chinese medicine, contains flavonoids and terpenoids. Its historical use for cognitive support has sparked modern research exploring its potential in conditions like Alzheimer's disease. Garlic, celebrated for its culinary uses, also boasts allicin, a sulfur-containing compound with antimicrobial properties. Throughout history, garlic has been utilized for its cardiovascular benefits and immune-boosting properties. Herbs hold cultural significance, reflecting the unique traditions and practices of diverse communities. Traditional Chinese Medicine emphasizes the balance of Qi through herbs like ginseng and astragalus [33, 34]. In Western herbalism, plants like chamomile and lavender are revered for their calming properties. The analysis of the aforementioned spices and herbs has revealed that they are all highly concentrated in antioxidants. Antioxidant substances are believed to shield cellular components from oxidative damage that arises during regular metabolic processes when ingested. Chronic illnesses, overly rigorous exercise, and exposure to pollutants in the environment can all lead to further oxidative damage. Oxidative reactions generate unstable radicals called free radicals that, if left unchecked, can react with and possibly change the structure and functionality of cellular proteins, cell membranes, RNA, DNA, lipoproteins, and carbohydrates. The antioxidant chemicals work by giving free radicals the electrons, this is especially crucial for long-term conditions like cardiovascular disease. In industrial cultures, cardiovascular disease is a leading cause of death for many people under the age of fifty. Atherosclerosis is a disorder of the vessels where the innermost layer thickens due to accumulation of fat and fibrous connective tissue, resulting in a disorder referred to as tightening of the artery walls. Atherosclerosis is the first stage of the illness and can cause hypertension and heart attacks [35]. The first signs of atherosclerosis, known as fatty streaks, are patches of yellow discoloration on the inside of the artery that do not extend into the interior cavity or obstruct blood flow. The sub-endothelial buildup of big foam cells containing intracellular lipid is what distinguishes the streaks. Derived from soft tissue and macrophages as foam cells are most likely the progenitors of fibrous plaques, which are raised lesions that's pale gray in color and have the potential to protrude into the artery wall and decrease the flow of blood across the vessel. When the fibrous plaque ruptures, material is released into the bloodstream and a blood clot forms, whereas hardening of the plaque causes the artery to become hard and hypertensive. Myocardial infarctions or strokes may result from localized occlusion of the artery or from transfer to remote locations. The high concentrations of antioxidants found in herbs and spices add to their therapeutic efficacy. These substances are essential to plants because they shield them from the sun's excessive energy intake during photosynthesis. The plant cell is shielded from oxidative damage and dangerous surplus energy is removed. Both oxygen species that react and reactive nitrogen compounds are generated during the oxidative phase of cellular metabolism. The radical hydroxyl ($OH-$), that could oxidize—that is, take an electron from practically a single molecule and harm the structures of cells and compounds, is the most reactive of all the free radicals. It additionally happens to be the most oxidizing. The antioxidant system's job is to make it easier for free radicals to donate electrons to one another, which lowers the chemical energy of hydroxyl radicals and other reactive oxygen or reactive nitrogen species. After that, the antioxidant molecule must be gradually decreased until the organic molecule is eventually freed as carbon dioxide or oxygen. Several antioxidant chemicals found in significant quantities in plants, including glutathione, ascorbic acid, polyphenols, tocopherols, and carotenoids, can combine chemically or non-enzymatically with an oxygen donor such a free radical. These substances found in herbs and spices give animals and people the necessary antioxidant component in their diets. Apart from the antioxidant chemicals' chemical non-enzymatic defense, an anti-oxidant mechanism made up of many enzymes known as the second stage enzymes is also present. These enzymes catalyze the transformation of hazardous metabolites into substances that are readily eliminated while also eliminating reactive oxygen species. Enzymes such as glutathione peroxidase and catalases remove hydrogen peroxide and organic peroxides, whereas superoxide dismutase eliminates superoxide radicals. NAD(P)H:quinine reductase, g-glutamyl

DOI: http://dx.doi.org/10.5772/intechopen.1004660

| S. no. | Herbs and spices | botanical name | Active constituents | Dose | Therapeutics uses | References |
|---|---|---|---|---|---|---|
| 1 | Garlic | *Allium sativum* L. | Alliin, allicin, diallyl sulfide, diallyl disulfide, and diallyl trisulfide | 2.4 g of *A. sativum* powder dose was found to reduce fasting glucose, 1.2 g of *A. sativum* powder dose was found for treatment for antidiabetic and 1.6 g for hypocholesterolemic treatment | Anti-diabetic, antihyperlipidemic, and anti-atherogenic properties | [36] |
| 2 | Turmeric | *Curcuma longa* L. | Curcumin, demethoxycurcumin, and bisdemethoxycurcumin | A daily dosage of 3.6 g of curcumin decreased inducible prostaglandin E2 synthesis by 62% and 57%, respectively. A daily oral dosage of 3.6 g of curcumin was suggested for the phase II assessment in the prevention or treatment of malignancies other than the gastrointestinal system | Rheumatoid arthritis, chronic anterior uveitis, conjunctivitis, skin cancer, small pox, chicken pox, wound healing, urinary tract infections, and liver ailments | [37–39] |
| 3 | Ginger | *Zingiber officinale var. rubrum* | 6-gingerol, 6-shogaol, and 6-paradol | Ginger extract intake at a level of 2 g/day for 28 days decreased telomerase reverse transcriptase activity | Anticancer, antiinflammatory, antioxidant, immunomodulatory, antiangiogenic, and antibacterial | [40, 41] |
| 4 | Rosemary | *Rosmarinus officinalis* L. | Rosmarinic acid | Rosemary powder (750 mg), the amount closest to regular culinary intake, exhibited beneficial impacts on the quickness of memory, which is necessary to efficiently recover data from both episodic and working memory | Aromatherapy | [42, 43] |
| 5 | Cinnamon | *Cinnamomum verum* J.Presl | Cinnamyl acetate | Taking 1.5 g of cinnamon for 12 weeks enhanced the antioxidant activity and lipid composition | Antioxidant | [44, 45] |

| S. no. | Herbs and spices | botanical name | Active constituents | Dose | Therapeutics uses | References |
|---|---|---|---|---|---|---|
| 6 | Thyme | *Thymus vulgaris* L. | Thymol, carvacrol | At dosages of 250, 500, and 750 mg/kg, thyme essential oil decreased inflammatory exudates and migratory cells | Treat bronchitis, whooping cough, and mucosal inflammation of the respiratory system | [46, 47] |
| 7 | Pepper | *Piper nigrum* L. | Ascorbic-acid, −carotene, piperine, camphene, carvacrol, eugenol, −terpinene | Lower dosages of piperine (<20 mg/kg/b.w.) have several therapeutic benefits, including anti-Alzheimer's disease (AD), antidepressant properties and Parkinson's disease (PD). | Antioxidant, antimicrobial, anticancer, anti-inflammatory, analgesic, antipyretic, hepatoprotective, bio-enhancing and enzyme inhibitory activities | [48, 49] |
| 8 | Clove | *Syzygium aromaticum* | Eugenol, β-caryophyllene, α-humulene, and eugenyl acetate | Clove has anti-inflammatory benefits in animal models at dosages of 0.05 and 0.20 mL/kg | Toothache, pain during dental work, dental plaque, hangover, indigestion, may help protect against cancer, can kill bacteria, may improve liver health | [50, 51] |

**Table 2.**
*Principal bioactive ingredients in spices that may have positive effects.*

cysteine synthetase, and additional members of the enzyme glutathione transferase family are also critical for antioxidant defense. The second stage of enzymes may also be induced by the breakdown byproducts of the sulfur-containing compounds found in Allium species. The second stage of enzymes can also be induced by the byproducts of breakdown of the compounds that contain sulfur from the Allium species, the anti-oxidant substances and the stage 2 enzymes function sequentially; antioxidant substances like quercetin can contribute an electron to a reactive oxygen compound or nitrogen species that reacts, and the free radical that results from this reaction may then stimulate the synthesis of the stage 2 enzymes' genes as shown in **Table 2**.

## 3. Traditional Herbalism in various countries

The history of herbal medicine dates back millennia, with civilizations such as ancient China, Egypt, India, and Greece harnessing the power of plants for medicinal purposes. Research shows that phytochemicals found in the spices turmeric, ginger, allspice, garlic and cinnamon might play a role in cancer prevention as shown in **Table 3**.

| S. no | Herbs and spices | Active constituents | Dose | Cancers types | References |
|---|---|---|---|---|---|
| 1 | Turmeric | Curcuminoids | A daily dose of 1.08 g of curcumin for from 10 to 30 days enhanced the health of colorectal cancer patients by raising the expression of P53 molecules in tumor cells | Colon cancer, lung cancer, breast cancer | [52–54] |
| 2 | Bay leaf | 1,8-cineole | 200 g of 1, 8-cineole has promise as a potent and secure colorectal cancer chemotherapy drug | Colorectal cancer | [55, 56] |
| 3 | Garlic | Diallyl trisulfide, allicin, allyl mercaptan diallyl disulfide, and diallyl sulphide, Garlicnin B1 | Consuming 5–20 mg of allicin daily is linked to a lower risk of lung, myeloma, gastric cancer, and colorectal cancer | Cancers of blood, breast, prostate, ovarian, and gastrointestine | [57–60] |
| 4 | Saffron | Crocin | At 150 mg/kg, crocin can prevent colon tumor development and angiogenesis in rat model | Anti-leukemic effects | [61, 62] |
| 5 | Mustard | Allyl isothiocyanate | Oral allyl isothiocyanate-affluent mustard seed powder at 71.5 mg/kg (with a sinigrin dosage of 9 μmol/kg) prevented bladder cancer development | Testicular cancer, ovarian cancer, cervical cancer | [63–66] |

**Table 3.**
*Herbs and spices uses for anticancer.*

The Ebers Papyrus, an ancient Egyptian medical document, details the use of herbs like aloe vera and garlic. Similarly, Ayurveda in India emphasizes the holistic approach to health, highlighting the importance of herbs like turmeric and neem. During the Middle Ages, herbal knowledge persisted through monastic gardens and herbal manuscripts. Hildegard von Bingen, a medieval herbalist, documented the uses of various herbs. The Renaissance witnessed a revival of herbalism, with herbal gardens flourishing and explorers bringing back new plant specimens from around the world. The nineteenth and twentieth centuries marked a shift toward scientific exploration of herbal constituents. The isolation of morphine from opium and quinine from cinchona bark exemplifies this era. In recent decades, advancements in technology have allowed for a deeper understanding of the molecular intricacies of herbs and their constituents. Asia boasts a rich heritage of herbal medicine deeply embedded in traditional healing systems such as Ayurveda, Traditional Chinese Medicine (TCM), and Kampo. The integration of ancient wisdom with modern research findings sheds light on the potential mechanisms of action and therapeutic benefits of herbs. Case studies from countries like Italy, Greece, and Morocco delve into how herbs like oregano, rosemary, and thyme are deeply woven into both culinary and medicinal practices. Examining the Mediterranean diet, renowned for its health benefits, this section explores how herbs play a dual role as flavor enhancers and therapeutic agents. The focus is on their potential in preventing and managing cardiovascular diseases, among others. Indigenous communities in Africa and the Americas have preserved unique herbal traditions [67–69].

## 4. Adverse events associated with herbs and spices

Although curcumin has been demonstrated to be useful in a variety of human illnesses with low toxicity, some researchers have identified unwanted side effects. Curcumin at dosages that varied from 0.45 to 3.6 g/day for 1–4 months was related to nausea and diarrhea, as well as an elevation in lactate dehydrogenase levels and blood alkaline phosphatase in human subjects. Thus, further research is needed to assess the long-term toxicity of curcumin before it is allowed for human usage [70, 71]. Higher dosages (≥20 mg/kg/b.w.) have detrimental effects on the body. Garlic has been linked to a variety of adverse impacts, including smelly breath, ulcero-necrotic lesions, body odor, skin blisters, flatulence, anaphylaxis, esophageal and abdominal pain, urticaria, angioedema, myocardial infarction, small intestinal obstruction, bleeding, contact dermatitis, asthma, and rhinitis. *Ginkgo biloba* extract is a natural medication used to improve memory and brain function. The National Toxicology Program (2013) found that the herb ginkgo biloba powder can induce liver and thyroid gland tumors in males and females mice and rats. Notani and Jayant [72] investigated the impact of nutrition on upper aerodigestive tract malignancies in male patients with oral, pharyngeal, esophageal, and laryngeal cancers. Red chili powder has been shown to may raise the incidence of aerodigestive track tumors by two to three times depending on the dose [72]. Some medical herbs and plants, including *Cannabis sativa* and *Papaver somniferum* cause hallucinations and contribute to neurotoxic syndromes [73].

## 5. Conclusion

As we conclude our exploration of herbs and spices in global disease management, it becomes evident that these botanical treasures are not just ingredients in the kitchen; they are bridges connecting diverse cultures, traditions, and the pursuit of health. Emphasizing the need for collaborative efforts between traditional healers, scientists, and healthcare professionals, the conclusion underscores the importance of incorporating herbal wisdom into mainstream healthcare practices. One goal has been to separate and identify specific compounds from specific spice-producing plants or spices so that the chemical may be sold as coming from a "natural" source. Studying the impact of individual plant-based components, however, has not shown the same positive results as using bulk extracts. It is far preferable to utilize herbs and spices for what they were meant to do, which was to improve the flavor of our meals and make sure that we had an enough intake of phytochemicals to be healthy. Future intervention research should evaluate the bioavailability of herbs and spices. As most reported activities are in vitro, it's important to examine aspects that may impede the clinical translation of spices. In-depth investigations are necessary to evaluate the efficacy and toxicity of spices, as well as their pharmacokinetic properties.

## Author details

Jose Prakash Dharmian[1*] and Prakash Ramakrishnan[2]

1 Department of Pharmaceutics, Crescent School of Pharmacy, B.S. Abdur Rahman Crescent Institute of Science and Technology, Chennai, Tamilnadu, India

2 Department of Pharmacology, Crescent School of Pharmacy, B.S. Abdur Rahman Crescent Institute of Science and Technology, Chennai, Tamilnadu, India

*Address all correspondence to: jose88prakash@gmail.com

## References

[1] Liu H, Bai Y, Huang C, Wang Y, Ji Y, Du Y, et al. Recent progress of electrospun herbal medicine nanofibers. Biomolecules. 2023;**13**(1):184

[2] Ren JL, Yang L, Qiu S, Zhang AH, Wang XJ. Efficacy evaluation, active ingredients, and multitarget exploration of herbal medicine. Trends in Endocrinology & Metabolism. 1 Mar 2023;**34**(3):146-157

[3] Lu Y, Luo Q, Jia X, Tam JP, Yang H, Shen Y, et al. Multidisciplinary strategies to enhance therapeutic effects of flavonoids from Epimedii folium: Integration of herbal medicine, enzyme engineering, and nanotechnology. Journal of Pharmaceutical Analysis. 2023;**13**(3):239-254

[4] Lindberg K, Martvall A, Lima MG, Franca CS. Herbal medicine promotion for a restorative bioeconomy in tropical forests: A reality check on the Brazilian Amazon. Forest Policy and Economics. 2023;**155**:103058

[5] Wang Z, Yu F, Hu F. Functional chitosan and its derivative-related drug delivery systems for nano-therapy: Recent advances. Pharmaceutics. 28 Feb 2024;**16**(3):337

[6] Pai V, Subraya CK, Holavana Halli Nanjundaiah AR, Kamath V, Kunhikatta V. Issues and challenges in pharmacovigilance of Herbal formulations. Current Drug Safety. 1 Feb 2024;**19**(1):19-23

[7] Sivakumar T, Deepa B. A critical review on antidiabetic potential of herbal plants and its their bioactive components. Journal of University of Shanghai for Science and Technology. 2023;**25**(01):303-314

[8] Sulaiman AA. Effect of single oral dose of proanthocyanidin on postprandial hyperglycemia in healthy rats: A comparative study with sitagliptin. Journal of Intercultural Ethnopharmacology. 2014;**3**(2):73

[9] Silva ML, Bernardo MA, Singh J, de Mesquita MF. Cinnamon as a complementary therapeutic approach for dysglycemia and dyslipidemia control in type 2 diabetes mellitus and its molecular mechanism of action: A review. Nutrients. 2022;**14**(13):2773

[10] Hafizur RM, Hameed A, Shukrana M, Raza SA, Chishti S, Kabir N, et al. Cinnamic acid exerts anti-diabetic activity by improving glucose tolerance in vivo and by stimulating insulin secretion in vitro. Phytomedicine. 2015;**22**(2):297-300

[11] Chandorkar N, Tambe S, Amin P, Madankar C. A systematic and comprehensive review on current understanding of the pharmacological actions, molecular mechanisms, and clinical implications of the genus Eucalyptus. Phytomedicine Plus. 2021;**1**(4):100089

[12] Al Kury LT, Abdoh A, Ikbariah K, Sadek B, Mahgoub M. In vitro and in vivo antidiabetic potential of monoterpenoids: An update. Molecules. 2021;**27**(1):182

[13] Pari L, Monisha P, Jalaludeen AM. Beneficial role of diosgenin on oxidative stress in aorta of streptozotocin induced diabetic rats. European journal of pharmacology. 15 Sep 2012;**691**(1-3):143-150

[14] Kumar RV, Sinha VR. A novel synergistic galactomannan-based unit dosage form for sustained release

of acarbose. AAPS PharmSciTech. 2012;**13**:262-275

[15] Anwar S, Desai S, Eidi M, Eidi A. Antidiabetic activities of fenugreek (*Trigonella foenum-graecum*) seeds. In: Nuts and Seeds in Health and Disease Prevention. 2011. pp. 469-478. DOI: 10.1016/b978-0-12-375688-6.10056-8

[16] Haeri MR, Limaki HK, White CJ, White KN. Non-insulin dependent anti-diabetic activity of (2S, 3R, 4S) 4-hydroxyisoleucine of fenugreek (*Trigonella foenum graecum*) in streptozotocin-induced type I diabetic rats. Phytomedicine. 2012;**19**(7):571-574

[17] Xu Y, Tang G, Zhang C, Wang N, Feng Y. Gallic acid and diabetes mellitus: Its association with oxidative stress. Molecules. 2021;**26**(23):7115

[18] Godi S, Agrawal A, Gupta V, Patil PV. Antioxidant and antiulcer activity of phytoconstituents isolated from *Cocinnia grandis* and *Diplocyclos palmatus* fruit. NVEO-Natural Volatiles & Essential Oils Journal| NVEO. 2024;**11**(01):14-37

[19] Otunola GA, Afolayan AJ. A review of the antidiabetic activities of ginger. In: Ginger Cultivation and Its Antimicrobial and Pharmacological Potentials. 19 Feb 2020. DOI: 10.5772/intechopen.88899

[20] Mostofa R, Ahmed S, Begum MM, Sohanur Rahman M, Begum T, Ahmed SU, et al. Evaluation of anti-inflammatory and gastric anti-ulcer activity of *Phyllanthus niruri* L.(Euphorbiaceae) leaves in experimental rats. BMC Complementary and Alternative Medicine. 2017;**17**:1-10

[21] Wan Zairullah WN, Fauzi NY, Jofrry SM, Fahrni ML, Ming LC, Choo CY. Herbs used for the management of hypertension: A systematic review. Current Traditional Medicine. 2024;**10**(3):35-49

[22] Lahlou S, Figueiredo AF, Magalhães PJ, Leal-Cardoso JH. Cardiovascular effects of 1, 8-cineole, a terpenoid oxide present in many plant essential oils, in normotensive rats. Canadian Journal of Physiology and Pharmacology. 2002;**80**(12):1125-1131

[23] Leherbauer I, Stappen I. Selected essential oils and their mechanisms for therapeutic use against public health disorders. An overview. Zeitschrift für Naturforschung C. 28 Jul 2020;**75**(7-8):205-223

[24] Song J, Zhao Y, Shan X, Luo Y, Hao N, Zhao L. Active ingredients of Chinese medicine with immunomodulatory properties: NF-κB pathway and Parkinson's disease. Brain Research. 2024;**1822**:148603

[25] Bertoncini-Silva C, Vlad A, Ricciarelli R, Giacomo Fassini P, Suen VM, Zingg JM. Enhancing the bioavailability and bioactivity of curcumin for disease prevention and treatment. Antioxidants. 8 Mar 2024;**13**(3):331

[26] Ho TJ, Ahmed T, Shibu MA, Lin YJ, Shih CY, Lin PY, et al. A prospective review of the health-promoting potential of Jing Si Herbal Tea. Tzu Chi Medical Journal. 2024;**36**(1):1-22

[27] Cui T, Liu W, Chen S, Yu C, Li Y, Zhang JY. Antihypertensive effects of allicin on spontaneously hypertensive rats via vasorelaxation and hydrogen sulfide mechanisms. Biomedicine & Pharmacotherapy. 2020;**128**:110240

[28] Bischoff-Kont I, Fürst R. Benefits of ginger and its constituent 6-shogaol in inhibiting inflammatory processes. Pharmaceuticals. 2021;**14**(6):571

[29] Gammeri L, Panzera C, Calapai F, Cicero N, Gangemi S. Asian herbal medicine and chronic urticaria: Which are the therapeutic perspectives? Natural Product Research. 2023;**37**(11):1917-1934

[30] Liu D, Hu Y, Wang D, Han H, Wang Y, Wang X, et al. Herbal medicines in the treatment of tinnitus: An updated review. Frontiers in Pharmacology. 2023;**13**:1037528

[31] Wu S, Wang C, Bai D, Chen N, Hu J, Zhang J. Perspectives of international multi-center clinical trials on traditional Chinese herbal medicine. Frontiers in Pharmacology. 2023;**14**:1195364

[32] Karahan F. Evaluation of trace element and heavy metal levels of some ethnobotanically important medicinal plants used as remedies in Southern Turkey in terms of human health risk. Biological Trace Element Research. Jan 2023;**201**(1):493-513

[33] Pushpangadan P, George V, Parambil Ijinu T, Ambika Chithra M. Biodiversity, bioprospecting, traditional knowledge, sustainable development and Value Added Products: A Review. Journal of Traditional Medicine and Clinical Naturopathy. 2018;**07**(01). DOI: 10.4172/2573-4555.1000256

[34] Babandi A, Yussuff YM, Yakasai HM, Ya'u M, Shehu D, Ibrahim S, et al. Determination of heavy metals contamination, risk prediction and antioxidant properties of anti-malarial herbal mixture sold in Kano state, Nigeria. Case Studies in Chemical and Environmental Engineering. 2024;**9**:100576

[35] Wang W, Li H, Shi Y, Zhou J, Khan GJ, Zhu J, et al. Targeted intervention of natural medicinal active ingredients and traditional Chinese medicine on epigenetic modification: Possible strategies for prevention and treatment of atherosclerosis. Phytomedicine. 2024;**122**:155139

[36] Houshmand B, Mahjour F, Dianat O. Antibacterial effect of different concentrations of garlic (Allium sativum) extract on dental plaque bacteria. Indian Journal of Dental Research. 1 Jan 2013;**24**(1):71-75

[37] Gogoi A, Munda S, Paw M, Begum T, Siddiqui MH, Gaafar AR, et al. Molecular genetic divergence analysis amongst high curcumin lines of Golden Crop (Curcuma longa L.) using SSR marker and use in trait-specific breeding. Scientific Reports. 11 Nov 2023;**13**(1):19690

[38] El-Saadony MT, Yang T, Korma SA, Sitohy M, El-Mageed A, Taia A, et al. Impacts of turmeric and its principal bioactive curcumin on human health: Pharmaceutical, medicinal, and food applications: A comprehensive review. Frontiers in Nutrition. 2023;**9**:1040259

[39] Gupta SC, Patchva S, Aggarwal BB. Therapeutic roles of curcumin: Lessons learned from clinical trials. The AAPS Journal. 2013;**15**:195-218

[40] Crichton M, Marshall S, Marx W, Isenring E, Lohning A. Therapeutic health effects of ginger (Zingiber officinale): Updated narrative review exploring the mechanisms of action. Nutrition Reviews. 1 Sep 2023;**81**(9):1213-1224

[41] Mohd Sahardi NF, Jaafar F, Tan JK, Mad Nordin MF, Makpol S. Elucidating the pharmacological properties of zingiber officinale roscoe (ginger) on muscle ageing by untargeted metabolomic profiling of human myoblasts. Nutrients. 25 Oct 2023;**15**(21):4520

[42] El Boukhari R, Fatimi A. A review of the patentability of rosemary-derived drugs and bioactive compounds. Drugs and Drug Candidates. 16 Mar 2023;**2**(1):172-188

[43] Rahbardar MG, Hosseinzadeh H. Therapeutic effects of rosemary (*Rosmarinus officinalis* L.) and its active constituents on nervous system disorders. Iranian Journal of Basic Medical Sciences. 2020;**23**(9):1100

[44] Nakhaee S, Kooshki A, Hormozi A, Akbari A, Mehrpour O, Farrokhfall K. Cinnamon and cognitive function: A systematic review of preclinical and clinical studies. Nutritional Neuroscience. 2024;**27**(2):132-146

[45] Borzoei A, Rafraf M, Niromanesh S, Farzadi L, Narimani F, Doostan F. Effects of cinnamon supplementation on antioxidant status and serum lipids in women with polycystic ovary syndrome. Journal of Traditional and Complementary Medicine. 2018;**8**(1):128-133

[46] Vassiliou E, Awoleye O, Davis A, Mishra S. Anti-inflammatory and antimicrobial properties of thyme oil and its main constituents. International Journal of Molecular Sciences. 2023;**24**(8):6936

[47] Hammoudi Halat D, Krayem M, Khaled S, Younes S. A focused insight into thyme: Biological, chemical, and therapeutic properties of an indigenous Mediterranean herb. Nutrients. 2022;**14**(10):2104

[48] Alyamoor AH, Mustafa YF. Bell pepper biowaste products as medicinally valuable nutraceuticals: a comprehensive review. Iraqi Journal of Pharmacy. 1 Dec 2023;**20**(2):152-167

[49] Balakrishnan R, Azam S, Kim IS, Choi DK. Neuroprotective effects of black pepper and its bioactive compounds in age-related neurological disorders. Aging and Disease. 2023;**14**(3):750

[50] Nait Irahal I, Darif D, Guenaou I, Hmimid F, Azzahra Lahlou F, Ez-zahra Ousaid F, et al. Therapeutic potential of clove essential oil in diabetes: Modulation of pro-inflammatory mediators, oxidative stress and metabolic enzyme activities. Chemistry & Biodiversity. 2023;**20**(3):e202201169

[51] Vicidomini C, Roviello V, Roviello GN. Molecular basis of the therapeutical potential of clove (*Syzygium aromaticum* L.) and clues to its anti-COVID-19 utility. Molecules. 2021;**26**(7):1880

[52] Iweala EJ, Uche ME, Dike ED, Etumnu LR, Dokunmu TM, Oluwapelumi AE, et al. Curcuma longa (Turmeric): Ethnomedicinal uses, phytochemistry, pharmacological activities and toxicity profiles-A review. Pharmacological Research-Modern Chinese Medicine. 22 Jan 2023:100222

[53] Puvvada N, Rajput S, Kumar BP, Mandal M, Pathak A. Exploring the fluorescence switching phenomenon of curcumin encapsulated niosomes: in vitro real time monitoring of curcumin release to cancer cells. RSC advances. 2013;**3**(8):2553-2557

[54] Pan D, Zheng M, Liu J, Sun Z, Shi X. Garlic extract participates in the proliferation and apoptosis of nonsmall cell lung cancer cells via endoplasmic reticulum stress pathway. Evidence-Based Complementary and Alternative Medicine. 6 Feb 2023;**2023**

[55] Son SY, Choi JH, Kim EB, Yin J, Seonu SY, Jin SY, et al. Chemopreventive activity of ellagitannins from acer pseudosieboldianum (Pax) Komarov leaves on prostate cancer cells. Plants. 2023;**12**(5):1047

[56] Gao S, Yang K, Nohara T, Ikeda T, Zhou JR, Yokomizo K, et al. Garlicnin B1, an active cyclic sulfide from garlic, exhibits potent anti-inflammatory and anti-tumor activities. Antioxidants. 2023;**12**(4):869

[57] Das A, Adhikari S, Deka D, Baildya N, Sahare P, Banerjee A, et al. An updated review on the role of nanoformulated phytochemicals in colorectal cancer. Medicina. 30 Mar 2023;**59**(4):685

[58] Oravetz K, Todea AV, Balacescu O, Cruceriu D, Rakosy-Tican E. Potential antitumor activity of garlic against colorectal cancer: Focus on the molecular mechanisms of action. European Journal of Nutrition. 2023:1-7.

[59] Mishra Y, Mishra V. Multifaceted roles of crocin, phytoconstituent of *Crocus sativus* Linn. In cancer treatment: An expanding horizon. South African Journal of Botany. 2023;**160**:456-468

[60] Yang X, Man D, Zhao P, Li X. Identification of the therapeutic mechanism of the saffron crocus on glioma through network pharmacology and bioinformatics analysis. Medical Oncology. 2023;**40**(10):296

[61] Shalwitz R, Day T, Ruehlmann AK, Julio L, Gordon S, Vandeuren A, et al. Treatment of sulfur mustard corneal injury by augmenting the DNA damage response (DDR): A novel approach. Journal of Pharmacology and Experimental Therapeutics. 2024;**388**(2):526-535

[62] Bakshi HA, Quinn GA, Nasef MM, Mishra V, Aljabali AA, El-Tanani M, et al. Crocin inhibits angiogenesis and metastasis in colon cancer via TNF-α/NF-kB/VEGF pathways. Cell. 2022;**11**(9):1502

[63] Srivastava S, Jayaswal N, Kumar S, Sharma PK, Behl T, Khalid A, et al. Unveiling the potential of proteomic and genetic signatures for precision therapeutics in lung cancer management. Cellular Signalling. 2024;**113**:110932

[64] Meena S, Chobhe KA, Golui D, Bana RS, Gawdiya S, Roy A. Assessing the impact of altered clay mineral and red mud derivatives on the characteristics of mustard (*B. Juncea*) and the soil arsenic content. International Journal of Plant & Soil Science. 2023;**35**(19):200-206

[65] Mansouri K, Rasoulpoor S, Daneshkhah A, Abolfathi S, Salari N, Mohammadi M, et al. Clinical effects of curcumin in enhancing cancer therapy: A systematic review. BMC Cancer. 2020;**20**:1-1

[66] Bhattacharya A, Li Y, Wade KL, Paonessa JD, Fahey JW, Zhang Y. Allyl isothiocyanate-rich mustard seed powder inhibits bladder cancer growth and muscle invasion. Carcinogenesis. 2010;**31**(12):2105-2110

[67] Ng JY, Anant S, Parakh ND. Characteristics of the research literature on herbal medicines corresponding with herbal supplements yielding the highest total sales: A bibliometric analysis. Advances in Integrative Medicine. 2023;**10**(2):64-79

[68] Dores AR, Peixoto M, Castro M, Sá C, Carvalho IP, Martins A, et al. Knowledge and beliefs about herb/supplement consumption and herb/supplement–drug interactions among the general population, including healthcare professionals and pharmacists: A systematic review and guidelines for a smart decision system. Nutrients. 2023;**15**(10):2298

[69] Chaughule RS, Barve RS. Role of herbal medicines in the treatment of infectious diseases. Vegetos. Feb 2024;**37**(1):41-51

[70] Venkataraman B, Almarzooqi S, Raj V, Bhongade BA, Patil RB, Subramanian VS, et al. Molecular docking identifies 1, 8-cineole (eucalyptol) as a novel PPARγ agonist that alleviates colon inflammation. International Journal of Molecular Sciences. 2023;**24**(7):6160

[71] Morbidoni L, Arterburn JM, Young V, Mullins D, Mulrow C, Lawrence V. Garlic: Its history and adverse effects. Journal of Herbal Pharmacotherapy. 2001;**1**(1): 63-83. DOI: 10.1080/j157v01n01_06

[72] Notani PN, Jayant K. Role of Diet in Upper Aerodigestive Tract 962 Cancers. 1987

[73] Carod FA. Neurological syndromes associated with the ingestion of 791 plants and fungi with a toxic component (II). Hallucinogenic fungi and 792 plants, mycotoxins and medicinal herbs. Revista de Neurologia. 2003;**36**(10):951

Chapter 9

# Medicinal Herbs

*Amrit Acharya, Gita Ghimire and Prabin Shakya*


## Abstract

This chapter explores the scientific research, supporting the multifaceted role of medicinal herbs and spices in our daily lives, focusing on their remarkable antimicrobial properties' relevance in modern health and food industry. These natural wonders have been used for centuries in traditional medicine and culinary traditions. The bioactive compounds confer their antimicrobial abilities and the traditional practices that harness their potential for health and healing. Moreover, these herbs and spices not only enhance flavor but also serve as natural preservatives, safeguarding products from microbial contamination, and the future trends in this culinary field. Balancing the health and culinary facets, examine the impact of integrating these natural wonders into our diets, fostering not only well-being but also culinary delight, bridging the realms of human health and the food industry. This chapter also highlights challenges, opportunities, application, and modern prospective of herbs and spices in health and food industry.


**Keywords:** medicinal herbs and spices, antimicrobial properties, bioactive compounds, natural preservatives, culinary field, human health

## 1. Introduction

Herbs and spices have been an integral part of human history, playing multifaceted roles in both traditional medicine and culinary practices including in pharmacy, cosmetology, to name a few. Ancient societies all across the world have been aware of and have made use of the therapeutic and flavor-enhancing capabilities of a wide variety of plants with the first documented evidence dating back to 1500 BC [1]. In the twenty-first century, a wide range of bioactive or secondary compounds derived from medical herbs and spices are employed in numerous industrial sectors, including food production to enhance flavor, supply vitamins and macro- and micro-nutrients, and prevent microbial food spoilage, medicine to treat a variety of illnesses, for cancer therapy and chemoprevention, and as a natural source of antimicrobials for treating infectious diseases, pharmacology, and cosmetology in dietary supplements, and due to the demand for preservative-free cosmetics, to lower the risk of synthetic preservative chemicals allergies. This chapter explores the enormous relevance of therapeutic herbs and spices by studying their different cultural, historical, and practical elements. Research has shown that natural compounds derived from herbal medicines, particularly traditional Chinese and Persian remedies, have antiviral properties against influenza viruses [2].

Researchers are merging modern science and traditional knowledge to investigate the bioactive compounds contained in herbs and spices and uncover potential medical uses for them including in medical and food preservation. Evidence-based medicine and traditional herbal wisdom together build a bridge between traditional medical practices and modern healthcare. In the food industry, enhancing the flavor and palatability of food with a blend of herbs and spices is a potential way to help customers lower their intake of salt, which in turn reduces their risk of hypertension and cardiovascular illnesses, as well as improve their acceptance of legumes [3].

Furthermore, in commercial food industry, several synthetic chemicals are used as food preservatives, and currently there is demand for natural components and chosen by consumers due to their health-associated benefits as preservatives in foods, including the need to develop alternative antimicrobials, due to which interest in herbs and spices, which have been used for millennia, has increased [4]. For millennia, spices and herbs have been used to taste, smell fragrance, and color dishes. Furthermore, in terms of ethnobotanical significance, locals use spices and plants in traditional medicine. Various research have revealed therapeutic benefits of certain spices and herbs, such as antibacterial, antioxidant, antidiabetic, and anti-inflammatory actions, among others. Often utilized for their antibacterial qualities against germs, they are employed to regulate microbes in food products [5]. Due to the natural antibacterial qualities of herbs and spices, which outperform many synthetic antimicrobial compounds used in the food business, the usage of natural preservatives has been rising steadily in recent years. According to earlier research [6], food products derived from plants, including extracts and essential oils, exhibit antimicrobial properties and can have their shelf life prolonged. For food preservation and product shelf life extension, researchers have been interested in herbs and spices for several decades. Spices and herbs may be utilized in some phases of food preparation. According to a number of studies [7], there is evidence that herbs and spices can be used in certain stages of food processing, such as maintaining and ensuring shelf life, preventing microbial degradation, and performing oxidation reactions. The food sector may find another use for the natural antibacterial chemicals found in herbs and spices, thanks to the development of some new technologies.

Herbs and spices are alchemists in the culinary arts, turning the ordinary materials into astounding gourmet experiences. These botanicals' aromatic components, essential oils, and diverse tastes have made them vital in kitchens throughout the world. Each herb and spice, from the spicy warmth of ginger to the delicate freshness of cilantro, adds a distinct flavor to foods, elevating them from plain nourishment to sensory joys.

The culinary use of herbs and spices is an ever-evolving art, influenced by geography, climate, and cultural preferences. In addition to enhancing taste, many of these botanicals also contribute to the preservation of food, historically crucial before the advent of modern refrigeration. Exploring the interplay between culinary traditions and the diverse array of herbs and spices offers a window into the dynamic evolution of global cuisine. In an organic living, traditional herbal remedies and prescriptions can be taken into consideration for the treatment and prevention of diseases.

Despite being a thriving subject of study from an economic and health perspective, it has received less attention. The need for natural and ethnic components for healthy living, along with shifting dietary preferences, has expanded the global market for traditional herbs, spices, and value-added goods. It's amazing how much more crucial good eating and lifestyle choices have become in this pandemic era. The topic of health maintenance has garnered significant attention globally. The usage of spices and plants with antibacterial, anti-inflammatory, and antioxidant qualities has grown globally [8, 9]. Therefore, given the current state of affairs and the increased relevance of herbs and spices, a thorough investigation of their nutraceutical and medical value as well as their potential for application in the creation of various value-added products should be conducted. Therefore, this chapter studies about some of the common herbs and spices, their traditional use, bioactive chemical components, and pharmacological properties and their application in terms of medical and in commercial food industries.

## 2. Common herbs and spices

Herbs and spices are categorized into a number of factors, based on the botanical origin, flavor character, and culinary use. Herbs (often leaves) and spices (derived from other parts of the plant) are distinguished by one popular classification scheme that is based on the plant component utilized. **Herbs parts such as: Leafy Herbs:** include basil (*Ocimum basilicum* L.), cilantro (*Coriandrum sativum*), and parsley (*Petroselinum crispum*). **Stem Herbs:** include rosemary (*Rosmarinus officinalis*) and thyme (*Thymus vulgaris*). **Seed Herbs:** include dill (*Anethum graveolens*) and fennel (*Foeniculum vulgare*). **Flower Herbs:** include chamomile (*Matricaria chamomilla*) and lavender (*Lavandula* spp.). Similarly, Spices parts can also be categorized into **Seed Spices:** include cumin (*Cuminum cyminum*), coriander (*Coriandrum sativum*), and mustard (*Brassica* spp.), **Fruit Spices**: include black pepper (*Piper nigrum*), cardamom (*Elettaria cardamomum*), and allspice (*Pimenta dioica*), **Root Spices**: include ginger (*Zingiber officinale*), turmeric (*Curcuma longa*), and garlic (*Allium sativum*), and **Bark Spices**: include cinnamon (*Cinnamomum verum*) and cassia (*Cinnamomum cassia*) [10–13].

Common herbs and spices play a significant role in both the herbal medicinal and food industries, offering a myriad of flavors, aromas, and potential health benefits. The dual roles of common herbs and spices not only enhance the sensory experience of food but also contribute to health and well-being. The integration of these botanicals into both traditional herbal medicine and modern culinary practices reflects a holistic approach to overall wellness.

**Turmeric** (*Curcuma longa*), a vibrant yellow spice, is well known for its culinary use in Indian cuisine and is rich in curcumin, which exhibits anti-inflammatory and antioxidant properties and immune-boosting properties. It is used in formulations to address health concerns like respiratory infections and digestive issues [14].

**Garlic** *(Allium sativum)* is a plant with allicin and is known for its antimicrobial properties, affecting both Gram-positive and Gram-negative bacteria. It also has antiviral properties, potentially aiding in respiratory infections. Garlic is used in herbal medicine for immune function, cardiovascular health, and respiratory well-being. Garlic, a versatile herb, is not only a staple in many cuisines but is also recognized for its potential cardiovascular benefits and antimicrobial properties [15].

**Rosemary** (*Rosmarinus officinalis*) is a fragrant herb widely used in Mediterranean cuisine and has been studied for its potential cognitive benefits. Rosmarinic acid, found in rosemary, may have neuroprotective effects, enhancing memory and cognitive function [13]. Its bioactive compounds, including rosmarinic acid and essential oils, have been proven effective against bacteria and fungi. Its therapeutic benefits extend to respiratory health, digestive issues, and skin conditions.

**Bakul** (*Mimusops elengi)* is known by its native name, Bakul, and this tiny to big tree may be found all throughout India. The plant has a significant role in the indigenous medical system, and different components of this plant are utilized to treat a range of systemic illnesses, including dental issues. It has demonstrated strong analgesic, antipyretic, and anti-inflammatory properties. For odontopathy, inflammation, and bleeding gums, *M. elengi* bark is used as a gargle due to its astringent and harsh properties. Toothbrushes are made from the sensitive stems [13].

**Neem** (*Azadirachta indica*) is possibly the most beneficial herb that is used in traditional medicine. Traditional medicine has employed every component of the tree as a home treatment for a variety of human illnesses. In India, the tree is still recognized as a "Village dispensary." The majority of the plant's components, including the seeds, fruits, leaves, bark, and roots, have been shown to contain substances with antibacterial, antiviral, antipyretic, anti-inflammatory, antiulcer, and antifungal qualities [1].

**Basil** (*Ocimum sanctum* L.; *Ocimum basilicum* L.) is a traditional, excellent medicinal plant that has been utilized for ages in Ayurvedic therapy. It has made a significant contribution to culinary flavor, to be used in cosmetics, and to be added in traditional medicine to relieve aches, inflammations, and coughs. It also adds a fragrant taste to a variety of foods, especially Italian fare like salads, pizzas, and pasta sauces. *O. basilicum* L. leaves have strong tasting ingredients and the essential oil extracted from the flowering plants and several chemotypes, such as methyl chavicol (estragole), linalool, eugenol, carvacrol, ursolic acid, rosmarinic acid, flavonoids, citronellol, and β-caryophyllene. There have been reports of antibacterial, anti-inflammatory, analgesic, antipyretic, antiulcer, antidiabetic, antioxidant, and anticancer properties in various plant sections [9].

**Gurjo** (*Tinospora cordifolia*) grows huge deciduous climbing shrubs that are native to the Indian Subcontinent. The herb is known as guduchi, giloy, or amrita in Ayurveda. The plant's extract is used as a treatment for a variety of illnesses, such as hepatitis and diabetes, throughout India and South Asia. The plant is particularly mentioned in many sections of the nation due to its usage in tribal or folk medicine. Numerous intriguing discoveries from in-depth phytochemical, pharmacological, and clinical studies on the medication have been reported [9].

**Ginger** (*Zingiber officinale*) is known for its distinctive flavor, and ginger has been used both culinarily and medicinally. Ginger is a common ingredient in both Asian and Indian cuisines because it gives the food a unique, cozy, and slightly spicy flavor. Its culinary adaptability includes both sweet and savory foods, such as gingerbread and ginger tea, as well as savory meals like stir-fries and curries. It contains bioactive ingredients, such as paradol, shogaol, and gingerol, which have anti-inflammatory and antioxidant properties [9].

**Clove** (*Syzygium aromaticum*) is among the most precious spices, having been used for numerous medical uses and as a food preservative for ages. This plant is one of the best sources of phenolic chemicals, including gallic acid, eugenol acetate, and eugenol, and it has a lot of potential uses in medicine, cosmetics, food, and agriculture. Clove has greater levels of antioxidant and antibacterial activity than many other therapeutic herbs [16].

**Cinnamon** (*Cinnamomum verum*) is widely used in baking and cooking, Cinnamon is not only flavorful but also has potential health benefits. Numerous studies indicate that cinnamon has antibacterial property and potential to help control blood sugar. It is one of the beloved spices with a warm and sweet flavor, and possesses notable antimicrobial properties attributed to its active components, including cinnamaldehyde. Scientific studies have highlighted cinnamon's effectiveness against a spectrum of bacteria and fungi, showcasing its potential as a natural antimicrobial agent. In herbal medicine, cinnamon has been traditionally utilized for its anti-inflammatory and antioxidant properties. Its applications extend to formulations aimed at addressing gastrointestinal issues, respiratory ailments, and even as an immune system booster. The synergistic combination of cinnamon with other herbs and spices enhances its therapeutic potential, contributing to its widespread use in traditional and holistic health practices. It is crucial to exercise caution and consult healthcare professionals, particularly in the context of individual health conditions and potential interactions with medications [17].

**Peppermint** (*Mentha piperita* L.) is a versatile herb used in both sweet and savory dishes. It is also known for its refreshing aroma and cooling flavor. Medicinally, peppermint oil is known for its calming effects on the gastrointestinal tract and has been studied for its potential in managing irritable bowel syndrome (IBS) symptoms. Peppermint is also known for its antimicrobial properties due to menthol and other essential oils. It's traditionally used in herbal medicine for digestive benefits, respiratory issues, and oral health. It's also used in culinary dishes. Its therapeutic potential is enhanced by its synergistic combination with other herbs and spices. Peppermint's versatility extends to culinary uses, where its invigorating flavor is often incorporated into teas, desserts, and savory dishes [18].

**Oregano** (*Origanum vulgare*), a flavorful herb commonly used in Mediterranean cuisine, possesses significant antimicrobial properties primarily attributed to its rich content of essential oils, including carvacrol and thymol. Scientific studies have demonstrated oregano's efficacy against various bacteria and fungi, highlighting its potential as a natural antimicrobial agent. In herbal medicine, oregano has been traditionally utilized for its antimicrobial and anti-inflammatory effects, making it a popular choice for addressing respiratory issues and promoting overall immune health. Oregano's versatile applications extend to culinary uses, where its robust flavor enhances a variety of dishes. The synergistic combination of oregano with other herbs and spices further amplifies its therapeutic potential, emphasizing its role in both traditional and holistic health practices [19].

**Fenugreek** (*Trigonella foenum-graecum* L.) is used both as a spice and as a medicinal herb. It is rich in fiber and has been studied for its potential benefits in blood sugar regulation and cholesterol management. Fenugreek is a maple-like herb with notable antimicrobial properties attributed to its bioactive compounds, like fenugreekine, diosgenin, and galactomannan. It has been traditionally used in herbal medicine for its digestive and anti-inflammatory properties and also in immune health formulations. Fenugreek seeds are used in culinary preparations and synergistically combined with other herbs and spices may enhance its therapeutic potential. More research is needed to fully understand its antimicrobial effects, antifertility, anticancer, and antiparasitic properties [20]. These herbs and spices exemplify the diverse ways in which nature's bounty contributes not only to the sensory pleasure of our meals but also to potential health and wellness benefits. Alkaloids, saponins, tannins, phenols, and several other secondary metabolites are medicinal benefits [9, 20].

**Thyme** (*Thymus vulgaris*), a fragrant herb with a robust flavor, is renowned for its potent antimicrobial properties, largely attributed to its essential oils, particularly thymol. Scientific studies have demonstrated thyme's effectiveness against various bacteria and fungi, underscoring its role as a natural antimicrobial agent. In herbal medicine, thyme has been traditionally used for its antiseptic and expectorant properties, making it a valuable remedy for respiratory issues such as coughs and bronchitis. Thyme's versatility extends to culinary applications, where it enhances the flavor of a variety of dishes. The synergistic combination of thyme with other herbs and spices enhances its therapeutic potential, emphasizing its role in both traditional and holistic health practices [9].

**Coriander** (*Coriandrum sativum* L.) is a fragrant plant that may be used in many ways. In cooking, both the leaves and the seeds are used. Traditional medical systems have investigated the possible health advantages of coriander for its benefits to the digestive system. Its antibacterial, antioxidant, and anti-inflammatory qualities have also been explored [21, 22]. Its bioactive compounds, including linalool and cineole, have been proven effective against bacteria and fungi. Coriander's aromatic qualities make it popular in culinary dishes worldwide [15]. The synergistic combination of coriander with other herbs and spices enhances its therapeutic potential, contributing to its role in traditional and holistic health practices [23].

**Cardamom** (*Elettaria cardamomum*) is widely used in Indian and Middle Eastern cuisines for its aromatic and a pleasant lemony flavor. Medicinally, cardamom has been investigated for its potential gastrointestinal benefits, including its carminative and anti-inflammatory effects [24]. Its high essential oil content, especially those of limonene, cineol, and terpinene, gives it significant antibacterial qualities [24, 25]. Cardamom's effectiveness against a range of bacteria and fungi has been demonstrated in scientific research, indicating its promise as a natural antibacterial agent. Cardamom has long been used in herbal medicine for its digestive properties and capacity to ease gastrointestinal problems. Cardamom has a wider range of uses in formulations meant to support digestive health because of its antibacterial qualities. Furthermore, because of its savory and aromatic nature, it is a widely used ingredient in food and drink preparations around the globe [24].

**Sage** (*Salvia officinalis*) is a fragrant herb with a distinct flavor and silvery-green leaves and is commonly used in Mediterranean cuisine. Medicinally, sage has been traditionally used for its cognitive-enhancing properties, and recent research suggests potential benefits in neurodegenerative conditions. It has antimicrobial properties due to its essential oils primarily thujone, camphor, and cineole. It's used in herbal medicine for oral health, sore throats, respiratory conditions, and culinary tastes [26].

**Fennel** (*Foeniculum vulgare*), a mild anise-like herb, has a sweet, licorice-like flavor and is used both as a herb and as a spice. Medicinally, fennel has been studied for its digestive and anti-inflammatory properties, and its seeds are commonly used in traditional medicine. Fennel has antimicrobial properties due to bioactive compounds like anethole and fenchone. Its traditional use in herbal medicine and its use in digestive and oral health formulations suggest potential benefits. Fennel's aromatic seeds are used in culinary preparations [27].

**Saffron** (*Crocus sativus*) is the powerful plant recognized for its unique flavor, color, and therapeutic qualities, which is produced from the saffron crocus or autumn crocus flower. Safranal, together with crocin and picrocrocin, is the active ingredient that gives it its medicinal properties. These substances support the anti-inflammatory, antioxidant, and neuroprotective qualities of saffron. Saffron, a traditional medicine, has been used for mood enhancement, depression and anxiety relief, and managing conditions like premenstrual syndrome (PMS). Its compounds modulate neurotransmitters like serotonin and may have anticancer properties. Moreover, saffron has demonstrated potential in the management of age-related macular degeneration (AMD), the main reason why older people have vision impairment. Saffron's antioxidant qualities might shield the retina from oxidative damage [20, 28].

**Rhodiola** (*Rhodiola rosea*), a traditional herbal remedy from the *Rhodiola rosea* plant, is known for its adaptogenic properties. Its primary active compounds are rosavin, salidroside, and p-tyrosol, which help the body adapt to stressors and maintain balance. Rhodiola is used to combat fatigue, enhance mental performance, and improve resilience to stress. It modulates the body's stress response by influencing stress hormone levels and supporting adrenal function, making it beneficial for individuals experiencing chronic stress or mental fatigue. Rhodiola also has potential benefits in improving mood and alleviating symptoms of mild to moderate depression. It also enhances physical performance and reduces exercise-induced fatigue, benefiting athletes and those engaged in strenuous activities. However, it is advisable to consult a healthcare professional before using Rhodiola [28].

**Paracress** (*Acmella oleracea* L.) is a tropical herbaceous plant that is commonly used as a culinary flavoring, especially flower head, and as an antiaging component in cosmetics and as a remedy for toothache and fever. Its extract is also recognized for its antibacterial, antifungal, anti-inflammatory activities, aphrodisiac, antinociceptive, diuretic, analgesic, local anesthetic, antioxidant, vasorelaxant, immunomodulatory, etc. [29, 30]. Previous research has revealed that the presence of alkamides is responsible for *A. oleracea*'s bioactivity [30]. According to research, the major active alkamide of this plant is spilanthol including triterpenoids, phytosterols, limonene, β-caryophyllene, etc. [9].

**Sweet flag** (*Acorus calamus* L.) is a tall perennial Acoraceae monocot plant found in wetlands. Long used as medicine, sweet flag's aromatic leaves and rhizomes have a spicy flavor. Dried and powdered rhizome is used to substitute nutmeg, ginger, and cinnamon in recipes. Rhizomes: Used to treat epilepsy, mental illnesses, chronic diarrhea, dysentery, abdominal discomfort, antimicrobial, antioxidant, insecticidal activities, anticonvulsant, and hypolipidemic properties, among other conditions; they also have antispasmodic, carminative, anthelmintic, aromatic, expectorant, nauseating, nervine, sedative, and stimulant effects. It has been stated that the plant contains glucoside, alkaloids, and essential oil that contains sesquiterpenes, calamen, clamenol, and calameon. Together with eugenol, pinene, and camphene, it also includes a bitter glycoside known as acorine [9, 31].

**Jimbu** (*Allium hypsistum, A. przewalskianum*) is indigenous to Nepal's north-central Himalayan areas. In Nepal, the aboveground components of these species are utilized as spices and remedies. Its dried leaves are used for flavoring and seasoning, as well as to treat coughs and colds, sore throats, stomach disorders, flues, and high altitude sickness. 1,2-bis(methylthio)ethene, 2,4-dimethyl thiophene, dimethyl disulfide and dimethyl trisulfide, phenolic compounds, flavonoids, and other phytochemicals are abundant in this plant [9, 32].

**Onion** (*Allium cepa* L.) is among the most frequently grown and well-liked vegetable crops in the world. With its unique flavor, the onion bulb holds the third position as the most important horticultural spice, and its economic worth is substantial. Several indigenous cultures have long used *A. cepa* for medicinal purposes, in addition to its culinary uses. Antibacterial, antioxidant, analgesic, anti-inflammatory, antidiabetic, hypolipidemic, antihypertensive, and immunoprotective properties are only a few of the many bioactive compounds and pharmacological activity found in *A. cepa*. Onions have flavonoids, organosulfur, phytosterols, and saponins among their major bioactive constituents [9].

**Aloe** (*Aloe vera* L.) is employed all throughout the world as a traditional medication. There are triangular, fleshy green leaves with white teeth on the margins of this succulent plant. Fresh leaf gel is employed in many formulations and cosmetic treatments, while aloe vera juice has long been used for its purgative qualities. Aloe vera's medicinal benefits are attributed to hundreds of nutritional and bioactive compounds, such as vitamins, enzymes, minerals, sugars, lignin, anthraquinones, saponins, salicylic acid, and amino acids. In conventional medical systems, it is used to treat burn damage, eczema, fever, inflammation, and cosmetics. Its secondary metabolites also have immune-boosting, anticancer, antidiabetic, antiaging, and antibacterial properties [9, 33].

**Black cardamom** (*Amomum subulatum Roxb.*) is a fragrant and therapeutic spice originating from the Eastern Himalayas. It is used to flavor and preserve many types of coffee, liquors, confections, and beverages. Large cardamom's main aroma-giving ingredient is volatile oil, whereas its main active compound is 1,8-cineole. Black or large cardamom has been shown to have allopathic, analgesic, anti-inflammatory, antibacterial, antioxidant, antiulcer, cardio-adaptogenic, and hypolipidemic properties, in addition to its numerous culinary and medicinal applications [9, 34].

**Hot pepper** (*Capsicum annuum*) fruits, generally known as chili, have been utilized since ancient times as food veggies, flavoring components, natural colorants, and traditional medicines. Chili includes significant amounts of pigments (such as chlorophyll, anthocyanin, and lutein) that may have health advantages; it also contains a variety of other exceptional health-promoting chemical substances, including vitamins, minerals, flavonoids, carotenoids, and capsaicinoids. And capsaicin, the primary active component responsible for these species' spicy taste, has been shown to have a favorable impact on health. Consumption of dietary chili and capsaicinoids, particularly capsaicin, is engaged in body weight reduction and their potential antiobesity benefits, urinary problems, antioxidants, antibacterial, anticancer, and analgesic capacity [9, 34, 35].

**Indian bay leaf** (*Cinnamomum tamala*) is a perennial Indian plant which is also known as Tejpat, and it is an important traditional medicinal plant found throughout South Asia. The plant has numerous synonyms and is known by various names depending on the location. Aromatic leaves are a key source of bioactive chemicals with extensive therapeutic, culinary, and medical applications. The plant contains a variety of bioactive chemicals, including saponins, phytosterols, fatty acids, monoterpenes, sesquiterpenes, geraniol, linolool, bornyl acetate, caryophyllene oxide, p-coumaric acid, vanillic acid, and others that have traditionally been used to treat skin rashes, bad breath, earaches, and arthritis. Indian bay leaves have antidiarrheic, antitumor, anti-inflammatory, antiarthritic, antitumor, antioxidant, cancer chemopreventive, gastroprotective, antibacterial, antipyretic, and anxiolytic properties, among other things [9, 34, 35].

These herbs and spices not only contribute to the diverse and rich tapestry of global cuisines but also offer potential health benefits, as evidenced by research in both traditional and modern contexts.

## 3. Mechanism of action of herbal medicines

The mechanism of action of herbal medicines can vary widely, as these remedies often contain complex mixtures of compounds that can affect the body in different ways [34–36]. Unlike conventional pharmaceutical drugs that typically contain a single active ingredient, herbal medicines often consist of a combination of compounds that may work synergistically. Some of the common mechanisms through which herbal medicines may exert their effects can be any one of the following;

i. Interact with cellular receptors to modulate signaling pathways and cellular responses.

ii. Affect the enzymes that participate in metabolic processes, including the metabolism of neurotransmitters or cellular respiration.

iii. Neutralize free radicals and protect cells from oxidative damage.

iv. Influence inflammation pathways.

v. Direct cell cycle progression and apoptosis, particularly in cancer prevention and treatment.

vi. Engage with DNA, preventing DNA damage and mutations.

vii. Impact the brain's neurotransmitter activity, which may have an impact on cognitive function.

viii. Alter hormone synthesis or interact with hormone receptors to impact hormonal homeostasis.

Overall, herbal medicines have potential health-promoting effects.

The herbs and spices consist of phytochemicals which are bioactive molecules found such as alkaloids, flavonoids, terpenoids, and glycosides. These substances have a variety of physiological impacts on the organism and are also critical to the effectiveness of herbal medicines and major constituents for the therapeutic properties. Phytochemicals function in a variety of ways, including interactions at the molecular, cellular, and physiological levels. The effects may vary depending on the specific phytochemical, its concentration, and the individual's health. Furthermore, synergistic interactions among numerous phytochemicals in a plant influence the cumulative therapeutic advantages of herbs and spices in human health and well-being. Several research have been conducted to study the possible health advantages of bioactive chemicals found in herbs and spices, such as anti-inflammatory, antioxidant, antibacterial, chemopreventive, and antimutagenic properties, among others. Herbs and spices include a variety of bioactive substances, including polyphenols, quinines, organosulfur compounds, alkaloids, polypeptides, and their derivatives, some of which are briefly described below [37–39]:

**Polyphenols** are secondary metabolite molecules, which come in a range of structural forms in plants, and are molecules that are engaged in metabolic processes. The number of phenol rings and one or more aromatic rings attached to hydroxyl groups in the structure of these molecules determine their classification. Plants include phenolic acids, flavonoids, tannins, stilbenes, lignans, and coumarins as their main class of polyphenols. Foods originating from plants, such as spices and herbs, are rich in polyphenols, with dried forms having a greater concentration than fresh forms. Antioxidant activity is just one of the many pharmacological properties of polyphenols that are widely recognized. Additionally, they possess neuroprotective, anti-inflammatory, and metabolic anticancer qualities [39–41].

**Terpenoids** are a kind of important phytochemical that is present in spices and herbs and is divided into several categories according to the quantity of isoprene units it contains. They are the main components of essential oils that are present in plants and are used in traditional medicine because of their aromatic qualities. The essential oils of many different plants, including herbs and spices, comprise monoterpenes, a kind of terpene containing two isoprene units. Plants include a variety of monoterpenes, such as borneol, camphor, citral, geraniol, lavandulol, thymol, α–terpineol, limonene, menthol, carvone, eucalyptol, and perillaldehyde. The anticancer, antibacterial, antifungal, antiviral, antihyperglycemic, analgesic, anti-inflammatory, and antiparasitic properties of these bioactive substances are explored. Carvone in caraway, menthol in peppermint, thymol in thyme, linalool in coriander, citral in lemongrass, β-caryophyllene in clove, α–zingiberene in ginger, and capsaicin in chili pepper are among the terpenoids found in herbs and spices [42].

**Saponins** are glycosides generated from plants that have diverse biological activities and surface-active or detergent qualities, as well as steroid and triterpene features. Plant groups include Agavaceae, Amaryllidaceae, Asparagaceae, Bromeliaceae, Dioscoreaceae, Liliaceae, Palmae, and Scrophulariaceae and are rich in saponins [43]. These compounds have a range of pharmacological activities, such as immunomodulatory, hepatoprotective, anti-inflammatory, and anticancer effects [44].

**Alkaloids** are another type of phytochemical with a nitrogen atom in the ring structure. Pyrollidine, pyrrolizidine, pyridine, quinoline, isoquinoline, imidazole, and other alkaloids are grouped into numerous categories. Morphine has analgesic properties, while ephedrine is useful for asthma [45].

Herbal phytochemicals can interact with various cellular receptors, including receptors on the cell membrane or within the cell. By binding to these receptors, they can modulate signaling pathways and influence cellular responses. For example, certain flavonoids can interact with estrogen receptors or cell surface receptors involved in inflammation. Similarly, they may also modulate the activity of enzymes involved in metabolic pathways. This can lead to the inhibition or activation of specific biochemical processes. Alkaloids, for instance, may interact with enzymes involved in neurotransmitter metabolism or cellular respiration. Many phytochemicals, such as flavonoids and polyphenols, exhibit antioxidant properties. They can neutralize free radicals and reactive oxygen species (ROS), protecting cells and tissues from oxidative damage. This antioxidant activity contributes to the overall health-promoting effects of herbal medicines. Some phytochemicals have anti-inflammatory properties, influencing pathways involved in inflammation. Terpenoids, for example, may modulate the production of inflammatory mediators and cytokines. Phytochemicals can influence cell cycle progression and apoptosis (programmed cell death). This is particularly relevant in the context of cancer prevention and treatment. Certain compounds may induce apoptosis in cancer cells or inhibit the proliferation of abnormal cells. Some phytochemicals can interact with DNA, affecting processes such as replication and repair. This interaction may contribute to the prevention of DNA damage and mutations. Certain alkaloids and polyphenols have been studied for their potential DNA-protective effects. Phytochemicals found in herbs can influence neurotransmitter activity in the brain. For example, alkaloids such as those found in *Ginkgo biloba* may impact neurotransmitter release and receptor function, potentially affecting cognitive function. Phytochemicals can influence hormonal balance by interacting with hormone receptors or affecting hormone synthesis. Isoflavones, found in soy and red clover, for example, have estrogenic activity and may influence hormone levels in the body. Similarly, Saffron's active components include saffron and crocin. They are crucial in regulating the brain's many neurotransmitter systems that include norepinephrine, glutamate, serotonin, and dopamine. They have also been shown to lessen depression. It was found that the saffron dosage was equally effective as fluoxetine and imipramine, two antidepressants, with no adverse effects [28, 34, 35].

## 4. Pharmacological activities of herbs and spices

The vast majority of spices and herbs used in culinary have medicinal properties. Pharmacological characteristics of many herbs and spices used in cooking include antibacterial, anti-inflammatory, antioxidant, antidiabetic, antihyperlipidemic, hepatoprotective, antipyretic, and digestive [9].

### 4.1 Antimicrobial properties

Microbial infections can cause a variety of ailments. Antimicrobials are substances that either kill or prevent the development of bacteria, viruses, and fungus. Phytochemicals have been found to have antimicrobial activity by mechanisms, such as cell membrane damage, inhibition of enzyme and toxin activity, suppression of virulence factors, or production of bacterial biofilm [37]. Several herbs and spices used in cooking include very strong bioactive chemicals, such as flavonoids, alkaloids, tannins, glycosides, and saponins that work as a defensive mechanism against various microbes, insects, and herbivores. Antimicrobial components present include allicin in garlic, cinnamaldehyde in cinnamon, and allyl isothiocyanate in mustard [46]. It has been demonstrated that cumin possesses exceptional antibacterial action against *Bacillus licheniformis*, *A. tumefaciens*, *B. subtilis*, *Pseudomonas oleovorans*, *Trichophyton rubrum*, *S. cerevisiae, and Saccharomyces pombe.* Additionally, bacteria, such as *Saccharomyces cerevisiae, Escherichia coli*, and *Bacillus subtilis*, have been demonstrated to be inhibited by it [47, 48].

Research has demonstrated that the alcoholic extract of cumin inhibits a variety of bacteria, including *Escherichia coli*, *Bacillus subtilis*, *Bacillus licheniformis*, *A. tumefaciens*, *B. subtilis*, *Pseudomonas oleovorans*, *Trichophyton rubrum*, *S. cerevisiae*, and *Saccharomyces pombe* [48]. Based on tests conducted on the aldehyde component of the oil containing the antimicrobial chemical compound cuminaldehyde, the entire cumin oil inhibits *Aspergillus flavus* and *Aspergillus niger* by over 90%, demonstrating the antifungal properties of cumin oil [49]. Cumin oil has been shown to have antifungal characteristics in a recent study; *Aspergillus flavus* and *Aspergillus niger* were suppressed by over 90% when the aldehyde portion of the oil containing the antibacterial chemical compound cuminaldehyde was analyzed [49]. The clove essential oil has a high concentration of eugenol and it has been proved to have diverse antimicrobial activity against *Listeria monocytogenes* strains [50], *Bacillus cereus* [51], and five strains of *Staphylococcus epidermidis* [52]. Clove also possesses fungicidal activity, and its chemical components, such as carvacrol and eugenol, are known to have fungicidal properties against *Candida albicans* and *Trichophyton mentagrophytes* [53]. Cinnamon's antimicrobial capacity has been evaluated against *Staphylococcus aureus*, attributed to the chemical component in cinnamon known as cinammaldehyde [54]. Cinnamaldehyde inhibition of bacterial growth can be caused via inhibition of cell wall synthesis, inhibition of cell membrane function, inhibition of protein synthesis, or inhibition of nucleic acid synthesis [54].

Six extracts of black pepper (*Piper nigrum*) and turmeric (*C. longa*) were found to be antibacterial against *B. subtilis*, *Bacillus megaterium*, *Bacillus sphaericus*, *Bacillus polymyxa*, *S. aureus*, *E. coli*, and 11 molds, including *Aspergillus luchuensis*, *A. flavus*, *Penicillium oxalicum*, *Rhizopus stolonifer*, *Scopulariopsis* sp., and *Mucor* sp. [55].

Similarly, Ajoene is a sulfur-containing chemical that is obtained from garlic and has broad-spectrum antibacterial action. It inhibits platelet aggregation and the growth of Gram-positive bacteria; *Streptomyces griseus, Mycobacterium smegmatis, Bacillus cereus, and Bacillus subtilis*. Since reduction by cysteine, which interacts with disulfide bonds, eliminated ajoene's antimicrobial action, it indicates that the disulfide bond in ajoene is important for the compound's antibacterial activity [21]. Basil oil has potential as an effective antibacterial agent to reduce *Salmonella enteritidis* in food. Similarly, the effectiveness of oregano essential oil (OEO) and two of its main constituents, thymol and carvacrol, against *Pseudomonas aeruginosa* and *Staphylococcus aureus* was studied [56].

According to earlier research, a fresh extract of *Nigella sativum* (black cumin) and *Allium sativum* (garlic) combined exhibited greater antibacterial sensitivity against urinary tract infection isolates than either extract alone or with other medications like cefalexin, cotrimoxazole, and nalidixic acid [57].

### 4.2 Antioxidant properties

Antioxidant chemicals included in herbal medicines have the ability to neutralize free radicals, stop oxidative damage to cellular structures, and boost the body's defense mechanisms against free radicals. These substances have the ability to chelate metal ions that can lead to the production of free radicals, like copper and iron. Additionally, they have the ability to activate the body's defense mechanisms against free radicals by upregulating key enzymes, including glutathione peroxidase (GPx), catalase (CAT), and superoxide dismutase (SOD). Certain herbal remedies have the ability to block the actions of enzymes including lipoxygenase, xanthine oxidase, and NADPH oxidase ((nicotinamide adenine dinucleotide phosphate) oxidase) that generate reactive oxygen species. They can fend off oxidative damage to cell membranes, and vitamin E and other lipid-soluble antioxidants can help keep them stable. Moreover, they might have anti-inflammatory properties, shield mitochondria from oxidative stress, and improve nitric oxide's bioavailability—a chemical that's essential for vascular health [9, 36, 47].

Antioxidant qualities are a well-known characteristic of some herbal remedies. They might aid in removing dangerous free radicals from the body, which are linked to oxidative stress and a number of illnesses. The potential health benefits of herbal medicines are mostly attributed to their antioxidant activity. Reactive oxygen species (ROS) and free radicals can have detrimental effects on the body that antioxidants can counteract or lessen. As byproducts of regular metabolic activities, these reactive chemicals can harm DNA, proteins, and cells, which can lead to aging and a number of chronic diseases: Herbs rich in catechins (green tea), berries (anthocyanins), turmeric (curcumin), and garlic (allicin) are a few examples of those having antioxidant activity. It's crucial to remember that many factors, like the particular chemicals present, their bioavailability, and individual metabolic differences, can affect how effective herbal antioxidants are. Cinnamon contains polyphenolic polymers, which may serve as antioxidants, increase the effects of insulin, and help manage diabetes and glucose intolerance [24]. Furthermore, even while foods and herbs high in antioxidants might improve general health, they should only be a small portion of a balanced diet and way of life that also includes other health-promoting practices [9, 36, 47].

### 4.3 Anti-inflammatory properties

Herbal medicines can provide a variety of benefits, including suppressing pro-inflammatory mediators, such as cytokines, prostaglandins, and leukotrienes, regulating inflammatory enzymes, and demonstrating antioxidant action. Some anti-inflammatory herbs are also antioxidants, meaning they neutralize free radicals and reduce oxidative stress, which can aid with inflammation. Herbal medicines can also have an effect on the immune system, influencing the balance of pro- and anti-inflammatory responses. Some herbal substances can disrupt inflammation-related signaling pathways, such as the mitogen-activated protein kinase (MAPK) pathway [22]. They can also inhibit the expression of adhesion molecules, decreasing the recruitment of inflammatory cells during inflammation. They can also help to maintain cell membranes and prevent the production of inflammatory mediators. Certain plant-based substances can be used to induce the synthesis of anti-inflammatory mediators, like interleukin-10 (IL-10), which can be used to reduce inflammation. All things considered, herbal remedies have a lot to offer in the way of reducing inflammation [58].

The pro-inflammatory substances such as hydrogen peroxide ($H_2O_2$) and tumor necrosis factor-α (TNF-α) were inhibited and protected by rosemary, sage, and thyme, which in turn prevented the release of interleukin-8 (IL-8) from peripheral blood lymphocytes (PBLs). The fact that these reductions were mostly only observed in PBLs exposed to $H_2O_2$ suggests that there may be more going on here than just the suppression of one pro-inflammatory mediator [58–60].

### 4.4 Food preservation and shelf life

Herbs and spices have been used for flavoring, coloring, scent, enhancing agents, and food preservation. As ancient as human civilization, food preservation uses a range of techniques to keep perishable foods safe and of high quality. Plants have traditionally been important for food preservation because they provide naturally occurring chemicals that stop oxidation and microbial development. Herbs and spices, such as basil, thyme, oregano, and rosemary, are powerful preservatives as well as gastronomic treats. They are extremely useful for pickling vegetables, flavoring meats, and extending the shelf life of different foods such as fruits, vegetables, and beverages, because of their antibacterial qualities and extending their freshness and nutritional values [61].

The use of herbs and spices as natural preservatives and medicinal agents has been the subject of increased investigation and the effects of certain spices and herbs' bioactive properties on health. Bioactive compounds found in herbs and spices have the power to lower or even completely prevent the risk of degenerative diseases, including diabetes, obesity, cancer, and heart disease. Herbs and spices have antibacterial properties that can be effectively utilized to prevent the formation of pathogenic microbes and the spoiling in food items [9].

These materials from plants provide a variety of bioactive substances that improve flavor, prevent spoiling, and prolong food shelf life [43]. For example, rosemary has the strong flavor and perfume, due to active ingredients like rosmarinic acid and carnosic acid and also has the ability to suppress bacterial and fungal development and impart a unique flavor to food. Rosemary's antibacterial capabilities make it a perfect addition to marinades for meats and poultry. Similarly, some well known for their fragrant application, thyme, oregano, and basil include chemicals like carvacrol and thymol that help repel microorganisms. These herbs provide their protective qualities to a wide range of culinary creations, guaranteeing taste and durability, from pickled veggies to savory sauces. Warm, earthy tones of cinnamon, enhanced with cinnamaldehyde, improve baked goods and help preserve fruits and spices by preventing the formation of dangerous germs. Using herbs and spices into different preservation methods not only enhances the flavor but also protects the quality and lifespan of our cooking, enhancing our culinary adventures and respecting the traditional knowledge of plant-based dishes. In summary, using plants to preserve food not only improves culinary experiences but also supports sustainability, health, and the preservation of cultural culinary traditions (**Table 1**) [13, 63].

| Herbs/spices | Application(s) | Preservative activity |
|---|---|---|
| Garlic | Culinary (flavoring), Medicinal (antibacterial, antiviral) | Antimicrobial (against bacteria, viruses) Antioxidant |
| Turmeric | Culinary (coloring, flavoring), Medicinal (anti-inflammatory) | Anti-inflammatory, Antioxidant, Antimicrobial |
| Cinnamon | Culinary (flavoring), Medicinal (anti-inflammatory) | Antimicrobial (against bacteria, fungi), Antioxidant |
| Rosemary | Culinary (flavoring), Medicinal (antioxidant) | Antioxidant, Antimicrobial (against bacteria) |
| Thyme | Culinary (flavoring), Medicinal (antibacterial) | Antimicrobial (against bacteria), Antioxidant |
| Oregano | Culinary (flavoring), Medicinal (antimicrobial) | Antimicrobial (against bacteria, fungi), Antioxidant |
| Basil ginger | Culinary (flavoring), Medicinal (anti-inflammatory), Culinary (flavoring), Medicinal (antinausea, anti-inflammatory) | Antimicrobial (against bacteria), Antioxidant Antimicrobial (against bacteria), Anti-inflammatory |
| Cloves | Culinary (flavoring), Medicinal (antimicrobial) | Antimicrobial (against bacteria, fungi), Antioxidant |
| Peppermint | Culinary (flavoring), Medicinal (digestive aid) | Antimicrobial (against bacteria, fungi), Antioxidant |
| Lemon balm | Culinary (flavoring), Medicinal (calming) | Antimicrobial (against bacteria, viruses), Antioxidant |
| Sage | Culinary (flavoring), Medicinal (anti-inflammatory, Memory enhancement) | Antimicrobial (inhibits microbial growth), Antioxidant, Flavor enhancer |
| Cilantro (Coriander) | Culinary (flavoring), Medicinal (anti-inflammatory, detoxification, digestive aid) | Preservative (inhibits microbial growth), Flavor enhancer, Antioxidants |
| Chili pepper | Culinary (heat and flavoring), Medicinal (pain relief, metabolism boost, anti-inflammatory) | Antimicrobial (against bacteria), adds heat and flavor enhancer, Antioxidants |
| Saffron | Culinary (flavoring), Medicinal (Antidepressant) | Antimicrobial (against bacteria, fungi), Flavor enhancer, Antioxidant |
| Rhodiola | Medicinal (adaptogen and Mood Enhancement, immune) | Antimicrobial, Antioxidants |

**Table 1.**
*Culinary and medicinal application and preservative activity of some herbs and spices [2, 9, 13, 28, 46, 62].*

## 5. Challenges and opportunities

The challenges and opportunities in the exploration of herbs and spices in human health and the food industry underline the need for a holistic and collaborative approach involving various stakeholders, including researchers, policymakers, industry professionals, and consumers. Embracing sustainability, quality control, and evidence-based practices will contribute to the responsible integration of herbs and spices into contemporary lifestyles. The use of herbal spices in human health has a number of challenges, including possible drug interactions, variability in the quantities of active compounds, and standardization problems. Variations in the content of herbal products, for instance, may impact their safety and effectiveness [59]. For instance, certain drugs may interact with St. John's Wort, a popular herbal cure, affecting its efficacy. Further supporting the need for caution are interactions between herbal spices such as St John's Wort *(Hypericum perforatum)* and certain pharmaceuticals [60]. For the safe and successful integration of herbal spices into healthcare, it is imperative to solve two critical challenges: ensuring uniform quality and comprehending possible interactions between herbs and drugs. For example, achieving consistent therapeutic results is difficult due to the variety in ginger's active ingredients. In order to guarantee the trustworthy and secure incorporation of herbal spices into medical procedures, it is imperative that these issues be resolved.

The exploration of herbs and spices in the realms of human health and the food industry presents both challenges and significant opportunities. One of the primary challenges is the necessity for rigorous scientific validation to substantiate health claims associated with these botanicals. The establishment of evidence-based guidelines for their safe and effective use requires well-designed clinical trials [59, 62]. Regulatory complexities, including issues related to labeling, health claims, and safety assessments, further pose challenges in integrating herbs and spices into mainstream healthcare and the food industry. Additionally, ensuring sustainable sourcing practices, consumer education, and addressing allergen considerations are vital aspects that require careful attention to balance the potential benefits with potential risks [59].

On the flip side, there are ample opportunities to harness the diverse properties of herbs and spices for innovative product development. The creation of functional foods, nutraceutical supplements, and natural preservatives using these botanicals opens up new avenues for the food industry. As the demand for natural and healthy alternatives rises, there is an opportunity to fortify foods with essential nutrients, promoting the development of products that align with modern nutritional trends. Furthermore, the global culinary diversity can be enriched by incorporating a wide array of herbs and spices, not only enhancing flavors but also contributing to a healthier and more varied diet [43, 64]. Collaborative research initiatives, personalized nutrition solutions, and the empowerment of local communities through sustainable cultivation practices underscore the potential for a holistic and integrative approach that benefits human health, the environment, and the economy. Finally, antibiotic resistance prompts a renewed effort to test medicinal herbs for potential antibacterial, antifungal, and antiparasitic properties [65].

## 6. Scope

The scope of herbs and spices in the context of new perspectives on human health and the food industry is extensive and continually growing. Herbs and spices offer opportunities for innovative product development, with the potential to create functional foods, nutraceutical supplements, and natural preservatives [62]. These botanicals also play a role in fortifying foods with essential nutrients, aligning with modern nutritional trends and consumer preferences. Also, spices are becoming a major ingredient in meals all across the world. Beyond only enhancing food's color, flavor, and scent, spices are also used to assist people maintain their health and beauty when they are suffering from chronic illnesses [64]. Furthermore, the incorporation of diverse herbs and spices in global cuisines not only enhances flavors but also contributes to a healthier and more varied diet [65]. For example, cinnamon may have antioxidant and anti-inflammatory qualities, which may help it to be used to treat chronic illnesses [65]. In the future, new spices may be investigated for their medicinal uses and bioactive substances, hence increasing the contribution of herbal spices to the promotion of human well-being. The scope extends to collaborative research initiatives, personalized nutrition solutions, and community empowerment through sustainable cultivation practices, highlighting the holistic and integrative potential of herbs and spices in benefiting human health, the environment, and local economies [66, 67].

## 7. Application

The application of herbs and spices in the context of new perspectives on human health and the food industry is multifaceted. Herbs and spices are utilized for innovative product development, contributing to the creation of functional foods, nutraceutical supplements, and natural preservatives [68, 69]. These botanicals also play a significant role in fortifying foods with essential nutrients, aligning with modern nutritional trends and consumer preferences. Moreover, the incorporation of a diverse array of herbs and spices in global cuisines enhances flavors and contributes to a healthier and more varied diet [62]. These applications demonstrate the versatility and potential of herbs and spices in addressing contemporary health and culinary challenges.

The various studies have highlighted the antimicrobial efficacy of numerous herbs and spices against a spectrum of pathogens, including bacteria, fungi, and viruses [10]. For example, "bioactive constituents" and "biological activities" are present in coriander (*Coriandrum sativum*). The essential oils derived from coriander seeds and herbs possess antibacterial, antioxidant, hypoglycemic, hypolipidemic, anxiolytic, analgesic, anti-inflammatory, anticonvulsant, and anticancer effects, as demonstrated by their innovative pharmacological and therapeutic uses [20, 21]. Similarly, garlic (*Allium sativum*) is renowned for its possible cardiovascular advantages, including blood pressure-lowering and lipid profile improvement [34]. In addition, the active compounds in garlic, such as allicin, exhibit potent antibacterial and antifungal effects [15]. Garlic produces the sulfur-containing compound ajoene that inhibits platelet aggregation generated by all known agonists. Ajoene will be used as a pharmaceutical to treat thrombosis and has demonstrated broad-spectrum antibacterial, antifungal, and antiprotozoal effect. The antibacterial activities of ajoene, which are efficient against Gram-positive and Gram-negative bacteria, such as *Xanthomonas maltophilia, Escherichia coli, Lactobacillus plantarum, Mycobacterium smegmatis, Bacillus subtilis,* and *Streptomyces griseus*, appear to be dependent on the disulfide link in the compound. Likewise, ajoene's antifungal activity had been proven against *Aspergillus niger, Candida albicans, Saccharomyces cerevisiae*, and *Paracoccidioides brasiliensis*. Moreover, Ajoene exhibited antiprotozoal properties against *Plasmodium berghei* and *Trypanosoma cruzi*. Similarly, the polyphenols in oregano, thyme, and rosemary have demonstrated antimicrobial activity against foodborne pathogens [70]. Turmeric, containing curcumin, has exhibited antibacterial and antiviral properties, suggesting its potential in supporting immune function [67]. The anti-inflammatory qualities of turmeric (*Curcuma longa*) make it a promising treatment for ailments including arthritis [71]. *S. lavandulaefolia* Vahl., known as Spanish Sage, have shown anticholinesterase, antioxidant, anti-inflammatory, estrogenic, and central nervous system (CNS)-depressive (sedative) properties for their potential to treat Alzheimer's disease (AD) [26]. These examples show how using herbal spices may improve health by providing natural solutions for a range of well-being issues. These findings underscore the potential of incorporating herbs and spices into daily diets as a natural and flavorful approach to promoting human health by harnessing their antimicrobial properties.

## 8. Human health impact and culinary benefits

The utilization of herbal medicine not only contributes to human health but also enhances culinary experiences by offering a plethora of flavors and nutritional benefits. Herbs have been an integral part of traditional medicine systems globally for centuries, and their potential positive impact on human health is supported by scientific research. For instance, rosemary, commonly used in Mediterranean cuisine, contains compounds with antioxidant properties, and studies suggest that it may have neuroprotective effects, potentially benefiting cognitive function [72]. Similarly, the culinary use of garlic has been associated with numerous health benefits, including cardiovascular protection and anti-inflammatory effects due to its sulfur-containing compounds [73]. Similarly, Clove spice is used to cure earaches, inflammation, pain, antipyretic, antifungal, antibacterial, and peptic ulcers. Moreover, it is said to have anti-inflammatory, antiviral, antimicrobial, anticancer, antidiabetic, and antiobesity properties. Analgesic, anticonvulsant, hepatoprotective, anthelmintic, memory recall, anti-stress, insecticidal, antimutagenic, and antiulcerogenic are some other uses. Including these herbs in regular meals improves general health, in addition to giving flavors more depth and richness [62].

The culinary benefits of herbal medicine extend beyond taste, as many herbs possess medicinal properties that can positively impact human health. For example, basil, a staple in Italian cuisine, has demonstrated antimicrobial and anti-inflammatory properties [74]. These dual roles of herbs, both as sources of flavor enhancement and potential health-promoting agents, highlight the synergy between culinary arts and traditional medicine [75, 76], providing individuals with an opportunity to enjoy delicious meals while supporting their health and well-being.

## 9. Future modern prospect

In the modern health and food industry, there is a growing recognition of the valuable contributions of herbs and spices in promoting wellness and enhancing culinary experiences. Herbal medicine, with its roots deeply embedded in traditional practices, historical, cultural, and ecological regions, has gained renewed attention as consumers seek natural and holistic approaches to health [66]. Herbs and spices are not only prized for their aromatic and flavorful qualities but are increasingly acknowledged for their potential health benefits [62]. Research has demonstrated the presence of bioactive compounds in various herbs and spices that possess antioxidant, anti-inflammatory, and antimicrobial properties, among others [16, 62]. For instance, turmeric, a spice commonly used in traditional Indian medicine, contains curcumin, recognized for its anti-inflammatory and antioxidant effects [75]. Similarly, the medicinal properties of herbs like cloves, ginger, and garlic are being explored for their potential cardiovascular and immune system benefits [62, 76, 77]. Recent evidence explore the tulsi (*Ocimum sanctum*), snakeroot (*Polygala senega*), licorice root (*Glycyrrhiza glabra*), clove (*Syzygium aromaticum*), slippery elm and marshmallow osha root (*Ligusticum porteri*), and sage leaf (*Salvia officinalis*) are the most significant expectorant herbs against flu and cold [2, 78]. Similarly, *Astragalus membranaceus, Glycyrrhiza uralensis, Saposhnikoviae divaricata, Rhizoma Atractylodis macrocephalae, Rhizoma phragmitis, Lonicerae japonicae flos, Forsythiae fructus, Atractylodis rhizoma, Platycodonis radix, Agastache rugosa, and Cyrtomium fortunei, Herba eupatorii, Ophiopogon japonicus, Scrophularia ningpoensis, Adeinophora stricta,* and *Dendrobium nobile* have been associated with the control of viral infection, immunological inflammatory responses, and hypoxic response, even during COVID-19 infection [2].

The incorporation of herbs and spices into both the culinary and wellness spheres reflects a paradigm shift in consumer preferences toward natural, plant-based solutions [13, 43]. As individuals become more health-conscious and seek preventive measures, the demand for herbal supplements, teas, and functional foods has surged. The food industry has responded by incorporating these botanical ingredients into a diverse range of products, from herbal teas to spice-infused snacks, catering to the modern consumer's desire for both flavor and functional health benefits [79]. This trend underscores the integration of traditional knowledge with contemporary science, paving the way for a harmonious synergy between herbal medicine and the health and food industry. Future investigations on the therapeutic qualities of herbs using various models, including as *in vitro*, *in vivo*, *ex vivo*, and clinical/human trials, should be the main emphasis and requirement of study [62].

## 10. Conclusions

In conclusion, there is strong evidence for the dual importance of herbs and spices in both human health and culinary applications due to their synergy, bioactive components, and antibacterial qualities, antioxidant, anti-inflammatory, and other attributes [13]. Using these bioactive-rich properties not only help preserve food but may also be used to treat microbial illnesses in people. Also, herbs and spices in food preparation not only improve the taste, aroma, color experience of meals but also recognize their possible health and nutritional advantages [63].

As regards the antibacterial and dietary qualities of these bioactive substances, polyphenol, flavonoids, terpenoids, saponins, coumarins, alkaloids, and tannins, more investigation is necessary to clarify their precise mechanisms of action and maximize their uses in the culinary and medical fields [13, 37–39]. Herbs and spices' transformation from kitchen staples to bioactive compounds combines ancient wisdom with modern science, promoting healthier, diverse food experiences and promoting holistic well-being [63].

*DOI: http://dx.doi.org/10.5772/intechopen.1004443*

## Author details

Amrit Acharya[1*], Gita Ghimire[2] and Prabin Shakya[3]

1 Pulchock Campus, Tribhuvan University, Lalitpur, Nepal

2 Padma Kanaya Multiple Campus, Tribhuvan University, Kathmandu, Nepal

3 MiRON Laboratory and Research Center, Kathmandu, Nepal

*Address all correspondence to: amrit.acharya@pcampus.edu.np

## References

[1] Mistry KS, Sanghvi Z, Parmar G, Shah S. The antimicrobial activity of *Azadirachta indica*, *Mimusops elengi*, *Tinospora cardifolia*, *Ocimum sanctum* and 2% chlorhexidine gluconate on common endodontic pathogens: An *in vitro* study. European Journal of Dentistry. 2014;**8**(2):172-177. DOI: 10.4103/1305-7456.130591

[2] Shahrajabian MH, Sun W, Cheng Q. Traditional herbal medicine for the prevention and treatment of cold and flu in the autumn of 2020, overlapped with covid-19. Natural Product Communications. 2020;**15**(8):1-7. DOI: 10.1177/1934578X20951431

[3] Dougkas A, Vannereux M, Giboreau A. The impact of herbs and spices on increasing the appreciation and intake of low-salt legume-based meals. Nutrients. 2019;**11**(12):2901. DOI: 10.3390/nu11122901

[4] Bondi M, Lauková A, de Niederhausern S, et al. Natural preservatives to improve food quality and safety. Journal of Food Quality. 2017;**2017**:1-3. DOI: 10.1155/2017/1090932

[5] Karakol P, Kapi E. Use of selected antioxidant-rich spices and herbs in foods. In: Waisundara V, editor. Antioxidants-Benefits, Sources, Mechanisms of Action [Internet]. London, UK: IntechOpen; 2021. DOI: 10.5772/intechopen.96136. Available from: www.intechopen.com

[6] Gyawali R, Ibrahim SA. Natural products as antimicrobial agents. Food Control. 2014;**46**:412-429. DOI: 10.1016/j.foodcont.2014.05.047

[7] Radha Krishnan K, Babuskin S, Azhagu Saravana Babu P, et al. Antimicrobial and antioxidant effects of spice extracts on the shelf life extension of raw chicken meat. International Journal of Food Microbiology. 2014;**171**:32-40. DOI: 10.1016/j.ijfoodmicro.2013.11.011

[8] Khadka D, Dhamala MK, Li F, et al. The use of medicinal plants to prevent COVID-19 in Nepal. Journal of Ethnobiology and Ethnomedicine. 2021;**17**:26. DOI: 10.1186/s13002-021-00449-w

[9] Khanal A, Devkota HP, Kaundinnyayana S, et al. Culinary herbs and spices in Nepal: A review of their traditional uses, chemical constituents, and pharmacological activities. Ethnobotany Research and Applications. 2021;**21**:1-18. DOI: 10.32859/era.21.40

[10] Mashabela MN, Ndhlovu PT, Mbeng WO. Herbs and spices' antimicrobial properties and possible use in the food sector. In: Ivanišová E, editor. Herbs and Spices - New Advances [Internet]. London, UK: IntechOpen; 2023. DOI: 10.1016/B978-0-323-90572-5.00008-1. Available from: https://www.intechopen.com/chapters/84365

[11] Parthasarathy VA, Chempakam B, Zachariah TJ. Chemistry of Spices. Wallingford, UK; Cambridge, MA: CABI Pub; 2008. 445 p

[12] Gottardi D, Bukvicki D, Prasad S, Tyagi AK. Beneficial effects of spices in food preservation and safety. Frontiers in Microbiology. 2016;7:1394. DOI: 10.3389/fmicb.2016.01394

[13] Opara EI, Chohan M. Culinary herbs and spices: Their bioactive properties, the contribution of polyphenols and the challenges in deducing their true

health benefits. International Journal of Molecular Sciences. 2014;**15**(10):19183-19202. DOI: 10.3390/ijms151019183

[14] Kocaadam B, Şanlier N. Curcumin, an active component of turmeric (*Curcuma longa*), and its effects on health. Critical Reviews in Food Science and Nutrition. 2017;**57**:2889-2895. DOI: 10408398.2015.1077195

[15] Reuter HD. *Allium sativum* and *Allium ursinum*: Part 2 pharmacology and medicinal application. Phytomedicine. 1995;**2**(1):73-91. DOI: 10.1016/S0944-7113(11)80052-8

[16] Cortés-Rojas DF, de Souza CR, Oliveira WP. Clove (*Syzygium aromaticum*): A precious spice. Asian Pacific Journal of Tropical Biomedicine. 2014;**4**(2):90-96. DOI: 10.1016/S2221-1691(14)60215-X

[17] Anderson RA, Broadhurst CL, Polansky MM, Schmidt WF, Khan A, Flanagan VP, et al. Isolation and characterization of polyphenol type-a polymers from cinnamon with insulin-like biological activity. Journal of Agricultural and Food Chemistry. 2004;**52**(1):65-70. DOI: 10.1021/jf034916b

[18] McKay DL, Blumberg JB. A review of the bioactivity and potential health benefits of peppermint tea (*Mentha piperita* L.). Phytotherapy Research. 2006;**20**(8):619-633. DOI: 10.1002/ptr.1936

[19] Lambert RJ, Skandamis PN, Coote PJ, Nychas GJ. A study of the minimum inhibitory concentration and mode of action of oregano essential oil, thymol and carvacrol. Journal of Applied Microbiology. 2001;**91**(3):453-462. DOI: 10.1046/j.1365-2672.2001.01428.x

[20] Visuvanathan T, Than LTL, Stanslas J, Chew SY, Vellasamy S. Revisiting Trigonella foenum-graecum L.: Pharmacology and therapeutic potentialities. Plants (Basel). 2022;**11**(11):1450. DOI: 10.3390/plants11111450

[21] Laribi B, Kouki K, M'Hamdi M, Bettaieb T. Coriander (*Coriandrum sativum* L.) and its bioactive constituents. Fitoterapia. 2015;**103**:9-26. DOI: 10.1016/j.fitote.2015.03.012

[22] Charles DJ. Coriander. In: Antioxidant Properties of Spices, Herbs and Other Sources. New York, NY: Springer New York; 2012. pp. 255-263. DOI: 10.1007/978-1-4614-4310-0_21

[23] Mandal S, Mandal M. Coriander (*Coriandrum sativum* L.) essential oil: Chemistry and biological activity. Asian Pacific Journal of Tropical Biomedicine. 2015;**5**:421-428. DOI: 10.1016/j.apjtb.2015.04.001

[24] Alam A, Rehman NU, Ansari MN, Palla AH. Effects of essential oils of elettaria cardamomum grown in India and Guatemala on gram-negative bacteria and gastrointestinal disorders. Molecules. 2021;**26**(9):2546. DOI: 10.3390/molecules26092546

[25] Alkhalifah EAR, Alobaid AA, Almajed MA, Alomair MK, Alabduladheem LS, Al-Subaie SF, et al. Cardamom extract alleviates the oxidative stress, inflammation and apoptosis induced during acetaminophen-induced hepatic toxicity via modulating Nrf2/HO-1/NQO-1 pathway. Current Issues in Molecular Biology. 2022;**44**(11):5390-5404. DOI: 10.3390/cimb44110365

[26] Perry NS, Bollen C, Perry EK, Ballard C. Salvia for dementia therapy: Review of pharmacological activity and pilot tolerability clinical trial. Pharmacology, Biochemistry, and Behavior. 2003;**75**(3):651-659. DOI: 10.1016/s0091-3057(03)00108-4

[27] Badgujar SB, Patel VV, Bandivdekar AH. Foeniculum vulgare mill: A review of its botany, phytochemistry, pharmacology, contemporary application, and toxicology. BioMed Research International. 2014;**2014**:842674. DOI: 10.1155/2014/842674

[28] Ahmed SP, Shahi MH, Afzal M, Ali A. Alternate medicinal approach for the treatment of depression/mood disorder. In: Herbal Medicines [Internet]. Academic press, Cambridge, UK: Elsevier; 2022. pp. 69-89. DOI: 10.1016/B978-0-323-90572-5.00008-1. Available from: https://linkinghub.elsevier.com/retrieve/pii/B9780323905725000081

[29] de Alcantara BN, Kobayashi YT, Barroso KF, da Silva ID, de Almeida MB, Barbosa WL. Pharmacognostic analyses and evaluation of the *in vitro* antimicrobial activity of Acmella oleracea (L.) RK Jansen (Jambu) floral extract and fractions. Journal of Medicinal Plants Research. 2015;**9**(4):91-96. DOI: 10.5897/JMPR2014.5680

[30] Moreno SC, Carvalho GA, Picanço MC, Morais EG, Pereira RM. Bioactivity of compounds from *Acmella oleracea* against *Tuta absoluta* (Meyrick) (Lepidoptera: Gelechiidae) and selectivity to two non-target species. Pest Management Science. 2012;**68**(3):386-393. DOI: 10.1002/ps.2274

[31] Imam H, Riaz Z, Azhar M, et al. Sweet flag (*Acorus calamus* Linn.): An incredible medicinal herb. International Journal of Green Pharmacy. 2013;7:288. DOI: 10.4103/0973-8258.122053

[32] Pandey S, Chataut G, Maharjan S, et al. Unveiling the potential of Jimbu (*Allium przewalskianum*): Bioactive compounds, antioxidant, and antimicrobial properties of a native Himalayan spice herb. International Journal of Applied Sciences and Biotechnology. 2023;**11**:171-180. DOI: 10.3126/ijasbt.v11i4.59326

[33] Khanal M, Lamichhane S, Bhattarai A, Kayastha BL, Labh SN. Extract of Aloe vera (Aloe barbadensis miller) enhances the growth, protein contents, and gastrosomatic index (GaSI) of common carp *Cyprinus carpio*. Journal of Nutrition and Metabolism. 2021;**2021**:1-14. DOI: 10.1155/2021/8029413

[34] Hsouna AB, Michalak M, Akacha BB, Dhifi W, Saad RB, Brini F, et al. Assessment of the phytochemical composition, antimicrobial activity and anti-inflammatory effects of Lobularia maritima extracts on lipopolysaccharide-stimulated RAW 264.7 cells and their capacity to extend the shelf life of raw minced beef. Journal of Functional Foods. 2022;**99**:105327. DOI: 10.1016/j.jff.2022.105327

[35] Berber D, Toksöz O, Türkmenoğlu İ, Sesal C. A review on antimicrobial activities of some culinary herbs and spices against *Staphylococcus aureus*. Frontiers in Life Sciences and Related Technologies. 2022;**3**(2):75-85. DOI: 10.51753/flsrt.1083093

[36] Rombolà L, Scuteri D, Marilisa S, Watanabe C, Morrone LA, Bagetta G, et al. Pharmacokinetic interactions between herbal medicines and drugs: Their mechanisms and clinical relevance. Life. 2020;**10**(7):106. DOI: 10.3390/life10070106

[37] Guldiken B, Ozkan G, Catalkaya G, Ceylan FD, Yalcinkaya IE, Capanoglu E. Phytochemicals of herbs and spices: Health versus toxicological effects. Food and Chemical Toxicology. 2018;**1**(119):37-49. DOI: 10.1016/j.fct.2018.05.050

[38] Kapoor LD. CRC Handbook of Ayurvedic Medicinal Plants [Internet]. 1st ed. Boca Raton, USA: CRC Press; 2018. DOI: 10.1201/9781351070997.

Available from: https://www.taylorfrancis.com/books/9781351079440

[39] Jiang TA. Health benefits of culinary herbs and spices. Journal of AOAC International. 2019;**102**(2):395-411. DOI: 10.5740/jaoacint.18-0418

[40] Adhikari Devkota A, Kurauchi Y, Yamada T, Katsuki H, Watanabe T, Devkota HP. Antineuroinflammatory activities of extract and polymethoxy flavonoids from immature fruit peels of citrus 'Hebesu'. Journal of Food Biochemistry. 2019;**43**(6):e12813. DOI: 10.1111/jfbc.12813

[41] Khan H, Sureda A, Belwal T, Çetinkaya S, Süntar İ, Tejada S, et al. Polyphenols in the treatment of autoimmune diseases. Autoimmunity Reviews. 2019;**18**(7):647-657. DOI: 10.1016/j.autrev.2019.05.001

[42] Ludwiczuk A, Skalicka-Woźniak K, Georgiev MI. Terpenoids. In: Pharmacognosy [Internet]. Academic press, Cambridge, UK: Elsevier; 2017. pp. 233-266. Available from: https://linkinghub.elsevier.com/retrieve/pii/B9780128021040000111

[43] Hussain SA, Panjagari NR, Singh RRB, Patil GR. Potential herbs and herbal nutraceuticals: Food applications and their interactions with food components. Critical Reviews in Food Science and Nutrition. 2015;**55**(1):94-122. DOI: 10.1080/10408398.2011.649148

[44] El Aziz MM, Ashour AS, Melad AG. A review on saponins from medicinal plants: Chemistry, isolation, and determination. Nanomedicine Research Journal. 2019;**8**(1):282-288. DOI: 10.15406/jnmr.2019.08.00199

[45] Lu JJ, Bao JL, Chen XP, Huang M, Wang YT. Alkaloids isolated from natural herbs as the anticancer agents. Evidence-Based Complementary and Alternative Medicine. 2012;**2012**:485042. DOI: 10.1155/2012/485042

[46] Al-Wabel NA, Fat'hi SM. Antimicrobial activities of spices and herbs. Worldwide research efforts in the fighting against microbial pathogens. In: Proceedings of the II International Conference on Antimicrobial Research - ICAR2012, 21-23 November 2012, Lisbon (Portugal). 2013. pp. 1-4. Available from: https://www.researchgate.net/publication/302931553

[47] Parham S, Kharazi AZ, Bakhsheshi-Rad HR, Nur H, Ismail AF, Sharif S, et al. Antioxidant, antimicrobial and antiviral properties of herbal materials. Antioxidants. 2020;**9**(12):1309. DOI: 10.3390/antiox9121309

[48] De M, De A, Mukhopadhvay R. Antimicrobial activity of *Cuminum cyminum* L. Ars Pharmaceutica. 2003;**44**:257-269. DOI: 10481/28192

[49] Pawar VC, Thaker VS. *In vitro* efficacy of 75 essential oils against *Aspergillus niger*. Mycoses. 2006;**49**(4):316-323. DOI: 10.1111/j.1439-0507.2006.01241.x

[50] Mytle N, Anderson GL, Doyle MP, Smith MA. Antimicrobial activity of clove (*Syzgium aromaticum*) oil in inhibiting listeria monocytogenes on chicken frankfurters. Food Control. 2006;**17**(2):102-107. DOI: 10.1016/j.foodcont.2004.09.008

[51] Ogunwande IA, Olawore NO, Ekundayo O, Walker TM, Schmidt JM, Setzer WN. Studies on the essential oils composition, antibacterial and cytotoxicity of *Eugenia uniflora* L. International Journal of Aromatherapy. 2005;**15**(3):147-152. DOI: 10.1016/j.ijat.2005.07.004

[52] Chaieb K, Hajlaoui H, Zmantar T, Kahla Nakbi AB, Rouabhia M,

Mahdouani K, et al. The chemical composition and biological activity of clove essential oil, *Eugenia caryophyllata* (*Syzigium aromaticum* L. Myrtaceae): A short review. Phytotherapy Research. 2007;**21**(6):501-506. DOI: 10.1002/ptr.2124

[53] Tampieri MP, Galuppi R, Macchioni F, Carelle MS, Falcioni L, Cioni PL, et al. The inhibition of *Candida albicans* by selected essential oils and their major components. Mycopathologia. 2005;**159**(3):339-345. DOI: 10.1007/s11046-003-4790-5

[54] Winias S, Retno A. Effect of cinnamaldehyde from cinnamon extract as a natural preservative alternative to the growth of *Staphylococcus aureus* bacteria. Indonesian Journal of Tropical and Infectious Disease. 2015;**2**(1):38-41. DOI: 10.20473/ijtid.v2i1.188

[55] Butt MS, Pasha I, Sultan MT, Randhawa MA, Saeed F, Ahmed W. Black pepper and health claims: A comprehensive treatise. Critical Reviews in Food Science and Nutrition. 2013;**53**(9):875-886. DOI: 10.1080/10408398.2011.571799

[56] Goncagul G, Ayaz E. Antimicrobial effect of garlic (*Allium sativum*). Recent Patents on Anti-Infective Drug Discovery. 2010;**5**(1):91-93. DOI: 10.2174/157489110790112536

[57] Tornuk F et al. Efficacy of various plant hydrosols as natural food sanitizers in reducing Escherichia coli O157: H7 and *Salmonella typhimurium* on fresh cut carrots and apples. International Journal of Food Microbiology. 2011;**148**(1):30-35. DOI: 10.1016/j.ijfoodmicro.2011.04.022

[58] Isbill J, Kandiah J, Kružliaková N. Opportunities for health promotion: Highlighting herbs and spices to improve immune support and well-being. Integrative Medicine (Encinitas). 2020;**19**(5):30-42

[59] Posadzki P, Watson LK, Alotaibi A, Ernst E. Prevalence of use of complementary and alternative medicine (CAM) by patients/consumers in the UK: Systematic review of surveys. Clinical Medicine. 2013;**13**(2):126-131. DOI: 10.7861/clinmedicine

[60] Izzo AA, Ernst E. Interactions between herbal medicines and prescribed drugs: A systematic review. Drugs. 2001;**61**(15):2163-2175. DOI: 10.2165/00003495-200161150-00002

[61] Hassani FV, Shirani K, Hosseinzadeh H. Rosemary (Rosmarinus officinalis) as a potential therapeutic plant in metabolic syndrome: A review. Naunyn-Schmiedeberg's Archives of Pharmacology. 2016;**389**(9):931-949. DOI: 10.1007/s00210-016-1256-0

[62] Otunola GA. Culinary spices in food and medicine: An overview of *Syzygium aromaticum* (L.) Merr. and L. M. Perry [Myrtaceae]. Frontiers in Pharmacology. 2022;**12**:793200. DOI: 10.3389/fphar.2021.793200

[63] Kumar Paswan V, Shekhar Singh C, Kukreja G, Shankar Bunkar D, Kumar Bhinchhar B. Health benefits and functional and medicinal properties of some common Indian spices. In: Shabir Ahmad R, editor. Herbs and Spices - New Processing Technologies [Internet]. London, UK: IntechOpen; 2021. Available from: https://www. intechopen.com/books/herbs-and-spices-new-processing-technologies

[64] Majumdar A. Advantages of spices and herbs for human health nutrition and recoveries. In: International Pepper and Spices Conference, Tobasco, Mexico, on 2022 Apr 26. 2022. Availble from: https://www.researchgate.net/publication/360561567

[65] Rao PV, Gan SH. Cinnamon: A multifaceted medicinal plant.

Evidence-based Complementary and Alternative Medicine. 2014;**2014**:1-12. DOI: 10.1155/2014/642942

[66] Kunwar RM, Shrestha KP, Bussmann RW. Traditional herbal medicine in far-West Nepal: A pharmacological appraisal. Journal of Ethnobiology and Ethnomedicine. 2010;**6**(1):35. DOI: 10.1186/1746-4269-6-35

[67] Kunwar RM, Mahat L, Acharya RP, Bussmann RW. Medicinal plants, traditional medicine, markets and management in far-West Nepal. Journal of Ethnobiology and Ethnomedicine. 2013;**9**(1):24. DOI: 10.1186/1746-4269-9-24

[68] Ried K, Travica N, Sali A. The effect of aged garlic extract on blood pressure and other cardiovascular risk factors in uncontrolled hypertensives: The AGE at heart trial. Integrated Blood Pressure Control. 2016;**9**:9-21. DOI: 10.2147/IBPC.S93335

[69] Prasad S, Aggarwal BB. Turmeric, the Golden spice: From traditional medicine to modern medicine. In: Benzie IFF, Wachtel-Galor S, editors. Herbal Medicine: Biomolecular and Clinical Aspects [Internet]. 2nd ed. Boca Raton (FL): CRC Press/Taylor & Francis; 2011. ISBN-13: 978-1-4398-0713-2

[70] Naganawa R, Iwata N, Ishikawa K, Fukuda H, Fujino T, Suzuki A. Inhibition of microbial growth by ajoene, a sulfur-containing compound derived from garlic. Applied and Environmental Microbiology. 1996;**62**(11):4238-4242. DOI: 10.1128/aem.62.11.4238-4242.1996

[71] Daily JW, Yang M, Park S. Efficacy of turmeric extracts and curcumin for alleviating the symptoms of joint arthritis: A systematic review and meta-analysis of randomized clinical trials. Journal of Medicinal Food. 2016;**19**(8):717-729. DOI: 10.1089/jmf.2016.3705

[72] Moss M, Cook J, Wesnes K, Duckett P. Aromas of rosemary and lavender essential oils differentially affect cognition and mood in healthy adults. The International Journal of Neuroscience. 2003;**113**(1):15-38. DOI: 10.1080/00207450390161903

[73] Morvaridzadeh M, Cohen AA, Heshmati J, Alami M, Berrougui H, Zoubdane N, et al. Effect of extra virgin olive oil on anthropometric indices, inflammatory and cardiometabolic markers: A systematic review and meta-analysis of randomized clinical trials. The Journal of Nutrition. 2023;**154**(1):95-120. DOI: 10.1016/j.tjnut.2023.10.028

[74] Rattanachaikunsopon P, Phumkhachorn P. Antimicrobial activity of basil (*Ocimum basilicum* L) oil against *Salmonella enteritidis in vitro* and in food. Bioscience, Biotechnology, and Biochemistry. 2010;**74**(6):1200-1204. DOI: 10.1271/bbb.90939

[75] Aggarwal BB, Sundaram C, Malani N, et al. The molecular targets and therapeutic uses of curcumin in health and disease. In: Aggarwal BB, Surh YJ, Shishodia S, editors. Advances in Experimental Medicine and Biology. Vol. 595. New York, USA: Springer; 2007. pp. 1-75. DOI: 10.1007/978-0-387-46401-5_1

[76] Bode AM, Dong Z. The amazing and mighty ginger. In: Benzie IFF, Wachtel-Galor S, editors. Herbal Medicine: Biomolecular and Clinical Aspects. 2nd ed. Boca Raton (FL): CRC Press/Taylor & Francis; 2011. Chapter 7

[77] Nantz MP, Rowe CA, Muller CE, Creasy RA, Stanilka JM, Percival SS.

Supplementation with aged garlic extract improves both NK and γδ-T cell function and reduces the severity of cold and flu symptoms: A randomized, double-blind, placebo-controlled nutrition intervention. Clinical Nutrition. 2012;**31**(3):337-344. DOI: 10.1016/j.clnu.2011.11.019

[78] Purkayastha S, Dahiya P. Phytochemical screening and antimicrobial activity of some medicinal plants against multi-drug resistant bacteria from clinical isolates. Indian Journal of Pharmaceutical Sciences. 2012;**74**(5):443. DOI: 10.4103/0250-474X.108420

[79] Brendler T, Brinckmann JA, Daoust M, He H, Masé G, Steffan K, et al. Suitability of botanical extracts as components of complex mixtures used in herbal tea infusions—Challenges and opportunities. Frontiers in Pharmacology. 2022;**7**(13):1013340. DOI: 10.3389/fphar.2022.1013340

Chapter 10

# Harnessing Nature's Bounty: Phytoinsecticides for a Healthier and Sustainable World

*Thirunavukkarasu Selvamuthukumaran,*
*Palanisamy Dhanapriya and Nusrat Iqbal*

## Abstract

Global warming fuels pest infestations, causing massive crop losses and economic damage. Chemical insecticides, though initially effective, come with a heavy toll: environmental harm, health risks, and resistance development. Their overuse creates a vicious cycle, leading to even more pesticide use and devastating consequences for beneficial insects, soil, water, and human health making our current agricultural practices unsustainable. Phyto-insecticides derived from plants are safer and more sustainable alternatives that boast a long history of use and diverse modes of action, making it harder for pests to develop resistance. They pose lower risks to human health and the environment and can be produced sustainably from renewable plant sources. While promising, phyto-insecticides face hurdles. Limited plant biomass, variable effectiveness, and unstable formulations hinder their commercialization. However, innovative solutions are emerging: (1) callus culture: this technique offers a sustainable way to mass-produce valuable secondary metabolites like Azadirachtin and Pyrethrin; (2) understanding insect interactions: Deciphering how these compounds interact with insects paves the way for effective utilization and formulation design; (3) nanotechnology: nanoparticles enhance stability, bioavailability, and targeted delivery, boosting efficacy and reducing environmental impact. Excitingly, trace amounts of phyto-insecticide residues may offer additional benefits. Some compounds, like Azadirachtin, Piperine etc., possess potential nutraceutical properties, promoting bone health, managing diabetes, and even fighting cancer. This opens the door to "*nutraresidiceuticals*," where food treated with phyto-insecticides might enhance consumer health. Phyto-insecticides hold immense potential as a sustainable and effective pest management strategy. By addressing challenges related to biomass, formulation, and understanding their modes of action, we can unlock their full potential for a healthier and more sustainable future. Additionally, exploring the potential "nutraresidiceutical" benefits opens up exciting new avenues for research and development.

**Keywords:** phyto-insecticides, sustainable agriculture, callus culture, nanotechnology, nutraceuticals, nutraresidiceuticals



IntechOpen

## 1. Introduction

Agriculture, the bedrock of human civilisation, faces a constant challenge: the relentless onslaught of pests. This battle for food security intensified as global warming resulted in insect pests' increased geographic distribution, survival, and invasion, altered interspecific interaction, and converted diverse ecosystems into fertile breeding grounds for these destructive insects [1]. While estimates suggest that less than 0.5% of insects are classified as pests, their impact on crops, livestock, and human health cannot be understated. Studies reveal a staggering 20–40% crop loss due to pest damage, translating to an alarming 70 billion USD annually [2, 3].

As soaring populations and catastrophes caused by world wars necessitated increased food production, Agriculture was compelled to take more intense pest control tactics during the late 1940s and 1950s. The initial success of synthetic chemical insecticides though fuelled the Green Revolution through high-yielding crop varieties, had unleashed a cascade of unforeseen consequences and had become one of the most controversial issues [4, 5].

The widespread and often indiscriminate use of these potent chemicals has wrought havoc on natural ecosystems and disrupted ecological balance. Pesticides, designed to eliminate specific insects, inadvertently select resistant individuals, leading to the rise of "superbugs" immune to their effects [6]. This phenomenon creates a vicious cycle of escalating pesticide use and ultimately exacerbates the problem. Furthermore, these chemicals wreak havoc on natural enemies and beneficial insects, leaving the door open for secondary pest outbreaks and further ecological damage [7–10]. The long-term consequences of this disruption are far-reaching, impacting soil health, biodiversity, and overall sustainability. The human cost of pesticide use is simply unacceptable, particularly for vulnerable populations in developing countries. Chemical residues on food crops pose a significant threat to consumers, with studies linking pesticide exposure to various health issues, including cancer, birth defects, and neurological disorders [11]. Finally, the environmental impact of chemical pesticides cannot be ignored. These toxic substances pollute water sources and contaminate air and soil [12, 13].

Several hundred pesticides of different chemical nature are currently used in agriculture all over the world [14]. The burgeoning global insecticide market, valued at 19.5 billion USD in 2022 and projected to reach 28.5 billion USD by 2027, highlights the increasing reliance on pest control measures, even now [15].

## 2. Exploring a sustainable alternative: botanical insecticides

### 2.1 A return to plant-based pest control: a long history and renewed interest

The long-term consequences for the planet are dire, jeopardizing the very foundations of our food security and environmental well-being. In the face of these challenges, the need for a paradigm shift in pest management is undeniable. Moving away from this unsustainable reliance on chemical solutions and embracing a more holistic approach that prioritizes environmental sustainability and human health is a must.

Further, as the concept of pest management was radically revised towards advocating the suppression of pest populations below levels capable of causing economic injury rather than total eradication, insecticides of plant origin which exerted coherent management over insect pests for many centuries got re-attention.

Secondary metabolites have been exploited longer for managing pests [16, 17]. Botanical insecticides have a long history; their use in India, China, and Egypt since thousands of years ago highlighted their efficiency in pest management. The availability of wider choices, a broader spectrum of action, and relative safety made them attractive alternatives for synthetic insecticides [18].

As most of the botanicals are relatively non-toxic to humans, animals, and natural enemies they readily fit into IPM protocol [19, 20]. They help in reducing pesticide usage, environmental contamination, and human and animal health hazards [21].

In recent years, attempts have been made to identify plants, including herbs and weeds and their novel phytochemicals, for their insecticidal property. Different types of plant preparations such as crude extracts, solvent extracts, and powders have been reported for their insecticidal activity. Numerous plant species have been identified as possessing pesticidal properties [22, 23]. They possess secondary metabolites like alkaloids, non-protein amino acids, steroids, phenols, flavonoids, glycosides, glucosinolates, quinones, tannins, and terpenoids, etc., which are responsible for the protective action against the insect pests [24].

Their diverse modes of action, repellence, feeding deterrence, insecticidal, and insect growth regulatory activities, could act additively or synergistically, making resistance development more difficult and a high degree of biodegradation makes them a preferred choice [25, 26].

### 2.2 Challenges hindering wider adoption: addressing the gaps

Although there is a rich source of plants that could be harnessed as insecticides, commercialization of botanicals has not gained ground. The market share of botanicals along with other biopesticides is a mere 2% [27]. While a few botanicals were commercially exploited like neem, rotenone, pyrethrum, and some essential oils [28], the non-availability of biomass and formulations, lack of standardization and quality control measures hinder their commercialization.

Furthermore, their sensitivity to environmental conditions like light, temperature, humidity, substrate pH, etc. make them more unstable necessitating the need to formulate them as stable phyto-insecticide formulations. Improving phyto-insecticide formulations through new approaches like encapsulation, emulsification, or complexation, aiming to increase shelf life and enhance efficacy by optimizing their delivery and uptake by target insects, results in novel product development [25, 29–31].

Hence, research efforts are focused towards novel approaches like the use of nanotechnology to encapsulate active ingredients as nanoparticles for improved stability, control-release, and targeted delivery and utilizing biodegradable polymers to create environmentally friendly formulations with improved release profiles.

While formulation plays a crucial role, other aspects also contribute to the limited utilization of botanical insecticides. These include:

Seasonal and ecotype variability: the efficacy of phyto-insecticides can vary significantly depending on the season, and plant source. This makes it challenging to develop standardized and widely applicable products [32].

Synthesis difficulties: isolating and purifying specific active compounds from plants can be complex and expensive, hindering their commercialization.

## 3. Overcoming barriers to phyto-insecticide commercialization: tackling biomass, formulation, and standardization issues

### 3.1 Harnessing the power of callus culture: a promising solution

Hence, the synthesis of active ingredients to overcome the non-availability of biomass is not a reliable and economical alternative. However, it could be overcome through plant tissue culture-based *in vitro* secondary metabolite production, like callus culture systems, which offers tremendous scope as the said issues can be surmounted easily. They produce secondary metabolites at a rate similar to or superior to *in vivo* systems [33].

The use of in vitro callus culture for synthesizing various secondary metabolites, including Azadirachtin, Pyrethrin, and others and the bioefficacy of these metabolites against insects highlights their prospective role in solving biomass, ecotype and seasonal variability-related issues.

The following examples of Successful Callus Culture for the development of Insecticidal Compounds highlight its scope.

#### *3.1.1 Azadirachtin*

Callus culture offers a consistent and controlled environment for the production of Azadirachtin content. Studies have shown that increased Azadirachtin production was achieved with bioreactors and cell retention techniques [34]. Some authors demonstrated increased extracellular Azadirachtin with specific nitrate: ammonium ratios [35]. A literature reported higher Azadirachtin in zygotic embryo cultures compared to leaf and ovary cultures [36].

#### *3.1.2 Pyrethrin*

Production of pyrethrin an insecticidal compound naturally found in Chrysanthemums, have been enhanced through callus culture. Few authors opined higher Pyrethrin concentrations in callus extracts compared to intact plant extracts and observed positive effects of auxin, cytokinin, and other hormones on Pyrethrin production [37, 38].

#### *3.1.3 Other secondary metabolites*

Callus culture has been used to produce various other valuable compounds:

Podophyllotoxin was extracted from callus and plant root cultures of *Podophyllum hexandrum* Royle [39]. Lignan production in callus, regenerated shoots, and field-grown plants of *Phyllanthus amarus* Schumach & Thonn, were compared and higher yields reported in regenerated shoots [40]. Various secondary metabolites (Alkaloids and other compounds) in callus cultures of *Physalis peruviana* L. were quantified [41]. Callus culture conditions were optimized for *Ginkgo biloba* L. to enhance flavonoid and terpene lactone synthesis [42]. A protocol for callus induction in *Centella asiatica* L. was developed and various secondary metabolites (Triterpenes and alkaloids) were identified in the callus extract [43].

#### *3.1.4 Bioefficacy of callus extracts*

Several studies reported strong antifeedant and insecticidal activity of Azadirachtin extracted from callus cultures against various insects [44–48]. A knockdown effect was demonstrated against *Tribolium spp*. using Pyrethrin extracted from *Tagetes erecta* L. callus culture [49]. The larvicidal activity was reported against *Anopheles stephensi* larvae using rotenoids extracted from *Cassia tora* L. callus culture [50]. Repellence and knockdown effects were observed against *Tribolium sp.* using callus extracts rich in pyrethrin from *C. cinerariaefolium* L. [51].

These studies highlight the potential of *in vitro* callus culture as a sustainable and efficient method for producing valuable secondary metabolites with insecticidal and antifeedant properties.

### 3.2 Understanding and optimizing the impact: deciphering the complex modes of action for effective utilization

Further, instead of mere screening for bio-efficacy isolation, identification and evaluation of the active components of the plant products will lead to commercialization of the product if found effective. This may also lead to the synthesis of the components for commercial use if active components are economically and biologically very effective.

Botanicals generally possess dose-dependent varieties of action starting from repellency and leading even to the death of insects. However, large-scale utilization of botanicals in pest management is obstructed by non-availability of formulations. This constraint may be surrounded by conducting detailed studies on the active constituents of the effective plant fractions and their influence on various physiological systems. In support of the above-mentioned point, the availability of plant-based insecticidal formulations derived from neem oil and pyrethrum, whose physiological influences are well studied is available worldwide [52].

Among all present-day living forms, insects are one of the most ancient creatures. These insects with a long evolutionary history and remarkable adaptability roamed the world for almost 350 million years, the reason behind being on top of the evolutionary ladder [53]. Needless to explain one can understand their capacity to adapt. Such superior adaptability is mainly the product of complex and adaptable physiological systems [54].

Hence, the strategic application of control strategies aiming at disrupting the adaptive capacity of the insects is thought over. Consequently, the use of plants, an organism that co-evolved along with insects with a diverse array of bioactive compounds with anti-herbivore properties is to be re-considered [55]. These natural products may be effective in reducing the capacity for adaptation of insects. Their interaction with dynamic physiological processes, if studied will necessarily throw light on the best possible use for efficient control of insect pests.

The interaction between secondary metabolites and insect physiology, and the specific physiological processes targeted, define the "mode of action" of these natural insecticides. Understanding the mode of action is crucial for effective pest management as it reveals essential information about speed of action, target spectrum, and environmental safety. Furthermore, such studies guide formulation design and provide leads for developing novel insecticides.

However, the intricate nature of secondary metabolites, encompassing both their structure and function, presents a significant challenge compared to synthetic insecticides. Specifically, the dose-dependent effects observed (e.g., Azadirachtin inducing diverse morphogenetic defects and mortality based on applied concentration) complicate the determination of the precise mode of action. This challenge is further compounded by the lack of standardized experimental protocols for validation.

The diverse modes of action employed by secondary metabolites against insects can be broadly categorized into:

1. Neurotoxic

2. Cytotoxic

3. Enzyme inhibitory

4. Metabolic disrupters

5. Interactions with biomolecules

Intriguingly, individual secondary metabolites can exhibit multiple modes of action depending on the dose and target site. This multifaceted nature, combined with the vast diversity of secondary metabolites, presents a significant challenge for deciphering their precise modes of action. The complexity is further amplified in plant extracts, where synergistic or antagonistic interactions between multiple compounds occur. Therefore, to effectively elucidate the mode of action of a plant extract's anti-insect activity, it is crucial to isolate and purify the key bioactive molecules, and investigate the individual and combined actions of these purified compounds on relevant targets using appropriate biochemical and physiological assays.

By adopting this stepwise approach, researchers can gain a deeper understanding of the complex mechanisms employed by secondary metabolites in their defense against insects. This knowledge can then be harnessed to develop more targeted and effective pest management strategies.

### 3.3 Developing stable and effective nanoformulations: key to unlocking potential

The term "nano" is derived from the Greek word meaning dwarf, representing sizes ranging from 1 to 100 nm. Nanotechnology, as termed by Norio Taniguchi, involves the manipulation of matter at the nanoscale to produce materials with unique properties. According to the International Organization for Standardization and the Organization for Economic Cooperation and Development, nanomaterials are defined as materials having external dimensions or internal structures at the nanoscale.

Agrochemical formulations incorporated as nanoparticles (NPs) or in nanoscale materials with diverse coatings offer a promising avenue for improving the solubility and permeability of active ingredients (AIs). This, in turn, can result in enhanced bioavailability, allowing for a reduction in the AI dosage. Moreover, these formulations enable controlled release and targeted biodistribution of AIs, contributing to more efficient and environmentally friendly agricultural practices [56]. They also offer other advantages such as improved dispersion and wettability, biodegradability in both soil and the environment, non-toxic nature, and photogenerative properties [57, 58].

Various nanocarriers, such as nanocapsules, nanospheres, micelles, nano gels, nanoemulsions, nanodispersions, and inorganic materials, alter the properties of phyto-insecticide's active ingredients and offer unique advantages like controlled release kinetics, improved solubility, lower dose-requirement, enhanced stability, and prolonged efficiency in pest control. Techniques like coacervation, nanoprecipitation, and microemulsion are commonly used for encapsulating phytochemicals within nanoformulations. These methods help achieve uniform particle size distribution and enhance the overall efficacy.

### *3.3.1 Nano-formulations for sustainable development*

Nano formulations offer multiple advantages. These include active ingredient degradation prevention and improved activity due to smaller particle size and larger surface area [59]. Notably, it is anticipated that the mechanism of action against target pests will be enhanced compared to bulk equivalents, given the increased interaction with target pests resulting from the small size of nanoparticles [60]. Additionally, nanoparticles have demonstrated consistent leaf cover and plant penetration, supporting their potential in pest control and agriculture [61]. These findings indicated reduced insecticide usage that led to sustainable plant protection protocols.

### *3.3.2 Key aspects related to nanoformulations of phytochemicals*

Improved bioavailability: nanoformulations can enhance the bioavailability of phytochemicals, ensuring that a greater proportion of the compound reaches its target site. This is particularly important for phytochemicals with low solubility or poor absorption.

Increased stability: phytochemicals are often sensitive to environmental factors, such as light, heat, and oxygen, which can lead to degradation. Nanoformulations provide a protective environment, increasing the stability of phytochemicals and prolonging their shelf life.

Targeted delivery: nanoformulations enable targeted delivery of phytochemicals to specific cells or tissues. This targeted approach can enhance efficacy while minimizing potential side effects on non-targeted cells.

Controlled release: controlled release is a key feature of nanoformulations. This allows for a sustained and controlled delivery of phytochemicals over time, ensuring a more consistent effect.

### *3.3.3 Challenges and safety considerations of nanoformulation*

Nanoformulations, which involve manipulating materials at the nanoscale, hold great promise in various fields such as medicine, agriculture, and electronics due to their unique properties. However, as with any new technology, some challenges need to be addressed to ensure their successful implementation.

Production scale-up: transitioning from small-scale laboratory production to large-scale manufacturing can be complex. Processes that work well in the lab may not be easily scalable or cost-effective when producing nanoformulations on a commercial scale. Engineers and scientists must develop efficient and reproducible manufacturing processes to meet the demands of mass production.

Regulatory considerations: nanoformulations may fall under specific regulatory frameworks that govern the production, distribution, and use of nanomaterials. Regulatory agencies require comprehensive data on the safety, efficacy, and environmental impact of these formulations before approving them for commercial use. Meeting regulatory requirements often involves extensive testing and documentation, which can be time-consuming and expensive.

Safety evaluations: nanoformulations raise concerns about potential health and environmental risks associated with exposure to nanomaterials. Before these formulations can be widely used, thorough safety evaluations are essential to understand their potential hazards and mitigate any associated risks. This includes assessing their toxicity, biodistribution, biodegradability, and long-term effects on living organisms and ecosystems.

Given the complexity and novelty of nanoformulations, ensuring their safety is crucial to gaining public acceptance and regulatory approval. By addressing these challenges and conducting rigorous safety evaluations, researchers and industries can harness the potential benefits of nanoformulations while minimizing potential risks to human health and the environment.

### *3.3.4 Types of nanoformulations*

#### *3.3.4.1 Nanoemulsions*

Nanoemulsions (NEs) are small droplets of one liquid dispersed in another, stabilized by an amphiphilic surfactant, commonly water in oil (W/O) or oil in water (O/W) [62]. Very small oil/water emulsion nanoscale droplets with diameters less than 100 nm create nanoemulsions. The nanoemulsion is designed to improve pest management by increasing pesticide solubility, bioavailability, stability, and wettability. With sizes around 100 nm, these NEs exhibit advantageous properties such as a large surface area, strong stability, clear appearance, and adjustable rheology [63].

##### *3.3.4.1.1 Nanoemulsion preparation techniques*

High-energy techniques like membrane emulsification, high-pressure homogenization, and sonication, as well as low-energy methods based on nonionic surfactant phase transitions or spontaneous emulsification, are employed for NE preparation. NEs possess unique characteristics, including a high elastic modulus, Laplace pressure, surface area to volume ratio, and small droplet size, requiring specialized equipment like ultrasonic generators or high-pressure homogenizers due to the energy-intensive nature of the process. However, low-energy methods, like spontaneous emulsification, offer a simpler approach without the need for expensive equipment.

##### *3.3.4.1.2 Applications*

The incorporation of these natural ingredients aims to enhance bioavailability, reduce negative side effects, minimize non-specific absorption, and enable precise targeting to specific cells. Targeting ligands, such as folate, are introduced on the surface of NEs to further improve their performance in specific applications. This targeted technology often involves nanocarrier functionalization, which can be achieved through surface modification [64, 65] and/or ligand grafting [66].

#### *3.3.4.2 Nanoencapsulation*

Nanoencapsulation involves enclosing pesticides within nano-sized materials, often polymers, to create nanocapsules that release the pesticide under specific environmental conditions [67]. It is employed in crop protection chemical formulations to achieve enhanced solubility, specificity, and stability at varying environmental conditions *viz*., pH and temperature [68]. Additionally, nanoencapsulation enables active substances' control release and precise targeting [69].

*3.3.4.2.1 Nanoencapsulation preparation techniques*

Creating efficient polymeric nanocapsules with various synthesis protocols including layer-by-layer deposition, double emulsification, emulsification-coacervation, solvent evaporation, nanoprecipitation, melt dispersion, emulsion polymerization, interfacial polymerization, interfacial deposition methods, solvent displacement technique, and emulsion-evaporation. These methods contribute to the utilization of this formulation technique in the development of chemically varying botanical AIs as phyto-insecticide formulations.

*3.3.4.2.2 Applications*

Their design enables them to withstand environmental processes such as leaching, evaporation, photolytic hydrolysis, and microbial destruction. Importantly, these formulations feature built-in switches for regulating the release and availability of pesticides over a specific duration, optimizing their efficiency. These facts highlight their utility in phyto-insecticide formulation development as their commercial success entirely depends on these parameters.

*3.3.4.3 Nanosuspension*

Nanosuspension technology offers significant improvements in the solubility, wettability, and activity of poorly soluble substances, with uniform particle sizes below 100 nm. The key outcomes arise from the increased surface area, rapid dissolving rate, and excellent penetrability. Notably, the nanosuspension approach reduces the required amount of surfactants and eliminates the need for organic solvents, making it an environmentally friendly and efficient method for phyto-insecticide formulation.

*3.3.4.3.1 Nanosuspension preparation techniques*

In contrast to traditional methods like micronization, which involves milling coarse powders to create suspension concentrations (SC) resulting in particle sizes around 13 μm, nanosuspension technology takes the process from micrometer- to nanometer-sized particles. Phyto-insecticides often have low solubility, and micronization may not achieve sufficient bioavailability. Nanosuspensions, typically in the size range of 200–500 nm, become the next step in enhancing the effectiveness of these poorly soluble AIs.

Two main methods for producing nanosuspensions are bottom-up (controlled precipitation/crystallization) and top-down technologies. In the bottom-up method, the AI is dissolved in an organic solvent and precipitated with an anti-solvent in the presence of a stabilizer, resulting in homogeneous particles with smaller sizes. This method may also generate amorphous nanoparticles with high dissolution rates. On the other hand, the top-down approach involves various methods such as media milling, high-pressure homogenization, and microfluidization to reduce particle size. Combination procedures that integrate pre-treatment with subsequent size-reduction phases are commonly used for producing more homogeneous nanoparticles. Overall, nanosuspension technology presents a versatile and effective means of addressing solubility challenges and enhancing phyto-insecticide formulation performance.

##### *3.3.4.3.2 Applications*

*Chrysanthemum coronarium* L. and *Azadirachta indica* A. Juss nanosuspensions exhibit versatility in targeting a spectrum of pests, including *Aedes aegypti*, *Culex quinquefasciatus*, *Myzus persicae*, *Plutella xylostella*, *Spodoptera litura*, *Tribolium castaneum*, and *Helicoverpa armigera* [70]. Similarly, Rotenone nanosuspensions target *Bursaphelenchus xylophilus* [71].

### *3.3.5 Nanoformulation characterization*

#### *3.3.5.1 Particle size and its distribution*

The characterization of particle size and distribution is pivotal for evaluating the efficacy and stability of nanoformulations. Small and uniform particle sizes contribute to an increased surface area, facilitating improved interactions with biological systems. Dynamic Light Scattering (DLS) is a widely employed technique for determining the size and distribution of nanoformulations. To achieve optimal efficiency, samples are diluted with deionized water before analysis to prevent multiple scattering caused by aggregation via electrostatic interaction. During the measurement process, the polydispersity index (PDI) is a critical parameter, with a value less than 0.5 considered acceptable for agricultural use, indicating good uniformity of particle diameter. Samples with higher PDI, signifying polydispersity, are deemed unsuitable for characterization using DLS [72].

The size of nanoemulsions is influenced by various factors, with surfactant concentration and packing parameters playing a significant role. Surfactant packing parameters, crucial for changes in surfactant curvature and finer nanoemulsion droplet formation, are highly affected by the ratio of hydrophobic and hydrophilic regions. The surfactant arrangement at the oil/water (O/W) boundary with low interfacial tension creates a bicontinuous microemulsion, resulting in smaller particles. Research studies consistently indicate that an increased ratio of surfactant to oil leads to smaller droplet sizes [73]. This scientific understanding contributes to the nuanced design and optimization of nanoemulsions for phyto-insecticide formulation development.

#### *3.3.5.2 Viscosity, zeta and pH measurement*

The electrophoretic properties, specifically the zeta potential, of nanoformulations are commonly measured using Zetasizer. The zeta potential, influenced by the surface properties around the particles, is indicative of nanoformulation stability, with pH value playing a crucial role. A negative zeta potential generates repulsive forces that surpass the attractive forces among droplets, preventing coagulation and coalescence in the emulsion. The stability of nanoformulation tends to decrease with an increase in oil concentration in the system [74–76]. Utilizing an Ostwald viscometer, viscosity values of nanoformulations influenced by the nature of surfactants, organic phase components, and oil viscosity can be measured. Phyto-insecticide nanoformulations typically exhibit low viscosity, given their classification as O/W type with high water loading. However, surfactant concentration can be a variable influencing the viscosity of nanoformulations [77]. This scientific understanding contributes to the meticulous design and optimization of nanoformulations, especially in the context of phyto-insecticide delivery systems.

#### *3.3.5.3 Morphology and stability study*

The shape and morphology of nanoformulations are elucidated through advanced imaging techniques such as atomic force microscopy (AFM), transmission electron microscopy (TEM), and cryogenic-field emission scanning electron microscopy (Cryo-FESEM). Commonly observed shapes for nanoformulations include spherical or core-shell structures, attributed to the formation of nano micelle clusters during the preparation process.

To assess stability, rigorous tests involve varying storage time and temperatures, typically spanning 0, 5, 10 days, and even up to 12 months, with temperatures ranging from 4 to 54°C. Stability is determined by monitoring sample appearance and measuring physicochemical properties, such as zeta potential and particle size, at predetermined intervals. Stable systems exhibit no changes like phase separation, creaming, flocculation, coalescence, or sedimentation. The zeta potential and particle size are compared before and after storage, with maintained values indicating stability.

Temperature plays a critical role in stability, with higher temperatures potentially inducing instability through particle movement and emulsifier dissolution into water, leading to particle aggregation. Factors like Ostwald ripening, occurring in the initial 5–10 days, contribute to stability challenges like flocculation and coalescence. The coarsening mechanism in nanoemulsions, influenced by factors like the oil phase fraction, emphasizes the role of appropriate surfactants and small droplet sizes in controlling stability. High curvature of nanoformulations prevents flocculation and coalescence due to Laplace pressure, but Ostwald ripening remains a concern for long-term storage. Strategies to counter Ostwald ripening involve increasing droplet elasticity and adding surfactants to reduce interfacial free energy, forming a protective barrier against coalescence [78–80].

#### *3.3.5.4 Retention and contact angle measurement*

Retention and contact angle measurements are crucial in understanding the interaction between nanoformulation and leaf surfaces. These metrics provide insights into the affinity of the phyto-insecticide liquid towards the leaves, impacting the efficiency of the nanoformulation. Increased adhesion of nanoformulations to leaves can enhance their effectiveness in pest control.

Retention is typically measured using methods like dipping and micro-weighing, providing a quantitative assessment of the amount of phyto-insecticide retained on the leaf surface. On the other hand, contact angle measurements involve using precision instruments equipped with a charged-coupled device (CCD) camera to evaluate the angle formed between the pesticide droplet and the leaf surface. A decreasing contact angle with an increasing agrochemical content indicates a lower interfacial tension of the active ingredient, facilitating effective pesticide diffusion on the plant surface [81].

### 3.4 Beyond pest control: exploring the health benefits of phyto-insecticide treated foods

Azadirachtin A, a compound from the neem tree, shows promising health benefits against bone health, cancer, and inflammation. In calvaria cells, it boosts bone formation (1–5 mg/kg BW) [82]. It also fights cervical cancer cells (135 μM) [83] and reduces tumor growth in buccal cancer (10–100 μg/kg BW) [84]. Additionally, Azadirachtin (120 mg/kg BW) reduces inflammation and pain in mice. This versatile compound holds the potential for diverse therapeutic applications [85].

Piperine, the active compound in black pepper, packs a punch in diverse areas. It inhibits HeLa cancer cell growth (20–100 μg/ml), tackles high blood sugar (20 mg/kg), and even fights obesity (40 mg/kg). Lower doses (1–10 mg/kg) bring down blood pressure and protect the heart. Piperine enhances cognitive and motor function in Parkinson's (10 mg/kg), combats seizures (10 mg/kg), and improves mobility in Alzheimer's (2 mg/kg). This spicy wonder holds promise for a range of health benefits [86].

The Pongamia tree, also known as *Millettia pinnata* L., boasts a treasure trove of health benefits. Its stem extract (10–40 mg/ml) packs an antibacterial punch, flower extracts (150 mg/kg) offer antioxidant protection against liver damage. Even leaves and bark possess antiplasmodial activity against malaria ($IC_{50}$ 9–43 μg/ml). Pongamia's effects extend to inflammation (70% ethanolic extract, 300–1000 mg/kg) and stomach ulcers (methanolic seed extract, 25 mg/kg). It even targets pancreatic cancer cells (leaf extract, 100 mg/ml) and helps manage blood sugar (aqueous extract, 300 mg/orally). This versatile tree truly stands as a testament to nature's healing power [87].

From fighting microbes (100 μg/ml) to shielding cells from damage (25 mg/kg), *Andrographis paniculata* (Burm.f.) Nees a powerful medicinal plant, is a natural wellness warrior. Its extracts stifle colon cancer cell growth (10 μg/mL), curb diabetes (50 mg/kg), and even block blood vessel formation in tumors (10 μg/mL). This versatile herb shines as a champion for diverse health concerns, offering nature's healing touch in a single potent package.

*Gloriosa superba* L. hides remarkable medicinal talents. Its leafy extracts, potent at just—2.97 mg/ml, thinning the blood and preventing unwanted clots [88]. Deeper down, the root tubers hold an antimicrobial arsenal. Aqueous, methanol, and petroleum ether extracts, wielded at 500–1000 μg/ml, vanquish both Gram-positive and Gram-negative bacteria. Colchicine, offers a natural balm for pain, soothing inflammation at a mere 2 mg/kg BW [89].

The versatile *Vitex negundo* L. offers a treasure trove of bioactivities. Leaf extracts exhibit promising effects against *Mycobacterium tuberculosis* (MIC $\leq$ 100 μg/ml), fungi (MIC 16–24 μg/ml), and inflammation (2.5–5.0 g/kg body weight). Ethanolic leaf extracts demonstrate pronounced analgesic activity (100–500 mg/kg) while alcoholic seed extracts protect the liver (250 mg/kg). Notably, *V. negundo* boasts remarkable anti-proliferative properties against diverse cancer cell lines, including hormone-dependent breast cancer (MCF-7), non-hormone-dependent breast cancer (MDA-MB-231), ovarian cancer (Caov-3), cervical cancer (HeLa), liver cancer (HepG2), and human foreskin fibroblast cells (Hs27) ($IC_{50}$ 65.38 μg/ml). These findings suggest *V. negundo* deserves further exploration for its potential therapeutic applications across various medical fields [90]

*Cleistanthus collinus* (Roxb.) Benth. ex Hook. f. unveils promising anti-cancer agents. Cleistanthus A, at 5 μg/ml, triggers lethal apoptosis and DNA strand breaks in cervical cancer cells (Si Ha), suggesting a potent cytotoxic effect. This effect manifests through inhibited DNA synthesis and amplified damage [91]. Cleistanthus B, showcasing selectivity, exhibits lower $GI_{50}$ values against 10 cancer cell lines compared to 5 normal lines. This selectivity, ranging from $1.6 \times 10^{-6}$ to $4 \times 10^{-5}$ M for tumor cells and $2 \times 10^{-5}$ M to $4.7 \times 10^{-4}$ M for normal cells, promises effective tumor suppression with minimal harm to healthy tissue [92]. Cleistanthus A and B both compounds holds their distinct potent cytotoxic and can be used in cancer therapy.

Eucalyptus oil, extracted from the leaves, reveals antimicrobial and anti-cancerous potential. At a potent concentration of 20 ml/L, it effectively inhibits the growth of common bacterial pathogens like *Staphylococcus aureus, Salmonella typhi, Bacillus subtilis,* and *Escherichia coli* [93]. Furthermore, its anti-cancerous activity shines against diverse cancer cell lines, demonstrated by its $IC_{50}$ values (half-maximal inhibitory concentration) of 4.75, 8.8, and 11.8 μg/mL against colorectal cancer (HCT-116), breast cancer (MCF-7), and liver cancer (HepG-2) cells, respectively [94].

Phyto insecticides render various insecticidal properties due to the presence of secondary metabolites such as steroids, tannins, alkaloids, terpenoids, flavonoids, coumarins, phenols etc., [95] also offer medicinal effects [96]. Synthetic insecticides become persistent pollutants (residues) and result in inevitable loss to human health and the environment [97, 98]. A pesticide residue is any specified substance in food, agricultural commodities, or animal feed resulting from the use of a pesticide [99]. In 2021, the European Food Safety Authority reported that 39.8% and 2.1% of food samples contain one or more residues in concentrations below or equal to permitted levels (MRLs) and the latter with residues exceeding the permitted levels [100].

Talukder et al. reported that, unlike conventional pesticides, botanical pesticides leave minimal or no residue, minimizing risks to human health, the environment, and ecosystems [101]. Whereas EPA exempt, botanical oils with a zero re-entry interval, zero pre- and post-harvest interval and maximum residue levels (MRLs) [102] and vulnerable to the degradation process [103].

### 3.5 Fate of phytoinsecticidal residues and potential for "nutraceuticalization"

What is the effect of residual phytoinsecticidal compounds present in treated produce? Recalling Paracelsus's principle, "the dose makes the poison," the *low concentrations* of active ingredients in these natural insecticides might paradoxically exhibit *medicinal properties*. This intriguing possibility suggests the potential transformation of treated food into "*nutraresidiceuticals*."

As discussed in Section 3.4, trace amounts of phytoinsecticide *residues* remaining in treated *food* could function as *pharmaceutics*, similar to *nutraceuticals*, potentially *enhancing consumer health*. Consequently, phytoinsecticidal treatment might offer a *unique and advantageous* additional dimension to their application.

## 4. Conclusion

Developing effective and stable phyto-insecticide formulations is a multifaceted challenge. By addressing the sensitivity of these natural compounds through innovative formulations and tackling biomass availability and variability difficulty in their synthesis through callus culture, we can unlock their potential for sustainable pest management. However further research is needed to optimize callus culture for various plants and scale production to decipher the intricate modes of action and to validate the potential health benefits of nutraresidueuticals through clinical trials.

The future of phytoinsecticides lies in a confluence of cutting-edge technologies and strategic development approaches. Genomics, transcriptomics, and metabolomics will unlock deeper knowledge of biosynthetic pathways and modes of action, paving the way for targeted optimization. Artificial intelligence and machine learning will accelerate the discovery of novel bioactive compounds through high-throughput screening of plant extracts. Synergistic effects and broader pest control can be achieved by developing combination formulations that integrate botanicals with other biocontrol agents. Exploring the potential of "nutraceuticalization" could offer additional health benefits from trace residues in treated food, opening exciting new research avenues.

By overcoming current challenges and harnessing these powerful tools, phytoinsecticides have the potential to revolutionize pest management, contributing to a more sustainable and healthier food production system for all.

## Author details

Thirunavukkarasu Selvamuthukumaran[1*], Palanisamy Dhanapriya[1] and Nusrat Iqbal[2]

1 Phyto-Insecticide Formulation Lab, Department of Entomology, Faculty of Agriculture, Annamalai University, Chidambaram, India

2 Formulation Division, Department of Chemicals and Petrochemicals, Institute of Pesticide Formulation Technology, Ministry of Chemicals and Fertilizers, Government of India, Sector 20, Udhyog Vihar, Gurugram, Haryana, India

*Address all correspondence to: entoselva@gmail.com

## References

[1] Skendžić S, Zovko M, Živković IP, Lešić V, Lemić D. The impact of climate change on agricultural insect pests. Insects. 2021;**12**:440. DOI: 10.3390/insects12050440

[2] OECD iLibrary. OECD-FAO Agricultural Outlook 2012-2021 [Internet]. 2012. Available from: OECD-FAO Agricultural Outlook 2012 | READ online (oecd-ilibrary.org) [Accessed: January 10, 2024]

[3] USDA. Researchers Helping Protect Crops from Pests [Internet]. 2023. Available from: https://www.nifa.usda.gov/about-nifa/blogs/researchers-helping-protect-crops-pests [Accessed: January 18, 2024]

[4] Carson R. Silent Spring. United States: Houghton Mifflin; 1962. 323 p

[5] Guedes RN, Smagghe G, Stark JD, Desneux N. Pesticide-induced stress in arthropod pests for optimized integrated pest management programs. Annual Review of Entomology. 2016;**61**:43-62. DOI: 10.1146/annurev-ento-010715-023646

[6] Zhu F, Lavine L, O'Neal S, Lavine M, Foss C, Walsh D. Insecticide resistance and management strategies in urban ecosystems. Insects. 2016;**7**:2. DOI: 10.3390/insects7010002

[7] Dutcher JD. A review of resurgence and replacement causing pest outbreaks in IPM. General Concepts in Integrated Pest and Disease Management. 2007;**1**:27-43. DOI: 10.1007/978-1-4020-6061-8

[8] Armes NJ, Jadhav DR, DeSouza KR. A survey of insecticide resistance in *Helicoverpa armigera* in the Indian subcontinent. Bulletin of Entomological Research. 1996;**86**(5):499-514. DOI: 10.1017/S0007485300039298

[9] Hashemi SM, Damalas CA. Farmers' perceptions of pesticide efficacy: Reflections on the importance of pest management practices adoption. Journal of Sustainable Agriculture. 2010;**35**:69-85. DOI: 10.1080/10440046.2011.530511

[10] Fang Y, Nie Z, Die Q, Tian Y, Liu F, He J, et al. Organochlorine pesticides in soil, air, and vegetation at and around a contaminated site in southwestern China: Concentration, transmission, and risk evaluation. Chemosphere. 2017;**178**:340-349. DOI: 10.1016/j.chemosphere.2017.02.151

[11] Sarkar S, Gil JD, Keeley J, Jansen K. The Use of Pesticides in Developing Countries and their Impact on Health and the Right to Food [Research Report]. Brussels, Belgium: European Union; 2021

[12] Rodríguez-Cruz MS, Jones JE, Bending GD. Field-scale study of the variability in pesticide biodegradation with soil depth and its relationship with soil characteristics. Soil Biology and Biochemistry. 2006;**38**:2910-2918. DOI: 10.1016/j.soilbio.2006.04.051

[13] Regnault-Roger C, Philogène BJ. Past and current prospects for the use of botanicals and plant allelochemicals in integrated pest management. Pharmaceutical Biology. 2008;**46**:41-52. DOI: 10.1080/13880200701729794

[14] Singh L, Choudhary SK, Singh PK. Pesticide concentration in water and sediment of River Ganga at selected sites in middle Ganga plain. International Journal of Environmental Sciences. 2012;**3**:260-274. DOI: 10.6088/ijes.2012030131026

[15] Statista. Market Value of Insecticides Worldwide in 2020 and 2022, with a Forecast for 2023 and 2027 [Internet]. 2023. Available from: Insecticide market value worldwide 2020-2027 | Statista

[16] Jacobson M, Crosby DG. Naturally Occurring Insecticides. New York, USA: Marcel Dekker. CABI Databases; 1971. 585 pp

[17] Maramorsch K. Recent Advances in Medicinal, Aromatic and Spice Crops. Vol. I. New Delhi: Today & Tomorrow's Printers and Publishers; 1991. 272 p

[18] Jayaraj S, Rabindra RJ. The local view on the role of plant protection in sustainable agriculture in India. In: Ciba Foundation Symposium 177-Crop Protection and Sustainable Agriculture: Crop Protection and Sustainable Agriculture: Ciba Foundation Symposium, Madras, India. 2007. pp. 168-190

[19] Ogunnika CB. Medicinal Plants: A Potential Agroforestry Components in Nigeria. InProceedings of the Humboldt Kellog/3rd Annual Conference of School of Agriculture and Agricultural Technology. Chichester, UK: John Wiley & Sons, Ltd; 2007. pp. 2-7

[20] Oyun MB, Agele SO. Testing of extracts from *Eucalyptus camaldulensis*, *Acacia auriculiformis Gliricidia sepium* for weed suppression in a maize farm. In: Proceeding of the Humboldt Kellog/5th SAAT Annual Conference of formulations of Medicinal Plants in Plant and Animal Production, Nigeria. 20 Apr 2009. pp. 89-93

[21] Grainge M, Ahmed S. Handbook of Plants with Pest-Control Properties. New Jersey: John Wiley & Sons Limited; 1988. 470 p

[22] Sahayaraj K, Raja SK. Effect of biopesticides on *Rhynocoris marginatus* (Fabricius)(Heteroptera: Reduvidae). Journal of Biological Control. 2003;**17**(1):43-45

[23] Kathuria V, Kaushik N. Comparison of different materials as *Helicoverpa armigera* (Hübner) oviposition substrates. International Journal of Tropical Insect Science. 2004;**24**:336-339. DOI: 10.1079/IJT200439

[24] Isman MB. Botanical insecticides in the twenty-first century—Fulfilling their promise? Annual Review of Entomology. 2020;**65**:233-249. DOI: 10.1146/annurev-ento-011019-025010

[25] Dougoud J, Toepfer S, Bateman M, Jenner WH. Efficacy of homemade botanical insecticides based on traditional knowledge. A review. Agronomy for Sustainable Development. 2019;**39**:1-22. DOI: 10.1007/s13593-019-0583-1

[26] Kedia A, Prakash B, Mishra PK, Singh P, Dubey NK. Botanicals as eco friendly biorational alternatives of synthetic pesticides against *Callosobruchus* spp. (Coleoptera: Bruchidae)—A review. Journal of Food Science and Technology. 2015;**52**:1239-1257. DOI: 10.1007/s13197-013-1167-8

[27] Hikal WM, Baeshen RS, Said-Al Ahl HA. Botanical insecticide as simple extractives for pest control. Cogent Biology. 2017;**3**:1404274

[28] Sinha B, Biswas I. Potential of bio-pesticides in Indian agriculture vis-a-vis rural development. India Science and Technology. 2009:340-342. DOI: 10.2139/ssrn.1472371

[29] Isman MB, Machial CM. Pesticides based on plant essential oils: From traditional practice to commercialization. Advances in Phytomedicine. 2006;**3**:29-44. DOI: 10.1016/S1572-557X(06)03002-9

[30] Riyajan SA, Sakdapipanich JT. Development of a controlled release neem capsule with a sodium alginate matrix, crosslinked by glutaraldehyde and coated with natural rubber. Polymer Bulletin. 2009;**63**:609-622. DOI: 10.1007/s00289-009-0126-z

[31] Ahmad SN, Islam MT, Abdullah DK, Omar D. Study of tannery phase diagram of rotenone *Derris elliptica* (Fabales: Fabaceae): A novel approach of insecticide formulation. Asian Journal of Chemistry. 2013;**25**:4718-4722

[32] Pavela R. History, presence and perspective of using plant extracts as commercial botanical insecticides and farm products for protection against insects–a review. Plant Protection Science. 2016;**52**:229-241

[33] Vijaya Sree N, Udayasri P, VVY AK, Ravi BB, Phani KY, Vijay VM. Advancements in the production of secondary metabolites. Journal of Natural Products. 2010;**3**:112-123

[34] Veitch GE, Beckmann E, Burke BJ, Boyer A, Ayats C, Ley SV. A relay route for the synthesis of azadirachtin. Angewandte Chemie International Edition. 2007;**46**:7777-7779. DOI: 10.1002/anie.200703028

[35] Sujanya S, Poornasri Devi B, Sai I. In vitro production of azadirachtin from cell suspension cultures of *Azadirachta indica*. Journal of Biosciences. 2008;**33**:113-120. DOI: 10.1007/s12038-008-0027-6

[36] Singh M, Chaturvedi R. Sustainable production of azadirachtin from differentiated in vitro cell lines of neem (*Azadirachta indica*). AoB Plants. 2013;**5**:plt034. DOI: 10.1093/aobpla/plt034

[37] Ahmed SA, Shawkat MS, Ibrahim KM. Production of Pyrethrins in tissue cultures of pyrethrum (*Chrysanthemum cinerariaefolium*). Al-Mustansiriyah Journal of Science. 2011;**22**(6):317-337

[38] Hitmi A, Sallanon H, Barthomeuf C. Effects of plant growth regulators on the growth and pyrethrin production by cell cultures of *Chrysanthemum cinerariaefolium*. Australian Journal of Botany. 2001;**49**:81-88. DOI: 10.1071/BT00008

[39] Ahmad R, Sharma VK, Rai AK, Shivananda RD, Shivananda BG. Production of lignans in callus culture of *Podophyllum hexandrum*. Tropical Journal of Pharmaceutical Research. 2007;**6**(4):803-808. DOI: 10.4314/tjpr.v6i4.14661

[40] Nitnaware KM, Naik DG, Nikam TD. Thidiazuron-induced shoot organogenesis and production of hepatoprotective lignan phyllanthin and hypophyllanthin in *Phyllanthus amarus*. Plant Cell, Tissue and Organ Culture (PCTOC). 2011;**104**:101-110. DOI: 10.1007/s11240-010-9796-3

[41] Lashin II, Elhaw MH. Evaluation of secondary metabolites in callus and tissues of *Physalis peruviana*. International Journal of Modern Botany. 2016;**6**:10-17. DOI: 10.5923/j.ijmb.20160601.03

[42] Cheng S, Zhang W, Nannan SU, Feng XU, Linling LI, Yongling LI, et al. Production of flavonoids and terpene lactones from optimized *Ginkgo biloba* tissue culture. Notulae Botanicae Horti Agrobotanici Cluj-Napoca. 2014;**42**:88-93. DOI: 10.15835/nbha4219393

[43] Rao S, Usha K, Arjun. Production of secondary metabolites from callus cultures of *Centella asiatica* (L.) Urban. Annals of Phytomedicine-An International Journal. 2015;**4**(1):74-78

[44] Kearney ML, Allan EJ, Hooker JE, Mordue AJ. Antifeedant effects of in vitro culture extracts of the neem tree, Azadirachta indica against the desert locust (*Schistocerca gregaria* (Forskål)). Plant Cell, Tissue and Organ Culture. 1994;**37**:67-71. DOI: 10.1007/BF00048119

[45] Zounos AK, Allan EJ, Mordue AJ. Bioactive compounds from neem tissue cultures and screening against insects. Pesticide Science. 1999;**55**(4):497-500. DOI: 10.1002/(SICI)1096-9063(199904)55:4%3C497::AID-PS946%3E3.0.CO;2-J

[46] Capataz Tafur J, Orozco Sánchez F, Vergara Ruiz R, Hoyos SR. Antifeedant effect of cell suspension extracts of *Azadirachta indica* on *Spodoptera frugiperda* JE Smith under laboratory conditions. Revista Facultad Nacional de Agronomía Medellín. 2007;**60**(1):3703-3715

[47] Rafiq M, Dahot MU, Naqvi SH, Mali M, Ali N. Efficacy of neem (*Azadirachta indica* A. Juss) callus and cells suspension extracts against three lepidopteron insects of cotton. Journal of Medicinal Plants Research. 2012;**6**:5344-5349. DOI: 10.5897/JMPR12.190

[48] Ashokhan S, Othman R, Abd Rahim MH, Karsani SA, Yaacob JS. Effect of plant growth regulators on coloured callus formation and accumulation of azadirachtin, an essential biopesticide in *Azadirachta indica*. Plants. 2020;**9**:352. DOI: 10.3390/plants9030352

[49] Sarin R. Insecticidal activity of callus culture of *Tagetes erecta*. Fitoterapia. 2004;**75**:62-64. DOI: 10.1016/j.fitote.2003.07.011

[50] Vats S. Larvicidal activity and in vitro regulation of rotenoids from *Cassia tora* L. 3 Biotech. 2018;**8**:13. DOI: 10.1007/s13205-017-1038-5

[51] Paramesha M, Manivannan S, Rao SA, Srikanth KS, Neelwarne B, Shetty NP. Augmentation of pyrethrins content in callus of *Chrysanthemum cinerariaefolium* and establishing its insecticidal activity by molecular docking of NavMS Sodium Channel Pore receptor. 3 Biotech. 2018;**8**:1-10. DOI: 10.1007/s13205-018-1387-8

[52] Perry AS, Yamamoto I, Ishaaya I, Perry R, Perry AS, Yamamoto I, et al. Botanical insecticides. Insecticides in Agriculture and Environment: Retrospects and Prospects. 1998:78-91. DOI: 10.1007/978-3-662-03656-3_13

[53] Current Biology. Insect evolution [Internet]. 2015. S0960-9822(15)00927-6.pdf (cell.com)

[54] Pullin AS. Physiological relationships between insect diapause and cold tolerance: Coevolution or coincidence. European Journal of Entomology. 1996;**93**:121-130

[55] Bruce TJ. Interplay between insects and plants: Dynamic and complex interactions that have coevolved over millions of years but act in milliseconds. Journal of Experimental Botany. 2015;**66**:455-465. DOI: 10.1093/jxb/eru391

[56] Rai VK, Mishra N, Yadav KS, Yadav NP. Nanoemulsion as pharmaceutical carrier for dermal and transdermal drug delivery: Formulation development, stability issues, basic considerations and applications. Journal of Controlled Release. 2018;**270**:203-225. DOI: 10.1016/j.jconrel.2017.11.049

[57] Kah M, Hofmann T. Nanopesticide research: Current trends and future priorities. Environment International. 2014;**63**:224-235. DOI: 10.1016/j.envint.2013.11.015

[58] Chauhan M, Reddy KP, Gopinath CS, Deka S. Copper cobalt sulfide nanosheets realizing a promising electrocatalytic oxygen evolution reaction. ACS Catalysis. 2017;7:5871-5879. DOI: 10.1021/acscatal.7b01831

[59] Balaji S, Mandal BK, Ranjan S, Dasgupta N, Chidambaram R. Nano-zirconia–evaluation of its antioxidant and anticancer activity. Journal of Photochemistry and Photobiology B: Biology. 2017;**170**:125-133. DOI: 10.1016/j.jphotobiol.2017.04.004

[60] Gupta G, Khan J, Singh NK. Advances and opportunities of nano materials in wastewater treatment. NanoWorld Journal. 2022;**2022**(8):S97-S103. DOI: 10.17756/nwj.2022-s1-018

[61] Moon H, Lertpatipanpong P, Hong Y, Kim CT, Baek SJ. Nano-encapsulated quercetin by soluble soybean polysaccharide/chitosan enhances anti-cancer, anti-inflammation, and anti-oxidant activities. Journal of Functional Foods. 2021;**87**:104756. DOI: 10.1016/j.jff.2021.104756

[62] Singh Y, Meher JG, Raval K, Khan FA, Chaurasia M, Jain NK, et al. Nanoemulsion: Concepts, development and applications in drug delivery. Journal of Controlled Release. 2017;**252**:28-49. DOI: 10.1016/j.jconrel.2017.03.008

[63] Anton N, Vandamme TF. Nano-emulsions and micro-emulsions: Clarifications of the critical differences. Pharmaceutical Research. 2011;**28**:978-985. DOI: 10.1007/s11095-010-0309-1

[64] Attia MF, Dieng SM, Collot M, Klymchenko AS, Bouillot C, Serra CA, et al. Functionalizing nanoemulsions with carboxylates: Impact on the biodistribution and pharmacokinetics in mice. Macromolecular Bioscience. 2017;7:1600471. DOI: 10.1002/mabi.201600471

[65] Liu Z, Gao J, Zhang G, Cheng Y, Zhang YW. From two-dimensional nano-sheets to roll-up structures: Expanding the family of nanoscroll. Nanotechnology. 2017;**28**:385704. DOI: 10.1088/1361-6528/aa7bf8

[66] Geng TY, Xu C, Xu GH. Albumin conjugated lipid nanoemulsion for site specific delivery of rapamycin at inflammatory site of spinal cord injury. International Journal of Clinical and Experimental Medicine. 2016;**9**:21028-21037

[67] Nuruzzaman MD, Rahman MM, Liu Y, Naidu R. Nanoencapsulation, nano-guard for pesticides: A new window for safe application. Journal of Agricultural and Food Chemistry. 2016;**64**:1447-1483. DOI: 10.1021/acs.jafc.5b05214

[68] Madhavi V, Reddy A, Madhavi G, Reddy NB. Nanoencapsulation of pesticides: Sustainable perspective in agriculture. AIP Conference Proceedings. 2020;**2280**:1. DOI: 10.1063/5.0018027

[69] Ebadollahi A, Valizadeh B, Panahandeh S, Mirhosseini H, Zolfaghari M, Changbunjong T. Nanoencapsulation of acetamiprid by sodium alginate and polyethylene glycol enhanced its insecticidal efficiency. Nanomaterials. 2022;**12**:2971. DOI: 10.3390/nano12172971

[70] Hazafa A, Jahan N, Zia MA, Rahman KU, Sagheer M, Naeem M. Evaluation and optimization of nanosuspensions of *Chrysanthemum coronarium* and *Azadirachta indica* using response surface methodology for pest management. Chemosphere. 2022;**292**:133411. DOI: 10.1016/j.chemosphere.2021.133411

[71] Lin HU, Han-hong XU, Ming-long LI. The characterization of aqueous nanosuspension of rotenone and the bioactivity against *Bursaphelenchus xylophilus*. Chinese Journal of Pesticide Science. 2005;7:171-175

[72] Díaz-Blancas V, Medina DI, Padilla-Ortega E, Bortolini-Zavala R, Olvera-Romero M, Luna-Bárcenas G. Nanoemulsion formulations of fungicide tebuconazole for agricultural applications. Molecules. 2016;**21**:1271. DOI: 10.3390/molecules21101271

[73] Anjali CH, Sharma Y, Mukherjee A, Chandrasekaran N. Neem oil (*Azadirachta indica*) nanoemulsion—a potent larvicidal agent against *Culex quinquefasciatus*. Pest Management Science. 2012;**68**:158-163. DOI: 10.1002/ps.2233

[74] Heydari M, Amirjani A, Bagheri M, Sharifian I, Sabahi Q. Eco-friendly pesticide based on peppermint oil nanoemulsion: Preparation, physicochemical properties, and its aphicidal activity against cotton aphid. Environmental Science and Pollution Research. 2020;**27**:6667-6679. DOI: 10.1007/s11356-019-07332-y

[75] Al-Assiuty BA, Nenaah GE, Ageba ME. Chemical profile, characterization and acaricidal activity of essential oils of three plant species and their nanoemulsions against *Tyrophagus putrescentiae*, a stored-food mite. Experimental and Applied Acarology. 2019;**79**:359-376. DOI: 10.1007/s10493-019-00432-x

[76] Badawy ME, Saad AF, Tayeb ES, Mohammed SA, Abd-Elnabi AD. Development and characterization of nanoemulsions of some insecticides by high energy technique for targeting delivery. Journal of Agricultural Research. 2019;**57**:15-23

[77] Yuliani S, Noveriza R. Effect of carrier oil and co-solvent on the formation of clove oil nanoemulsion by phase inversion technique. In: IOP Conference Series: Earth and Environmental Science. Vol. 309, No. 1. Kuta, Bali, Indonesia: IOP Publishing; 2019. p. 012036

[78] Guan W, Tang L, Wang Y, Cui H. Fabrication of an effective avermectin nanoemulsion using a cleavable succinic ester emulsifier. Journal of Agricultural and Food Chemistry. 2018;**66**:7568-7576. DOI: 10.1021/acs.jafc.8b01388

[79] Koroleva M, Nagovitsina T, Yurtov E. Nanoemulsions stabilized by non-ionic surfactants: Stability and degradation mechanisms. Physical Chemistry Chemical Physics. 2018;**20**:10369-10377. DOI: 10.1039/c7cp07626f

[80] Patel RP, Joshi JR. An overview on nanoemulsion: A novel approach. International Journal of Pharmaceutical Sciences and Research. 2012;**3**:4640-4650. DOI: 10.13040/IJPSR.0975-8232.3(12).4640-50

[81] Qian K, Zhao H, He L. Preparation of fenpropathrin nanoemulsion by phase inversion temperature method and its study on performance. International Journal of Nanoscience. 2018;**17**:1850001. DOI: 10.1142/S0219581X18500011

[82] Islas JF, Acosta E, Zuca G, Delgado-Gallegos JL, Moreno-Treviño MG, Escalante B, et al. An overview of Neem (*Azadirachta indica*) and its potential impact on health. Journal of Functional Foods. 2020;**74**:104171. DOI: 10.1016/j.jff.2020.104171

[83] Priyadarsini RV, Murugan RS, Sripriya P, Karunagaran D, Nagini S. The neem limonoids azadirachtin and nimbolide induce cell cycle arrest and mitochondria-mediated apoptosis in human cervical cancer (HeLa) cells. Free Radical Research. 2010;**44**:624-634. DOI: 10.3109/10715761003692503

[84] Patel SM, Venkata KC, Bhattacharyya P, Sethi G, Bishayee A. Potential of neem (*Azadirachta indica* L.) for prevention and treatment of oncologic diseases. Seminars in Cancer Biology. 2016;**40**:100-115

[85] Soares D, Godin A, Menezes R, Nogueira R, Brito A, Melo I, et al. Anti-inflammatory and antinociceptive activities of azadirachtin in mice. Planta Medica. 2014;**80**:630-636. DOI: 10.1055/s-0034-1368507

[86] Haq IU, Imran M, Nadeem M, Tufail T, Gondal TA, Mubarak MS. Piperine: A review of its biological effects. Phytotherapy Research. 2021;**35**:680-700. DOI: 10.1002/ptr.6855

[87] Al Muqarrabun LM, Ahmat N, Ruzaina SA, Ismail NH, Sahidin I. Medicinal uses, phytochemistry and pharmacology of *Pongamia pinnata* (L.) Pierre: A review. Journal of Ethnopharmacology. 2013;**150**:395-420. DOI: 10.1016/j.jep.2014.05.028

[88] Ashokkumar K. *Gloriosa superba* (L.): A brief review of its phytochemical properties and pharmacology. International Journal of Pharmacognosy and Phytochemical Research. 2015;7:1190-1193

[89] Jana S, Shekhawat GS. Critical review on medicinally potent plant species: *Gloriosa superba*. Fitoterapia. 2011;**82**:293-301. DOI: 10.1016/j.fitote.2010.11.008

[90] Zheng CJ, Li HQ, Ren SC, Xu CL, Rahman K, Qin LP, et al. Phytochemical and pharmacological profile of *Vitex negundo*. Phytotherapy Research. 2015;**29**:633-647. DOI: 10.1002/ptr.5303

[91] Pradheepkumar CP, Panneerselvam N, Shanmugam G. Cleistanthin A causes DNA strand breaks and induces apoptosis in cultured cells. Mutation Research/Genetic Toxicology and Environmental Mutagenesis. 2000;**464**:185-193. DOI: 10.1016/s1383-5718(99)00179-5

[92] Pradheep Kumar CP, Panneerselvam N, Rajesh S, Shanmugam G. Cytotoxic and genotoxic effects of cleistanthin B in normal and tumour cells. Mutagenesis. 1996;**11**(6):553-557. DOI: 10.1093/mutage/11.6.553

[93] Tian Y, Dong F, Zhou X, Yang X. Repellent, insecticidal and antimicrobial activities of leaf essential oils from three eucalyptus species. Chemistry & Biodiversity. 2020;**17**:e1900580. DOI: 10.1002/cbdv.201900580

[94] Salehi B, Sharifi-Rad J, Quispe C, Llaique H, Villalobos M, Smeriglio A, et al. Insights into Eucalyptus genus chemical constituents, biological activities and health-promoting effects. Trends in Food Science & Technology. 2019;**91**:609-624. DOI: 10.1016/j.tifs.2019.08.003

[95] Ahmad W, Singh S, Kumar S. Phytochemical screening and antimicrobial study of *Euphorbia hirta* extracts. Journal of Medicinal Plants Studies. 2017;**5**:183-186

[96] Srijita D. Biopesticides: An ecofriendly approach for pest control. World Journal of Pharmacy and Pharmaceutical Sciences (WJPPS). 2015;**4**:250-265

[97] Gilden RC, Huffling K, Sattler B. Pesticides and health risks. Journal of Obstetric, Gynecologic & Neonatal Nursing. 2010;**39**:103-110. DOI: 10.1111/j.1552-6909.2009.01092

[98] Stockholm Convention. Stockholm Convention on Persistent Organic Pollutants [Internet]. 2001. Available

from: UNEP-POPS-COP-CONVTEXT-FULL.English.PDF

[99] FAO. Pesticide Registration Toolkit [Internet]. 2024. Available from: Residue definitions [a07-02-01a] | Pesticide Registration Toolkit | Food and Agriculture Organization of the United Nations (fao.org)

[100] EFSA. Pesticides in Food: Latest Data Published [Internet]. 2023. Available from: Pesticides in food: latest data published | EFSA (europa.eu)

[101] Talukder FA, Islam MS, Hossain MS, Rahman MA, Alam MN. Toxicity effects of botanicals and synthetic insecticides on *Tribolium castaneum* (Herbst) and *Rhyzopertha dominica* (F.). Bangladesh Journal of Environmental Sciences. 2004;**10**:365-371

[102] Keyplex. Botanicals Are Safe, Effective and Better Than Before [Internet]. 2015. Available from: Botanicals Are Safe, Effective and Better Than Before - KeyPlex Plant Nutrition

[103] Soković M, Glamočlija J, Marin PD, Brkić D, Van Griensven LJ. Antibacterial effects of the essential oils of commonly consumed medicinal herbs using an in vitro model. Molecules. 2010;**15**:7532-7546. DOI: 10.3390/molecules15117532

Section 3

# Medicine

Chapter 11

# Dietary Plant Flavone Cynaroside and Its Biological Significance

*Sabina Gayibova, Eva Ivanisova and Ulugbek Gayibov*

## Abstract

Flavonoids, the most diverse group of natural polyphenolics, are secondary plant metabolites that play a crucial role in human health protection. Two main classes—flavonols and flavones—comprise the main body of flavonoids with antioxidant properties and high biological activity, proven both *in vitro* and *in vivo*. Purified samples of flavones represent special interest. One of them, luteolin-7-glucoside (cynaroside), has attracted increasing attention as a potential agent possessing a number of biological activities. The current understanding of cynaroside bioactivities is outlined in this chapter, along with research gaps and potential future directions for this flavonoid's study.

**Keywords:** antioxidant, anticancer, cardio-protection, cynaroside, flavone, hepatoprotection

## 1. Introduction

Flavonoids assigned as low molecular weight secondary metabolism phytochemicals perform various biological properties both manifested *in vitro* and *in vivo* [1]. Extensive range of studies strongly suggest that long-term consumption of diets rich in plant flavonoids offers protection against development of cancers [2], cardiovascular diseases [3], diabetes [4], neurodegenerative diseases [5], and so on. By now, the mechanisms of flavonoids action have been proposed that include the influence of signaling processes, flowing in living systems, due to the specific interaction with regulatory proteins [6], the modification of eicosanoid biosynthesis [7], the prevention of platelet aggregation [8], the promotion of relaxation of cardiovascular smooth muscle cells [9], and so forth.

Flavones and flavone-derived compounds are a large group of flavonoids with a great number of properties and presented in vegetables, fruits, and aromatic plants [10]. The term "flavone" was used for the first time in 1895 by von Kostanecki and Tambor who were pioneers in the structural work of this particular class of flavonoids [11].

Cynaroside—which is 7-O-glucoside of luteolin (other names are luteolin-7-O-β-D-glucopyranoside or luteolin-7-O-β- Dglucopyranoside-5,3′,4′-trihydroxyflavone, Luteoloside [12], luteolin-7-glucoside, nephrocizin [13])—is a natural flavone,

one of the bioactive compounds purified from various genuses of plant families: Apiaceae [14], Lamiaceae [15], Asteráceae [16], and so on (**Figure 1** and **Table 1**).

The name "cynaroside" originates from the latin term Cynara denoting the name of the genus of artichokes where this flavone was firstly isolated in 1955 and defined to be glycoside in 1958 by Michaud (J.) and Masquelier (J.) correspondingly [56, 57]. Since that, interest to cynaroside has been growing considerably. For instance, due to

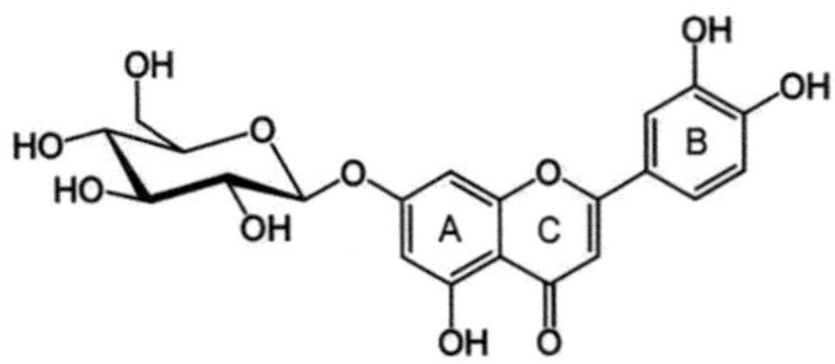


**Figure 1.**
*Chemical structure of cynaroside.*

| Family | Species |
|---|---|
| Apiaceae | *Pimpinella anisum* L. [10], *Daucus carota* [17], *Angelica keiskei* [18], *Anthriscus sylvestris* (L.) *Hoffm.* [19], *Dystaenia takeshimana* [20], *Ferula varia* [21] |
| Asteraceae | *Cynara scolymus* [22], *Ixeris chinensis* [22], *Chrysanthemum morfolium Ramat* [23], *Glossogyne tenuifolia* [24], *Tanacetum parthenium* [25], *Taraxacum officinale* [25], *Scorzonera cana var. jacquiniana* [26], *Scorzonera cinerea* [26], *Scorzonera eriophora* [26], *Scorzonera incisa* [26], *Scorzonera parviflora* [26], *Dendranthema morifolium* [27], *Chrysanthemum indicum Linnén.* [28], *Chrysanthemum morifolium* [28], *Achillea millefolium* L. [29], *Artemisia rupestris* [30], *Carduus crispus* [31], *Bidens tripartite* [32], *Cirsium canum* (L.) [32], *Smallanthus sonchifolius* [33] |
| Balsaminaceae | *Impatiens textori* [33] |
| Caprifoliaceae | *Lonicera japonica* [34], *Lonicera maackii* [35] |
| Fabaceae | *Leucaena leucocephala* [36], *Vicia pannonica var. purpurascens* [37] |
| Juncaceae | *Juncus jerardi Lois.*, *Luzula silvatica*, *Juncus ariculatus*, *Juncus conglameratus*, *Juncus filiformis*, *Juncus tenuis* [38] |
| Lamiaceae | *Salvia officinalis* [39], *Scutellaria immaculata* and *Scutellaria ramosissima* [39], *Lycopus europaeus* L. [40], *Elsholtzia blanda (Benth.) Benth* [41], *Phlomis younghusbandii* [42], *Melittis melissophyllum* L. [43] |
| Myrtaceae | *Pimenta racemosa* [44] |
| Oleaceae | *Ligustrum lucidum Ait* [45], *Phillyrea latifolia* [46], *Ligustrum delavayanum* and *L. vulgare* [47] |
| Polygonaceae | *Polygonum orientale* [48], *Rumex hastatus* [15] |
| Pteridaceae | *Pteris multifida* [49] |
| Rosaceae | *Crataegus monogyna* [50], *Agrimonia pilosa Ledeb* [51], *Prunus mume* [52] |
| Rubiaceae | *Ophiorrhiza mungos Linn* [52] |
| Salicaceae | *Salix matsudana* [53] |
| Scrophulariaceae | *Verbascum densiflorum* and *V. Phlomoides* [54], *Verbascum nubicum* [55] |

**Table 1.**
*Natural sources of cynaroside.*

electronic database PubMed, search for the term "cynaroside, luteolin-7-O-glucoside, nephrocizin" shows growing trend in publications (**Figure 2**).

Several lines of evidence have emerged in recent years, suggesting that cynaroside, despite its nontoxic nature, can minimize hepatotoxicity [58]; repair the antioxidant system [59]; normalize energy, lipid, and carbohydrate metabolism; and enhance bile flow [60]. Therefore, scientific interest in flavones as therapeutic agents is rapidly increasing. The most interesting biological activities of cynaroside are summarized in **Figure 3**.

The purpose of this chapter was to provide an overview of the distribution and biological properties of cynaroside. In this paper, we concentrate our attention on the most common knowledge about the biological effects of plant flavone cynaroside in the context of relevance to influencing human physiology and therefore interesting from a general perspective of improving human well-being.

The electronic databases included PubMed, and Google Scholar was initially used to find articles (published as early as 1955 up to manuscript submission). More than 100 articles were found in applied databases and manually screened sources. Reported publications focus on the cynaroside identification in plants, isolation, its biological and pharmacological properties *in vivo* and *in vitro*, but any articles summarizing or analyzing all this information were not observed. As was mentioned above, the search terms included "cynaroside" as well as "luteolin-7-glucoside" and

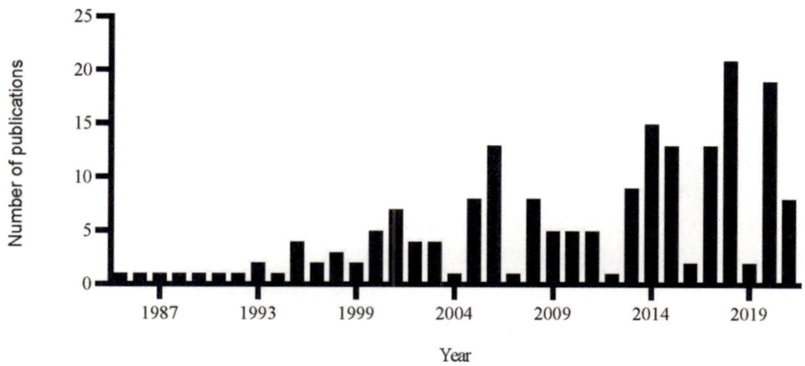


**Figure 2.**
*Trend of research in the field of cynaroside represented by PubMed database.*

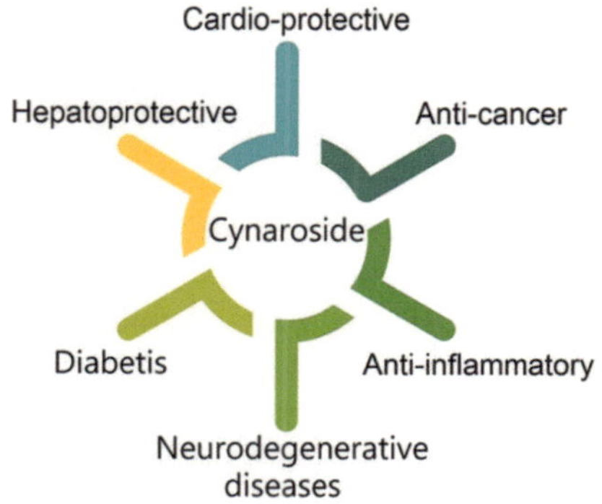


**Figure 3.**
*Some biological activities of cynaroside.*

"nephrocizin" (the mostly used alternative terms for "cynaroside"). Firstly, the search was limited to English-Language articles. Citation lists of retrieved articles were screened manually to ensure sensitivity of the search strategy. References of the included papers were then hand searched to identify in potential relevant studies in other languages (only abstract was used in Russian, French, and Chinese articles). Only published articles were included. No protocol was developed for this review.

## 2. Potential health benefits

Even though there is some data on LD50 of the plant extracts abundant on cynaroside [55], acute toxicity data on pure cynaroside are not available. In general, *in vivo* experiments carried on rats showed that the administration of cynaroside to six-week-old male Wistar rats for 1 week did not change animals' food and beverage consumption, as well as animal body weights when compared to the control [1]. Thus, the rats treated with pure cynaroside up to a dose of 30 mg/kg did not demonstrate any signs of hepatotoxicity [61].

## 3. Hepatoprotective effect

The proposed hepatoprotective effect of cynaroside was due to the great importance of Cynara (artischocke) abundant in the flavonoid in folk medicine against liver complaints.

*In vivo* carbon tetrachloride ($CCl_4$)-induced toxicity model revealed that pretreatment of experimental rats with cynaroside suppressed the elevation of alanine aminotransferase (ALT, GPT, and glutamic-pyruvic transaminase), aspartate transaminase (AST, GOT, and glutamic oxaloacetic transaminase), malondialdehyde (MDA), and 8-Oxo-2′-deoxyguanosine (8-OHdG) and inhibited the reduction of glutathione (GSH) in a dose-dependent manner [61]. *In vitro* experiments with fresh hepatocytes, primary hepatocyte cultures, and immortalized cell lines also showed the protective effect of cynaroside from cell damage induced by carbon tetrachloride ($CCl_4$) [61], as well as brombenzene [62] and t-BHP [63]. Interestingly, Adzet et al. [64] observed specific GOT-recovering effect on isolated hepatocytes preincubated with cynaroside.

However, Quisheng et al. [60] *in vivo* demonstrated dose-dependent decline of both AST and ALT levels in blood serum. It is well accepted that increased enzyme levels are indicative of a tissue injury: the attenuated increase of serum AST and ALT indicates a direct protective role in hepatocytes; while AST is not specific to the liver, any elevation of the ALT is a direct indication of a liver injury. In the meanwhile, some authors mark that ALT and AST have relatively long half-lives ($T\frac{1}{2}$) (literature estimates approximately 17 and 47 hours, respectively) [65, 66] and thus do not reflect immediate changes in liver injury or recovery [67]. Therefore it could be worth considering that in the work of Adzet et al., the period of incubation of cynaroside with toxicated cells did not last more than 1 hour, while in the work of Quisheng et al. [60], all treated animals were sacrificed 24 hours after receiving the administration of the LUTG or hepatotoxin. Thus, it might be worth considering the preliminary incubation time as one important factor in cynaroside hepatoprotection.

Cynaroide *in vitro* also recovered significant (up to 81%) GSH level in tBHP-induced HepG2 cells after 5 hours of incubation [63], which was even higher than by luteolin (53%) and twice higher than quercetin (40%), and *in vivo* cynaroside was able to attenuate the $CCl_4$-induced reduction of GSH in a dose-dependent manner [61, 63]. On the other hand, cynaroside did not have any effect on the recovery level of GST either on cell lines *in vitro* or on cells isolated from rats that were subjected to brombenzene *in vivo* [62]; the effect of cynaroside on its own in normal conditions was not shown by the authors.

It is known that GSH conjugates can be formed directly or be catalyzed by GST [68]. All the toxins mentioned above ($CCl_4$, brombenzene, t-BHP) are known to initiate the formation of free reactive species. Thus, $CCl_4$ intoxication metabolized by CYP2E1 is forced by the prevention of lipid peroxidation by decreasing the level of LPO and reducing the level of MDA. Wang et al. [69] found that cynaroside affected some isoforms of $P_{450}$, namely, isoforms CYP2C9, 3A4, and 1A2; however, no effect was found on 2E1 isoform. That lets to propose that in the case of $CCl_4$ intoxication, cynaroside does not have an effect of toxin metabolization, but protects the lipids through its antiradical properties against $CCl_3$ metabolite.

The results of Park et al. [62] indicate a protective effect of cynaroside in reducing the degree of liver lipid peroxidation caused by the hepatotoxin bromobenzene *in vivo* in rats. Thus, cynaroside at a concentration of 10 mg/kg almost completely neutralized the process of lipid peroxidation caused by hepatoxin and restored the level of the enzyme epoxide hydrolase (an enzyme that deactivates bromobenzene). Bromobenzene is converted by $P_{450}$ enzymes to its reactive, toxic metabolite bromobenzene 3,4-epoxide that is metabolized to a nontoxic bromobenzene 3,4-dihydrodiol by either epoxide hydrolase or glutathione S-transferase. Cynaroside even not effecting GST was nevertheless shown to recover the level of epoxide hydrolase, suggesting that cynaroside might improve the level of hydrolase and also act as a radical scavenger.

There are two pathways by which tBHP is metabolized; both of them induce oxidative stress. The first, provided by cytochrome P450, leads to the production of peroxyl and alkoxyl radicals [70]. These radicals initiate lipoperoxidation of membrane phospholipids with subsequent alterations to membrane fluidity and permeability. The other pathway employs glutathione peroxidase. tBHP is detoxified to tert-butanol, and GSH is depleted by oxidation to its disulfide form (GSSG) [71].

Results showed that cynaroside protects from AST and ALT leakage and also strongly suggest the ability of cynaroside in protecting hepatocyte against membrane fragility. Recent findings have demonstrated that the carbon ring B is deeply immersed into lipid bilayer. A ring is adjusted toward the aqueous environment; the glycosylated form (in position 7) is dispersed over the lipid bilayer, thus providing the defense for double bonds from oxidation [72].

## 4. Cardio-protective effect

It has been shown that cynaroside in pharmacological concentrations has a positive inotropic and vasodilator effect. On the models of isolated hearts of rabbits and guinea pigs, perfused using the Langendorff technique, it was shown that cynaroside increases pressure in the left ventricle, accelerates the process of blood ejection into the arterial system, and enhances the total and relative coronary blood flow and, consequently, the supply oxygen for nonischemic myocardial damage [73, 74].

Also, in the last decade, the processes of apoptosis in cardiovascular pathologies have been intensively studied, and more and more researchers are assigning a leading role to apoptosis in the pathogenesis of various cardiovascular diseases.

Xiao Sun et al. [70] demonstrated that cynaroside exhibits cytoprotective properties in relation to the H9C2 cardiac myocyte cell line against $H_2O_2$-induced apoptosis. Cardiomyocyte cells exposed to $H_2O_2$ suffered severe damage accompanied by apoptosis. Preincubation of cells with cynaroside reduced the degree of ROS generation and inhibition of caspase activation both in the mitochondrial pathway and through the so-called death receptors by enhancing the endogenous antioxidant activity of superoxide dismutase, glutathione peroxidase, and catalase, thereby inhibiting the formation of intracellular ROS. In addition, cynaroside supported normal mitochondrial function by regulating the expression of Bcl-2 protein (B-cell lymphoma protein 2, apoptosis regulator protein Bcl-2), as well as the expression of protein kinases JNK (c-Jun N-terminal kinase, mitogen-activated protein kinase) and p53 (cell cycle regulating transcription factor). The results of microscopic studies showed that after incubation with cynaroside, there was a dose-dependent restoration of the morphological structures of cardiomyocytes damaged by $H_2O_2$, such as a dense membrane packing compared to a loose membrane layer, normal physiological spindle shape versus cell swelling, round shape of the nucleus in comparison with vacuolar degeneration, and so forth. An analysis of generalized experimental data also indicates some possible mechanisms of the protective modulating action of cynaroside in certain pathological conditions of the heart, which indicates an urgent need for further in-depth study of the mechanisms of the action of cynaroside and also to consider it as the basis for the creation of potentially effective and safe cardioprotective agents that can make a significant contribution to the treatment and prevention of a number of cardiovascular diseases.

## 5. Anticancer effect

Evidences are accumulating that pure cynaroside possesses antiproliferative and cytotoxic effects on the tested cancer cell lines with minimal toxicity toward the normal (VERO) cells, indicating cancer cell-specific cytotoxic (apoptotic) effect. Various researchers mentioned that cynaroside does not cause cytotoxicity in relation to normal cells (human neonatal epidermal keratinocytes (HEKn) [75], VERO [76], non-tumorigenic BEAS-2B [77], and Huh7 [78].

Luteolin-7-glucoside in more or less extent has been shown to inhibit a proliferation of a series of human cancer cell lines: breast (MCF-7 [76, 77, 79, 80], MDA-MB-231 [80], A549 [76, 80], H292 [81], renal [77], colorectal (COLO 320 DM [76], gastric (AGS [76], liver (HepG-2 [78–80], Hep 3B [80, 82], LO2 [83], SNU-449 [82], Huh-7 [82], SMMC-7721 [82], MHCCLM3 [82] and MHCC97-H [82], cervical (HeLa [79], melanoma (UACC-62 [77], and some others (mouse fibroblast cell line L929 [84]. No significant effect was observed on macrophage RAW264.7 cells, Ehrlich ascites carcinoma [36], and Rhabdomyosarcoma [12].

## 6. Anti-inflammatory activity

Palombo et al. [75] in their study on a model of psoriasis showed the ability of luteolin-7-glucoside to regulate proliferative responses, as well as differentiation of cultured human keratinocytes treated with interleukin 22 (IL-22). It was found that local administration of cynaroside leads to a noticeable reduction in acanthosis. Thus, treatment with cynaroside led to a decrease in the expression of proliferation markers, an increase in the production of differentiation markers, and, in general, to a phenotypic improvement.

Other *in vivo* tests have shown that luteolin-7-O-glucoside is responsible for wound healing activity. It has also been found to have significant anti-inflammatory, antioxidant, anti-hyaluronidase and anti-collagenase activities. Phases in the healing processes (inflammation, proliferation, and remodeling) were observed and recorded in the experimental groups to varying degrees. All tissues treated with cynaroside ointments showed good healing processes with the potential for faster reepithelialization and high collagen concentrations compared to other groups tested. The anti-inflammatory activity of cynaroside was mainly demonstrated by the inhibition of nuclear factor-kappa B (NF-kappa B), cyclooxygenase-2 (COX-2), 5-lipoxygenase (5-LOX), and inducible NO synthase (iNOS) [85, 86].

## 7. Radical scavenging activity

An analysis of the scientific literature has shown that cynaroside has significant antiradical (ARA) and antioxidant (AOA) properties in relation to free radicals and ROS molecules (superoxide anion ($O_2^{\bullet-}$) [80], hydroxyl radical ($HO^{\bullet}$) [87], peroxyl radicals (ROO–) [23], and nitric oxide (NO) [85]. ARA of cynaroside using the DPPH method is quite widely presented in various works [23, 88]. It has been shown that cynaroside has a fairly high efficiency in the reduction of DPPH molecules (IC50 value varies in different studies: from 13.90 ± 1.46 μM to 277.3 ± 14.9 mmol/mol DPPH. In addition, it was found that cynaroside dose-dependently inhibits the formation of superoxide anion induced by various stimuli, such as N-formylmethionyl-leucyl-phenylalanine, human polymorphonuclear neutrophils and peripheral blood mononuclear cells, phorbol-12-myristate-13-acetate (PMA) and arachidonic acid (AA) in human neutrophils, superoxide anion produced by the xanthine/xanthine oxidase system, and so on [89]. Hsu et al. [80] showed that cynaroside prevents DNA dissociation caused by hydroxyl radical. Cynaroside at a maximum concentration of 1.8 μg/ml provided 64.0 (3.9%) and 83.1 (2.9%) protection of supercoiled DNA against nonspecific and site-specific oxidation caused by hydroxyl radical, respectively. In comparison, 4 μg/ml Trolox (a water-soluble vitamin E analog) provided 83.5 (0.5%) and 42.6 (2.5%) protection. Cynaroside also effectively slows down lipid peroxidation caused by peroxyl radical as well as reduces the concentration of nitric oxide (NO) free radicals produced by endothelial nitric oxide synthase (eNOS) under pathological conditions [85].

## 8. Antibacterial activity

Isolated cynaroside also exhibits antimicrobial activity, especially against gram-negative bacteria (*Escherichia coli*, *Streptococcus pyogenes*, and *Micrococcus luteus* with minimal inhibition zone of 0.5 mm), suppresses the formation of biofilm of gram-positive *Pseudomonas aeruginosa* and *Staphylococcus aureus*, and increases the frequency of mutations that reduce the resistance of *Salmonella typhimurium* to ciprofloxacin. Cynaroside also exhibits high activity against yeast fungi *Candida albicans*, *Candida lusitaniae*, *Saccharomyces cerevisiae*, and *Saccharomyces carlsbergensis* and molds *Aspergillus niger*, *Penicillium oxalicum*, *Mucor mucedo*, and *Cladosporium cucumerinum* [19].

## 9. Anti-coronavirus activity

The World Health Organization (WHO) welcomes the use of traditional medicines, including against COVID-19, assuming they must undergo the same serious testing for effectiveness and safety as any other new medicines. Moezzi [90] proposed cynaroside as a promising agent against SARS-nCoV2 by docking the flavonoid on the active site of the main peptidase.

## 10. Conclusion

The current chapter efforts to coordinate with current studies to identify more accurate biomarkers of the risks for nutrition-related diseases and should lead to dietary recommendations and the formulation of new food products. The food and nutritional supplement industry is very interested in the development and promotion of flavone-rich products as a result of the current evidence for flavones' protective effects against diseases. An integration of the results of past and future experiments in various disciplines, including biochemistry, cell biology, physiology, pathophysiology, epidemiology, and food chemistry, will be needed to identify the most effective properties of cynaroside and to determine the optimal levels of intake for better health.

## Acknowledgements

This study was supported by fund by the EU Erasmus Mundus Partnership (Action 2) CASIA project.

## Conflict of interest

The authors declare no conflict of interest.

## Author details

Sabina Gayibova[1*], Eva Ivanisova[2] and Ulugbek Gayibov[1]

1 Laboratory of Plant Cytoprotectors, Institute of Bioorganic Chemistry Named After A.S. Sadykov, Academy of Sciences of the Republic of Uzbekistan, Tashkent, Uzbekistan

2 Institute of Food Sciences, Faculty of Biotechnology and Food Sciences, Slovak University of Agriculture, Nitra, Slovakia

*Address all correspondence to: gayibova.sabina@gmail.com

## References

[1] Sá C, Oliveira AR, Machado C, Azevedo M, Pereira-Wilson C. Effects on liver lipid metabolism of the naturally occurring dietary flavone Luteolin-7-glucoside. Evidence-Based Complementary and Alternative Medicine. 2015;**2015**:e647832. Available from: https://www.hindawi.com/journals/ecam/2015/647832/

[2] Batra P, Sharma AK. Anti-cancer potential of flavonoids: Recent trends and future perspectives. 3 Biotech. 2013;**3**(6):439-459

[3] McCullough ML, Peterson JJ, Patel R, Jacques PF, Shah R, Dwyer JT. Flavonoid intake and cardiovascular disease mortality in a prospective cohort of US adults. The American Journal of Clinical Nutrition. 2012;**95**(2):454-464. Available from: https://www.ncbi.nlm.nih.gov/pmc/articles/PMC3260072/pdf/ajcn9520454.pdf

[4] Testa R, Bonfigli AR, Genovese S, De Nigris V, Ceriello A. The possible role of flavonoids in the prevention of diabetic complications. Nutrients. 2016;**8**(5):1-13. Available from: https://www.ncbi.nlm.nih.gov/pmc/articles/PMC4882722/

[5] Costa SL, Silva VDA, dos Santos SC, Santos CC, Paris I, Muñoz P, et al. Impact of plant-derived flavonoids on neurodegenerative diseases. Neurotoxicity Research. 2016;**30**(1):41-52

[6] Jana K, Yin X, Schiffer RB, Chen JJ, Pandey A, Stocco DM, et al. Chrysin, a natural flavonoid enhances steroidogenesis and steroidogenic acute regulatory protein gene expression in mouse Leydig cells. Journal of Endocrinology. 2008;**197**(2):315-323

[7] Kim HP, Son KH, Chang HW, Kang SS. Anti-inflammatory plant flavonoids and cellular action mechanisms. Journal of Pharmacological Sciences. 2004;**96**(3):229-245

[8] Ja G, Ml L, Castillo J, Benavente-Garcia O, Vicente V, Rivera J. Flavonoids inhibit platelet function through binding to the thromboxane A2 receptor. Journal of Thrombosis and Haemostasis. 2005;**3**(2):369-376

[9] Horáková Ľ. Flavonoids in prevention of diseases with respect to modulation of Ca-pump function. Interdisciplinary Toxicology. 2011;**4**(3):45-60

[10] Kumar S, Pandey AK. Chemistry and biological activities of flavonoids: An overview. The Scientific World Journal. 2013;**2013**(162750):1-16

[11] von Kostanecki S, Tambor J. Ueber die Constitution des Fisetins. Berichte der Deutschen Chemischen Gesellschaft. 1895;**28**:2302-2309

[12] Cao Z, Ding Y, Ke Z, Cao L, Li N, Ding G, et al. Luteoloside acts as 3C protease inhibitor of enterovirus 71 in vitro. PLoS One. 2016;**11**(2):e0148693

[13] Lin Y, Chen TY, Tseng HW, Lee MH, Chen ST. Chemical and biological evaluation of nephrocizin in protecting nerve growth factor-differentiated PC12 cells by 6-hydroxydopamine-induced neurotoxicity. Phytochemistry. 2012;**84**:102-115

[14] Choudhary M, Kumar V, Malhotra H, Singh S. Medicinal plants with potential antiarthritic activity. Journal of Intercultural Ethnopharmacology. 2015;**4**(2):147-179. DOI: 10.5455/jice.20150313021918

[15] Sahreen S, Khan MR, Khan RA. Comprehensive assessment of

phenolics and antiradical potential of Rumex hastatus D. Don. Roots. BMC Complementary and Alternative Medicine. 2014;**14**:47. DOI: 10.1186/1472-6882-14-47

[16] Ahmed SA, Kamel EM. Cytotoxic activities of flavonoids from Centaurea scoparia. The Scientific World Journal. 2014;**2014**:274207. DOI: 10.1155/2014/274207

[17] Shojaii A, Abdollahi FM. Review of pharmacological properties and chemical constituents of Pimpinella anisum. ISRN Pharmaceutics. 2012;**2012**:1-8. Available from: http://www.ncbi.nlm.nih.gov/pmc/articles/PMC3405664/

[18] Brooks JS, Williams EH, Feeny P. Quantification of contact oviposition stimulants for black swallowtail butterfly, Papilio polyxenes, on the leaf surfaces of wild carrot, Daucus carota. Journal of Chemical Ecology. 1996;**22**(12):2341-2357

[19] Žemlička L, Fodran P, Lukeš V, Vagánek A, Slováková M, Staško A, et al. Physicochemical and biological properties of luteolin-7-O-β-d-glucoside (cynaroside) isolated from *Anthriscus sylvestris* (L.) Hoffm. Monatshefte für Chemie—Chemical Monthly. 2014;**145**(8):1307-1318

[20] Kim JS, Kim JC, Shim SH, Lee EJ, Jin W, Bae K, et al. Chemical constituents of the root of Dystaenia takeshimana and their anti-inflammatory activity. Archives of Pharmacal Research. 2006;**29**(8):617-623. DOI: 10.1007/BF02968244

[21] Mamatkhanova M, Khalilov R, Syrov V, Mamatkhanov A, Kotenko L, Satimov G, et al. Technology for cinaroside production from the aerial part of Ferula varia and evaluation of its hypoazotemic activity. Pharmaceutical Chemistry Journal. 2009;**43**(3):160-162

[22] Wang J, Lv D, Liang X, Zhao M, Zhang S. Study on chemical constituents of the Ixeris chinensis. Journal of Chinese Medicinal Material. 2011;**34**(11):1706-1708

[23] Hu B, Jiang H, Yang J, Zeng S. Determination of luteolin and luteolin-7-beta-D-glucoside in chrysanthemum morfolium Ramat. From different collection time by RP-HPLC. Journal of Zhejiang University Medical Sciences. 2004;**31**(1):29-32

[24] Wu MJ, Wang L, Ding HY, Weng CY, Yen JH. Glossogyne tenuifolia acts to inhibit inflammatory mediator production in a macrophage cell line by downregulating LPS-induced NF-κB. Journal of Biomedical Science. 2004;**11**(2):186-199

[25] Williams CA, Hoult JRS, Harborne JB, Greenham J, Eagles J. A biologically active lipophilic flavonol from Tanacetum parthenium. Phytochemistry. 1995;**38**(1):267-270

[26] Williams CA, Goldstone F, Greenham J. Flavonoids, cinnamic acids and coumarins from the different tissues and medicinal preparations of Taraxacum officinale. Phytochemistry. 1996;**42**(1):121-127

[27] Küpeli Akkol E, Bahadır Acıkara Ö, Süntar I, Ergene B, Saltan ÇG. Ethnopharmacological evaluation of some Scorzonera species: In vivo anti-inflammatory and antinociceptive effects. Journal of Ethnopharmacology. 2012;**140**(2):261-270

[28] Wu XL, Li CW, Chen HM, Su ZQ, Zhao XN, Chen JN, et al. Anti-inflammatory effect of supercritical-carbon dioxide fluid extract from

flowers and buds of *Chrysanthemum indicum* Linnén. Evidence-Based Complementary and Alternative Medicine. 2013;**2013**:1-13

[29] Vitalini S, Beretta G, Iriti M, Orsenigo S, Basilico N, Dall Acqua S, et al. Phenolic compounds from *Achillea millefolium* L. and their bioactivity. Acta Biochimica Polonica. 2011;**58**(2):203-209

[30] Shuai S. Flavonoids of Erigeron canadensis. China Journal of Chinese Materia Medica. 2012;**37**(19):2902-2905

[31] Jeong DM, Jung HA, Choi JS. Comparative antioxidant activity and HPLC profiles of some selected Korean thistles. Archives of Pharmacal Research. 2008;**31**(1):28-33

[32] Wolniak M, Tomczykowa M, Tomczyk M, Gudej J, Wawer I. Antioxidant activity of extracts and flavonoids from *Bidens tripartita*. Acta Poloniae Pharmaceutica. 2008;**64**(5):441-447

[33] Russo D, Malafronte N, Frescura D, Imbrenda G, Faraone I, Milella L, et al. Antioxidant activities and quali-quantitative analysis of different *Smallanthus sonchifolius* [(Poepp. and Endl.) H. Robinson] landrace extracts. Natural Product Research. 2014;**29**(17):1673-1677

[34] Iwaoka E, Oku H, Iinuma M, Ishiguro K. Allergy-preventive effects of the flowers of *Impatiens textori*. Biological & Pharmaceutical Bulletin. 2010;**33**(4):714-716

[35] Cipollini D, Stevenson R, Enright S, Eyles A, Bonello P. Phenolic metabolites in leaves of the invasive shrub, *Lonicera maackii*, and their potential phytotoxic and anti-herbivore effects. Journal of Chemical Ecology. 2008;**34**(2):144-152

[36] Hassan RA, Tawfik WA, Abou-Setta LM. The flavonoid constitunts of *Leucaena leucocephala*. Growing in Egypt, and their biological activity. African Journal of Traditional Complementary and Alternative Medicine. 2014;**11**(1)67-72

[37] Kozan E, Anul SA, Tatlı İİ. In vitro anthelmintic effect of *Vicia pannonica* var. *purpurascens* on trichostrongylosis in sheep. Experimental Parasitology. 2013;**134**(3):299-303

[38] Stabursvik A. Luteolin-7-glucoside as a characteristic component of the plant genera Juncus and Luzula (family Juncaceae). Acta Chemica Scandinavica. 1968;**22**:2371-2373

[39] Lima CF, Valentao PCR, Andrade PB, Seabra RM, Fernandes-Ferreira M, Pereira-Wilson C. Water and methanolic extracts of Salvia officinalis protect HepG2 cells from t-BHP induced oxidative damage. Chemico-Biological Interactions. 2007;**167**(2):107-115

[40] Wojciechowski H, Gumbinger HG, Vahlensieck U, Winterhoff H, Nahrstedt A, Kemper FH. Analysis of the components of *Lycopus europaeus* L. in body fluids during metabolism studies comparison of capillary electrophoresis and high-performance liquid chromatography. Journal of Chromatography A. 1995;**717**(1-2):261-270

[41] Zhang W. Separation and identification of flavonoid glucosides from *Elsholtzia blanda* (Benth.) Benth. China Journal of Chinese Materia Medica. 1999;**24**(2):96-96

[42] Gao Y, Rui-chao L, Wang GL, Zhao H, Gao Y, Ciren B. Studies on the chemical constituents of *Phlomis younghusbandii*. Journal of Chinese Medicinal Materials. 2007;**30**(10):1239

[43] Skrzypczak-Pietraszek E, Pietraszek J. Seasonal changes of flavonoid content in *Melittis melissophyllum* L. (Lamiaceae). Chemistry & Biodiversity. 2014;**11**(4):562-570

[44] Moharram FA, Al-Gendy AA, El-Shenawy SM, Ibrahim BM, Zarka MA. Phenolic profile, anti-inflammatory, antinociceptive, anti-ulcerogenic and hepatoprotective activities of *Pimenta racemosa* leaves. BMC Complementary and Alternative Medicine. 2018;**18**(1):1-15

[45] Wang J, Hou G. Chemical constituents of the flower of *Ligustrum lucidum* Ait. Journal of Chinese Medicinal Material. 1990;**15**(3):168-170, 191

[46] Yazıcı-Tütüniş S, Gürel-Gurevin E, Üstünova S, Demirci-Tansel C, Meriçli F. Possible effects of *Phillyrea latifolia* on weight loss in rats fed a high-energy diet. Pharmaceutical Biology. 2016;**54**(10):1991-1997

[47] Nagy M, Križková L, Mučaji P, Kontšeková Z, Šeršeň F, Krajčovič J. Antimutagenic activity and radical scavenging activity of water infusions and phenolics from Ligustrum plants leaves. Molecules. 2009;**14**(1):509-518

[48] Huang Y, Zhang P, He F, Zheng L, Wang Y, Wu J. Simultaneous determination of four bioactive flavonoids from *Polygonum orientale* L. in dog plasma by UPLC–ESI-MS/MS and application of the technique to pharmacokinetic studies. Journal of Chromatography B. 2014;**957**:96-104

[49] Wang W, Wang Z, Zhou Y. Studies on chemical constituents from *Pteris multifida*. Journal of Chinese Medicinal Materials. 2008;**31**(8):1165-1167

[50] Nabavi S, Habtemariam S, Ahmed T, Sureda A, Daglia M, Sobarzo-Sánchez E, et al. Polyphenolic composition of *Crataegus monogyna* Jacq.: From chemistry to medical applications. Nutrients. 2015;**7**(9):7708-7728. Available from: https://www.ncbi.nlm.nih.gov/pmc/articles/PMC4586556/

[51] Zhu L, Tan J, Wang B, He R, Liu Y, Zheng C. Antioxidant activities of aqueous extract from *Agrimonia pilosa* Ledeb and its fractions. Chemistry & Biodiversity. 2009;**6**(10):1716-1726

[52] Shin EJ, Hur HJ, Sung MJ, Park JH, Yang HJ, Kim MS, et al. Ethanol extract of the *Prunus mume* fruits stimulates glucose uptake by regulating PPAR-γ in C2C12 myotubes and ameliorates glucose intolerance and fat accumulation in mice fed a high-fat diet. Food Chemistry. 2013;**141**(4):4115-4121

[53] Zhang J, Zhang YN, Han L. Studies on chemical constituents of leaves of Salix matsudana Koidz and their influence on lipolysis. China Journal of Chinese Materia Medica. 2000;**25**(9):538-541

[54] Klimek B, Olszewska MA, Tokar M. Simultaneous determination of flavonoids and phenylethanoids in the flowers of *Verbascum densiflorum* and *V. phlomoides* by high-performance liquid chromatography. Phytochemical Analysis. 2009;**21**(2):150-156

[55] El Gizawy H, Hussein MA, Abdel-Sattar E. Biological activities, isolated compounds and HPLC profile of *Verbascum nubicum*. Pharmaceutical Biology. 2019;**57**(1):485-497

[56] Michaud J, Masquelier J. Essai d'identification du cynaroside. Bulletin-Societe de Pharmacie de Bordeaux. 1958;**97**(2):77-81

[57] Masquelier J, Michaud J. Note sur les constituants polyphénoliques du *Cynara*

*scolymus* L. Bulletin-Societe de Pharmacie de Bordeaux. 1955;**95**(2):65-67

[58] Tien YH, Chen B, Hsu G-SW, Lin W, Huang JH, Lu Y. Hepatoprotective and anti-oxidant activities of *Glossogyne tenuifolia* against acetaminophen-induced hepatotoxicity in mice. The American Journal of Chinese Medicine. 2014;**42**(06):1385-1398

[59] Song YS, Park CM. Luteolin and luteolin-7-O-glucoside strengthen antioxidative potential through the modulation of Nrf2/MAPK mediated HO-1 signaling cascade in RAW 264.7 cells. Food and Chemical Toxicology. 2014;**65**:70-75

[60] Gebhardt R. Inhibition of cholesterol biosynthesis in HepG2 cells by artichoke extracts is reinforced by glucosidase pretreatment. Phytotherapy Research. 2002;**16**(4):368-372. DOI: 10.1002/ptr.960

[61] Qiusheng Z, Xiling S, Xubo MS, Changhai W. Protective effects of luteolin-7-glucoside against liver injury caused by carbon tetrachloride in rats. Die Pharmazie. 2004;**59**(4):286-289

[62] Park JC, Park JG, Kim HJ, Hur JM, Lee JH, Sung NJ, et al. Effects of extract from *Angelica keiskei* and its component, cynaroside, on the hepatic bromobenzene-metabolizing enzyme system in rats. Phytotherapy Research. 2002;**16**(S1):24-27

[63] Lima CF, Fernandes-Ferreira M, Pereira-Wilson C. Phenolic compounds protect HepG2 cells from oxidative damage: Relevance of glutathione levels. Life Sciences. 2006;**79**(21):2056-2068

[64] Adzet T, Camarasa J, Laguna JC. Hepatoprotective activity of polyphenolic compounds from *Cynara scolymus* against CCl4 toxicity in isolated rat hepatocytes. Journal of Natural Products. 1987;**50**(4):612-617

[65] Hayes PC, Bouchier IA, Beckett GJ. Glutathione S-transferase in humans in health and disease. Gut. 1991;**32**:813-818

[66] Knapen MF, Peters WH, Mulder TP, Steegers EA. A marker for hepatocellular damage. The Lancet. 2000;**355**(9213):1463-1464. Available from: https://www.thelancet.com/journals/lancet/article/PIIS0140-6736(05)74662-6/fulltext

[67] Maina I, Rule JA, Wians FH, Poirier M, Grant L, Lee WM. α-Glutathione S-transferase: A new biomarker for liver injury? Journal of Applied Laboratory Medicine. 2016;**1**(2):119-128. DOI: 10.1373/jalm.2016.020412

[68] Ishikawa T, Ali-Osman F. Glutathione-associated cis-diamminedichloroplatinum (II) metabolism and ATP-dependent efflux from leukemia cells. Molecular characterization of glutathione-platinum complex and its biological significance. Journal of Biological Chemistry. 1993;**268**(27):20116-20125

[69] Wang L, Ma X, Wang J, Li C. In vitro inhibitory effects of Cynaroside on human liver cytochrome P450 enzymes. Pharmacology. 2019;**104**(5-6):296-302

[70] Sun X, Sun G, Wang M, Xiao J, Sun X. Protective effects of cynaroside against H2O2-induced apoptosis in H9c2 cardiomyoblasts. Journal of Cellular Biochemistry. 2011;**112**(8):2019-2029

[71] Martín-Cordero C, López-Lazaro M, Piñero J, Ortiz T, Cortés F, Ayuso MJ. Glucosylated isoflavones as DNA topoisomerase II poisons. Journal of Enzyme Inhibition. 2000;**15**(5):455-460

[72] Kapral-Piotrowska J, Strawa JW, Jakimiuk K, Wiater A, Tomczyk M, Gruszecki WI, et al. Investigation of the membrane localization and interaction of selected flavonoids by NMR and FTIR spectroscopy. International Journal of Molecular Sciences. 2023;**24**(20):15275. DOI: 10.3390/ijms242015275

[73] Rump AFE, Schüßler M, Açar D, Cordes A, Theisohn M, Rösen R, et al. Functional and antiischemic effects of luteolin-7-glucoside in isolated rabbit hearts. General Pharmacology-the Vascular System. 1994;**25**(6):1137-1142

[74] Rump AFE, Schüssler M, Acar D, Cordes A, Ratke R, Theisohn M, et al. Effects of different inotropes with antioxidant properties on acute regional myocardial ischemia in isolated rabbit hearts. General Pharmacology: The Vascular System. 1995;**26**(3):603-611

[75] Palombo R, Savini I, Avigliano L, Madonna S, Cavani A, Albanesi C, et al. Luteolin-7-glucoside inhibits IL-22/STAT3 pathway, reducing proliferation, acanthosis, and inflammation in keratinocytes and in mouse psoriatic model. Cell Death & Disease. 2016;**7**(8):e2344. Available from: https://pubmed.ncbi.nlm.nih.gov/27537526/

[76] Baskar AA, Ignacimuthu S, Michael GP, Al NK. Cancer chemopreventive potential of Luteolin-7-O-glucoside isolated from *Ophiorrhiza mungos* Linn. Nutrition and Cancer. 2010;**1-9**:130-138

[77] Gálvez M, Martín-Cordero C, López-Lázaro M, Cortés F, Ayuso MJ. Cytotoxic effect of Plantago spp. on cancer cell lines. Journal of Ethnopharmacology. 2003;**88**(2-3):125-130. DOI: 10.1016/s0378-8741(03)00192-2

[78] Hwang YJ, Lee E, Kim HR, Hwang K. Molecular mechanisms of luteolin-7-O-glucoside-induced growth inhibition on human liver cancer cells: G2/M cell cycle arrest and caspase-independent apoptotic signaling pathways. Journal of Biochemistry and Molecular Biology. 2013;**46**(12):611-616

[79] Mamadalieva NZ, Herrmann F, El-Readi MZ, Tahrani A, Hamoud R, Egamberdieva D, et al. Flavonoids in *Scutellaria immaculata* and *S. ramosissima* (Lamiaceae) and their biological activity. Journal of Pharmacy and Pharmacology. 2011;**63**(10):1346-1357

[80] Hsu HF, Houng JY, Chang CL, Wu CC, Chang FR, Wu YC. Antioxidant activity, cytotoxicity, and DNA information of *Glossogyne tenuifolia*. Journal of Agricultural and Food Chemistry. 2005;**53**(15):6117-6125

[81] Zhou M, Shen S, Zhao X, Gong X. Luteoloside induces G0/G1 arrest and pro-death autophagy through the ROS-mediated AKT/mTOR/p70S6K signalling pathway in human non-small cell lung cancer cell lines. Biochemical and Biophysical Research Communications. 2017;**494**(1-2):263-269

[82] Fan SH, Wang Y, Lu J, Zheng YL, Wu DM, Li MQ, et al. Luteoloside suppresses proliferation and metastasis of hepatocellular carcinoma cells by inhibition of NLRP3 inflammasome. PLoS One. 2014;**9**(2):e89961-e89961

[83] Shao J, Wang C, Li L, Liang H, Dai J, Ling X, et al. Luteoloside inhibits proliferation and promotes intrinsic and extrinsic pathway-mediated apoptosis involving MAPK and mTOR signaling pathways in human cervical cancer cells. International Journal of Molecular Sciences. 2018;**19**(6):1664

[84] Emad F, Khalafalah AK, El Sayed MA, Mohamed AEH, Stadler M,

Helaly SE. Three new polyacetylene glycosides (PAGs) from the aerial part of *Launaea capitata* (Asteraceae) with anti-biofilm activity against *Staphylococcus aureus*. Fitoterapia. 2020;**143**:104548

[85] Lee SA, Moon S, Han SH, Hwang EJ, Park BR, Kim JS, et al. Chondroprotective effects of aqueous extract of *Anthriscus sylvestris* leaves on osteoarthritis in vitro and in vivo through MAPKs and NF-κB signaling inhibition. Biomedicine and Pharmacotherapy. 2018;**103**:1202-1211

[86] Szymański M, Dudek-Makuch M, Witkowska-Banaszczak E, et al. Comparison of the chemical composition and antioxidant activity of essential oils from the leaves and flowers of *Sambucus nigra*. Pharmaceutical Chemistry Journal. 2020;**54**:496-503. DOI: 10.1007/s11094-020-02228-5

[87] Grzegorczyk-Karolak I, Wysokińska H, Olas B. Studies on the antioxidant properties of extracts from the roots and shoots of two Scutellaria species in human blood plasma. Acta Biochimica Polonica. 2015;**62**(2):253-258

[88] Zou Y, Zhang M, Zhang T, Wu J, Wang J, Liu K, et al. Antioxidant and anti-inflammatory activities of Cynaroside from *Elsholtiza bodinieri*. Natural Product Communications. 2018;**13**(11):1934578X1801301-1

[89] Lu J, Feng X, Sun Q, Lu H, Manabe M, Sugahara K, et al. Effect of six flavonoid compounds from *Ixeris sonchifolia* on stimulus-induced superoxide generation and tyrosyl phosphorylation in human neutrophils. Clinica Chimica Acta. 2002;**316**(1-2):95-99

[90] Moezzi MS. Comprehensive in silico screening of flavonoids against SARS-CoV-2 main protease. Journal of Biomolecular Structure and Dynamics. 2022;**1-14**:9448-9461

Chapter 12

# Improvement of Cognitive Function by Wasabi Component "Hexaraphane"

*Isao Okunishi*

## Abstract

Hexaraphane (6-methylsulfinylhexyl isothiocyanate; 6-MSITC) is an isothiocyanate present in the rhizomes and roots of wasabi (*Eutrema japonicum* (Miq.) Kiudz.). It is known to induce detoxifying and antioxidant enzymes by activating the Keap1-Nrf2 system, ameliorating oxidative damage in the body. Hexaraphane was shown to inhibit brain damage and improve dementia symptoms in Alzheimer's model mice. Moreover, in two randomized controlled trials conducted on middle-aged and elderly subjects, the extract powder "Wasabi sulfinyl™" containing 0.8% hexaraphane improved memory, attention, and judgment. In a clinical study of fifteen patients with chronic fatigue syndrome, treatment with Wasabi sulfinyl ™ for 3 months improved brain fog and other symptoms.

**Keywords:** wasabi, hexaraphane, clinical study, brain function, dementia, brain fog

## 1. Introduction

Wasabi (*Eutrema japonicum* (Miq.) Kiudz.) is a plant species belonging to the Cruciferae family, native to Japan. It is used primarily as a condiment for its distinctive pungent and stimulating flavor. It is sometimes referred to as *Wasabia japonicum*, a synonym for *Eutrema. E. yunnanense*, also belonging to the genus *Eutrema*, grows in the wild in mainland China and is similar in form to *E. japonicum*, but lacks its characteristic pungent flavor. Genetic analyses indicated its separation from its Japanese counterpart approximately 5 million years ago, and the Japanese plant subsequently acquired the distinctive pungent flavor [1].

Since ancient times, wasabi has grown in mountain streams throughout Japan and has been used as a medicinal herb for centuries because of its characteristic pungent flavor (**Figures 1** and **2**). The earliest written record is "Wasabi" written on a wooden strip unearthed from a ruin in the village of Asuka, Nara Prefecture, dating from the Aska period [2]. As the area was a medicinal herb garden, it is believed to have been used as a medicinal herb since ancient times.

"Wasabi" is also mentioned in "Honzo Wamyo," Japan's oldest encyclopedia of medicinal plants 1100 years ago. According to the dictionary of medicinal food published more than 300 years ago, "Gleanings from the Materia Medica," it disperses depression, generates sweating, dispels wind (shallow evil poison as the cause of the

**Figure 1.**
*Wasabi cultivation. The wasabi fields in Izu City, Shizuoka Prefecture, Japan. The wasabi fields stretch from the upper reaches to the lower reaches, and the cold but abundant clear water nurtures the wasabi. Source: Kinjirushi Co., Ltd.*

**Figure 2.**
*Wasabi. Wasabi grows slowly for more than a year until the rhizome reaches about 15 centimeters. It has a distinctive flavor, sweetness, and consistency when ground. Source: Kinjirushi Co., Ltd.*

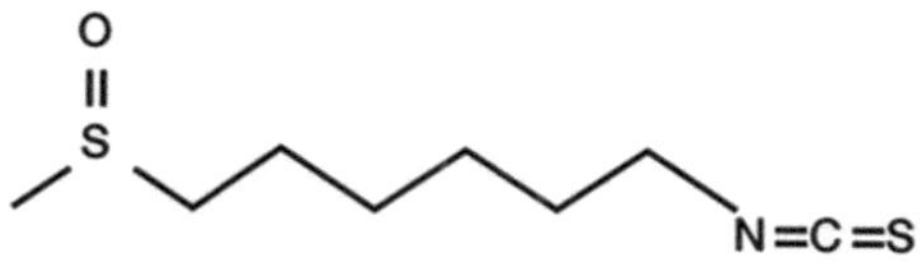

**Figure 3.**
*Hexaraphane. It is found in trace amounts in the rhizomes of wasabi. It is a nonvolatile amphipathic substance with a slightly astringent flavor, but no pungent taste. 6-methylsulfinylhexyl isothiocyanate; molecular weight: 205.3, $C_8H_{15}NOS_2$, CAS number: 4430-35-7. Source: Kinjirushi Co., Ltd.*

disease), purges dampness (the first of the five tinctures as the cause of the disease), eliminates stagnation (pain-causing), and removes plumpness (plumpness, one of the five stagnations, refers to the spleen stagnation). It is the best of the seven flavors of colic herbs. It is also used to treat biofilms and is recognized as a medicinal plant [3]. In addition to its appetite-boosting properties, modern catalogs of medicinal plants state that the rhizomes of wasabi, when ground and thinly applied to a cloth and applied to the areas affected by rheumatism, neuralgia, and tonsillitis, can be effective in relieving pain and that the sap of the rhizomes is effective in preventing fish and horsemeat poisoning [4]. Finely chopped leaves can also be used as a moisturizing bath additive by placing them into a bathtub in bags [5].

Although wasabi has been used in Japan since ancient times, its medicinal properties have little scientific basis and have been regarded only folklore. Recent studies have elucidated the active ingredients in wasabi and explored their functions. These functional ingredients have been used to promote health and beauty. In particular, hexaraphane (6-methylsulfinylhexyl isothiocyanate, 6-MSITC) (**Figure 3**), which is relatively abundant in wasabi rhizomes and roots, has been reported to exhibit antioxidant, anticarcinogenic, and detoxifying enzyme-inducing effects. Moreover, it was reported to inhibit platelet aggregation and ameliorate dementia. Here, I would like to focus on the clinical data on the effect of hexaraphane in improving brain function. I would also like to discuss some of the other functional ingredients in wasabi.

## 2. Functional components of wasabi

The extracts of leaves, stems, and roots of wasabi were analyzed by MS chromatography, and 42 major peaks were detected, including glucosinolates, phenylpropanoid glycosides, flavone glycosides, and hydroxycinnamic acids [6]. Some of these constituents had the strongest anti-inflammatory, antimicrobial, and cytotoxic effects compared to others.

Wasabi is known to contain many isothiocyanates, responsible for its characteristic pungent flavor. In addition, its leaves contain many polyphenols, such as isosaponarin (4'-O-glucosyl-6-C-glucosyl apigenin) and isovitexin. In addition, several new components with various functional properties are being elucidated [7].

Wasabi contains at least 21 kinds of isothiocyanates [8, 9]. Among them, the most abundant isothiocyanate is allyl isothiocyanate (AITC), which is the main component responsible for the pungent flavor. Hexaraphane is another isothiocyanate that has recently attracted attention as a functional ingredient. It has been shown to induce detoxification of metabolic enzymes [10] and exhibit anticancer [11], anti-inflammatory [12], antidiabetic [13], and antiallergenic/anti-atopic [14] effects. Recent studies have demonstrated its ability to activate hair papilla cells and improve dementia [15].

In addition, 6-methylthiohexyl isothiocyanate (6-MTITC) is a volatile component with the characteristic green note of wasabi. It is mainly used as a flavoring agent in processed food, but it has also been reported to have functional properties including antimicrobial and anticarcinogenic effects [16], detoxification and antioxidant effects [17], anticarcinogenic effects [18], and antiallergic effects [12].

To date, wasabi leaves have been partially used as food, but mostly discarded. However, studies have gradually identified the components of wasabi leaves and their functional properties, such as anti-influenza virus [19], anti-obesity [20, 21], and anticancer [22] effects.

Experimental and clinical studies have demonstrated the effects of wasabi leaves on the skin. For example, wasabi leaf extract was shown to decrease tyrosinase activity, melanin content, and formation of advanced glycation end products in vitro and to improve skin condition in a clinical trial [23].

Further research on the functional ingredients may enable more efficient use of wasabi leaves.

### 2.1 Allyl isothiocyanate (AITC)

Studies have reported the antimicrobial properties of AITC. Its pungent flavor is believed to have antimicrobial properties. It is a relatively volatile ingredient with a low molecular weight of 99.15 and a boiling point of 152°C. Its antimicrobial activity is also very effective in gas-phase exposures. Its minimum growth inhibitory concentrations against bacteria, yeasts, and molds ranged between 16 and 110 ng/mL [24]. Industrially, AITC-based antimicrobial products for lunch boxes and rice storage container and antimicrobial deodorizers for car air conditioners are available.

Due to its highly irritant nature, an acute toxicity $LD_{50}$ of 310 mg/kg has been reported in mice [25]. Moreover, it causes irritation of skin and mucous membranes on contact [26]. In addition, there are reports of suspected mutagenicity and genotoxicity [27]. Therefore, due caution should be exercised when using it as a functional material.

### 2.2 Hexaraphane

Hexaraphane is a nonvolatile isothiocyanate, mainly found in the rhizome and root, but at low levels of 200–500 μg/g [28]. The dietary intake ranges from 2.5 to 5.0 g per meal, with an estimated hexaraphane content of 0.5 to 2.5 mg per meal.

The $LD_{50}$ values for hexaraphane in rats ranged from 338 to 451 mg/kg, but the $LD_{50}$ values for the natural and synthetic extracts differed. Hexaraphane is detected in the blood 2 hours after ingestion, and approximately 50% of the ingested amount is excreted in the urine within 24 hours [29]. Animal studies have shown that sulforaphane (4-methylsulfinylbutyl isothiocyanate; 4-MSITC), an analog of hexaraphane, crosses the blood-brain barrier 15 minutes after ingestion [30].

Therefore, hexaraphane is considered to be an effective functional ingredient because it is also rapidly absorbed by the body. It is important to note that the most commonly used products are room temperature tubular ones. However, these processed products often contain no rhizomes or only a small amount of rhizomes and therefore very little hexaraphane [28]. Therefore, the consumption of ground rhizomes or the use of products or supplements primarily containing rhizomes is essential to utilize the benefits of hexaraphane.

Hexaraphane acts on the transcription factor Nrf2 (NF-E2-related factor 2)/ Keap1 (Kelch-like ECH-associated protein 1) system and promotes the transcriptional expression of antioxidant response elements. It is known to increase the expression of phase 2 enzymes such as GST and NAD(P)H quinone reductase (NQO1), and its GST-inducing activity was reported to be the highest among 20 vegetables [31].

Hexaraphane has been reported to inhibit the production of reactive oxygen species (ROS) by neutrophils stimulated with phorbol myristate acetate (PMA) [32]. A study suggested that hexaraphane does not directly eliminate the ROS produced by stimulation, but inhibits the ROS production by neutrophils by acting on the site of ROS production. Studies have shown that benzyl isothiocyanate (BITC), a typical papaya isothiocyanate, modifies cytochrome b558 [33] of the NADPH oxidase complex on leukocyte membranes. Hexaraphane is also believed to have the same mechanism of action.

In a study, rat basophil-like cells RBL-2H3 were stimulated with specific IgE antibodies, and the released histamine and leukotrienes were measured by ELISA. The addition of hexaraphane [12] significantly inhibited the release of histamine and leukotrienes.

In addition, we investigated the effects of hexaraphane on dermal papilla cells (DPCs) and showed that hexaraphane activated hair papilla cells and upregulated vascular endothelial growth factor and adenosine receptors. This suggests that hexaraphane may promote the growth of hair [34].

It is important to note here that wasabi is primarily sold in tube form as processed, but horseradish is used as a substitute for it. In other words, many products are made by coloring dried horseradish powder, kneading it with water, and then putting it into tube form, which contains almost none of wasabi's functional ingredients, including hexaraphane. Some wasabi products are made using the stems and leaves, but as the wasabi rhizomes and roots contain large amounts of hexaraphane, only processed products containing large amounts of these parts have these functional ingredients in adequate amounts.

### 2.3 Isosaponarin

Wasabi plants accumulate C-glucosyl flavones, including isosaponarin, in their leaves [7, 35]. Isosaponarin is one of the major flavonoids in wasabi and has been rarely reported from other species in the Plant Kingdom (**Figure 4**) [36].

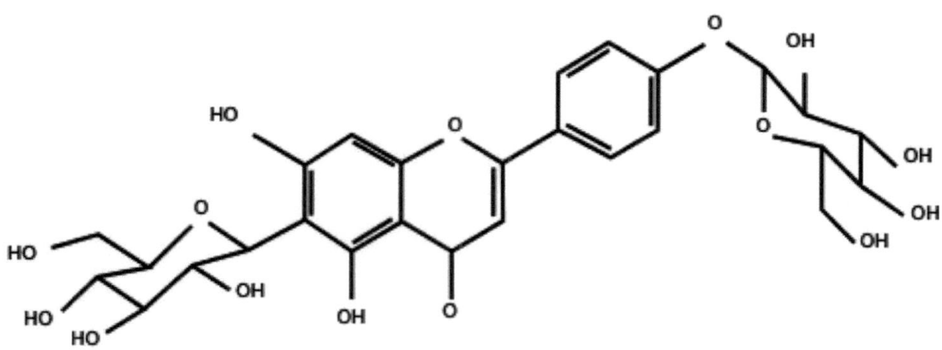


**Figure 4.**
*Isosaponarin. It is a unique flavonoid with C-bond found in the leaves of wasabi and has the structure of apigenin with two sugars attached. An enzymatic reaction removes the sugar and converts it to isovitexin. Molecular weight: 594.5, $C_{27}H_{30}O_{15}$, CAS number: 19416–87-6. Source: Kinjirushi Co., Ltd.*

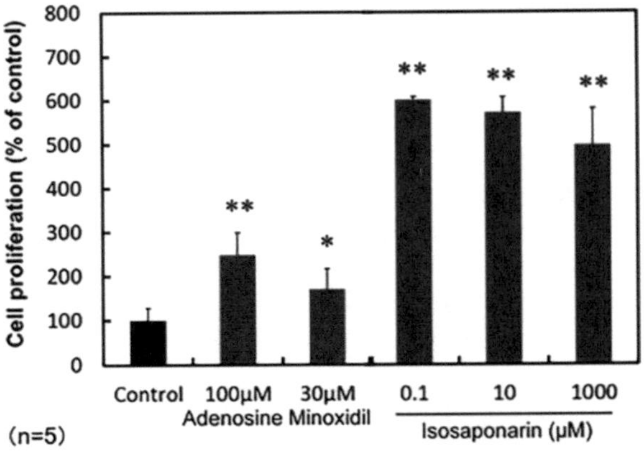


**Figure 5.**
*Effect of isosaponarin on the proliferation of DPCs. The effect of isosaponarin on amounts and activity of mitochondria in DPCs. DPCs were cultured for 72 hours with the indicated concentrations of isosaponarin, solvent control (DMSO), and positive controls (adenosine and minoxidil). The amounts of mitochondria were measured by Mito Tracker staining and further divided by the number of nuclei obtained by Hoechist staining to evaluate the number per cell. The data represent mean ± SD of five replicates. * and ** indicates significant differences versus control ($p < 0.05$ and $p < 0.01$ respectively; t-test). Source: Author.*

Isosaponarin and other flavonoids in wasabi leaves were shown to promote collagen production [37] and activate hair papilla cells.

Isosaponarin was found to increase collagen production in fibroblasts and enhance expression of transforming growth factor βII-type receptor (TβR-II) and prolyl 4-hydroxylase (P4H). These results suggested that isosaponarin promotes collagen synthesis by increasing TβR-II and P4H [37].

We also reported the effect of isosaponarin on DPCs. This was tested by adding wasabi leaf extract and isosaponarin to the culture of DPCs and assessing the cellular activity by the water-soluble tetrazolium method. Minoxidil and adenosine, which are used as ingredients in hair growth products, were used as positive controls. The results showed activation of DPCs at a low concentration of 0.1 μM, as well as an increase in vascular endothelial growth factor and other substances (**Figure 5**) [38]. Although further validation is needed, it is conceivable that isosaponarin could be used as a material to promote hair regrowth and inhibit hair loss. Indeed, formulations containing isosaponarin are now available as hair care materials.

## 3. Cognitive function improvement by hexaraphane

### 3.1 Animal studies

Hexaraphane is known to express detoxification and antioxidant genes by activating the Keap1/Nrf2 system. Activation of Keap1/Nrf2 increases antioxidant enzymes such as HO-1, thereby inhibiting oxidative damage to neurons [39]. In the study by Morroni et al., the administration of hexaraphane to Parkinson's disease model mice was found to improve symptoms [15].

Uruno et al. conducted animal studies on Alzheimer's Disease model mice. The Keap1-Nrf2 system was shown to protect neurons in the hypothalamus against oxidative damage [40]. They also demonstrated that Nrf2 induction improved the impairment in AD model mice in the passive avoidance test (PAT) and suppressed the elevation of the oxidative stress marker 8-OHdG. Administration of hexaraphane improved the impaired cognition of AD model mice in the PAT [41].

These studies suggest that hexaraphane may inhibit oxidative damage in the brain, preventing the onset of dementia or alleviating symptoms.

### 3.2 Safety

The characteristic pungent component of wasabi is AITC, which is approximately five times more abundant than hexaraphane [28, 42]. Therefore, a wasabi extract prepared to obtain hexaraphane would contain large amounts of AITC and its original glycoside, sinigrin. Ingestion of this extract may cause abdominal pain due to their high stimulant properties. We developed a special manufacturing process to remove AITC and sinigrin and developed "Wasabi Sulfinyl™ (WS)," an extract containing 0.8% hexaraphane and almost no AITC or sinigrin. Four-week overdose studies of WS confirmed the safety, with intake ranging from 500–2000 mg (containing 4–16 mg of hexaraphane) [43, 44]. No adverse effects were observed in these overdose trials demonstrating the safety of WS.

### 3.3 Randomized controlled trial

Cognitive function is known to be influenced by diet [45–47] and lifestyle. Intake of vegetables, fruits, and spices was shown to improve cognitive function [48–51]. Hexaraphane has antioxidant and anti-inflammatory properties and has been shown to improve memory and other functions in a mouse model of dementia. Thus, hexaraphane is expected to improve cognitive function. Below, we discuss several clinical trials on this subject.

#### *3.3.1 Neurocognitive functions in cognitively intact middle-aged and older adults*

Based on the antioxidant effects of hexaraphane and the results of animal studies, we conducted a randomized controlled trial for 8 weeks (UMIN000027407) [52]. Fifty cognitively intact elderly adults with memory complaints between the ages of 45 and 69 years were enrolled in this randomized double-blind, placebo-controlled trial. Subjects were randomized to receive 100 mg WS or placebo daily for 8 weeks. The effect on neurocognitive functions was assessed using a battery of four neuropsychological tests at baseline and after 4 and 8 weeks of treatment.

All subjects completed the study although four were excluded from the efficacy analysis due to adverse events or protocol violations. Analysis of the fully evaluable population indicated no significant differences between WS and placebo groups on any outcome measure. Because of the relatively small sample size, the unexpectedly high scores in the placebo group, the fact that all subjects were cognitively intact, and the fact that the eight-week treatment period was too short, it may not have been sensitive enough to discern the relatively subtle differences between the two groups. However, there were trends favoring WS in almost all parameters of the Group version of the Stroop Color Word Test (G-SCWT), of which the magnitude of improvement in correct responses in the Stroop interference task and the interference rate reached near-significant levels ($P = 0.083$ and $0.091$, respectively) at the end of 8-week treatment.

Exercise has been suggested to contribute to improved cognitive function [53, 54]. Therefore, we performed a subgroup analysis stratified according to their exercise habits or lack. A subset of nonexercisers exhibited significantly better performance in both the reverse-Stroop control task and the Stroop interference task of the G-SWCT, which assesses attention and judgment, at week 8. The Stroop test is divided into four levels of difficulty, and in this study, there was a significant difference between difficulty level 1 and difficulty level 4 (**Figure 6**). In difficulty level 1, an improvement in performance was observed after 8 weeks even with placebo, suggesting a learning effect; however, in difficulty level 4, there was no improvement in performance after a period of time with placebo, and only the WS intake group showed an improvement in performance. This suggests that the effect of WS is not just to maintain function by preventing impairment of brain function, but also to improve function. However, due to the small number of subjects and the stratified analysis of the results, more robust studies are required to obtain definitive evidence. Of note, no WS-induced side effects were observed in the study.

### *3.3.2 Memory functioning in healthy adults aged 60 years and older*

Nouchi et al. assessed cognitive function in 72 healthy subjects (19 males, 53 females; mean age: 65.4 years [range, 60–69 years]; mean MMSE score: 28.5) who received a capsule containing either 100 mg WS (0.8 mg hexaraphane/day) or a placebo for 12 weeks (UMIN000032694) [55].

A wide range of cognitive functions (e.g., executive function, episodic memory (immediate and delayed), processing speed, working memory, attention, reasoning, short-term memory, and visuospatial cognitive function) were assessed before and after the intervention.

There were no significant differences between the two groups baseline characteristics with respect to age (WS; 65.9 ± 3.9 years, placebo; 64.9 ± 3.6 years), education (WS; 12.5 ± 0.8, placebo; 12.9 ± 1.1), Mini-Mental State examination (WS; 28.5 ± 1.5, placebo; 28.8 ± 1.6), Frontal Assessment Battery at bedside (WS; 15.4 ± 1.9, placebo; 15.4 ± 1.9), Japanese Reading Ability Test (WS; 20.3 ± 3.9, placebo; 20.8 ± 3.6), and Geriatric Depression Scale (WS; 2.4 ± 1.4, placebo; 2.1 ± 1.1). And, the duration of supplement intake also did not differ between the two groups.

The WS group showed a significant improvement in working memory, measured by digit span backward ($p = 0.000$), and in episodic memory performance, measured by logical memory (LM) immediately ($p = 0.00$), LM delayed ($p = 0.03$), and the face and second name test ($p = 0.00$) (**Figure 7**). However, they did not find significant

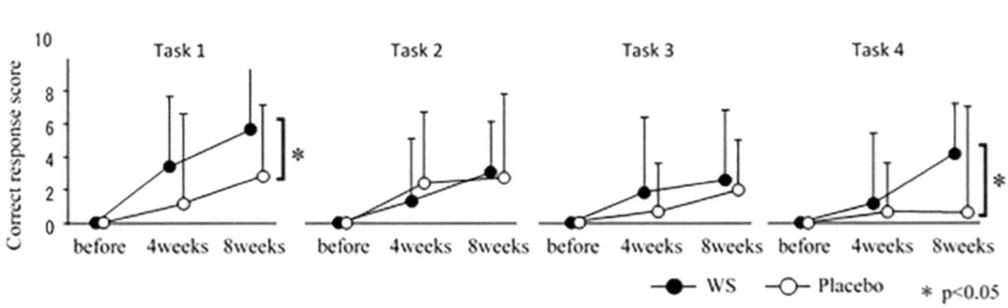


**Figure 6.**
*The result of cognitive performance for the G-SCWT tasks 1 to 4 in the nonexercizer subgroup. Subjects who had a habit to perform exercise for 30 minutes or longer for more than twice a week were included in the exerciser subgroup, and the remaining subjects were included in the nonexercizer subgroup. Data presented as mean ± SD. Comparison of the mean change from baseline for independent samples. Source: Author.*

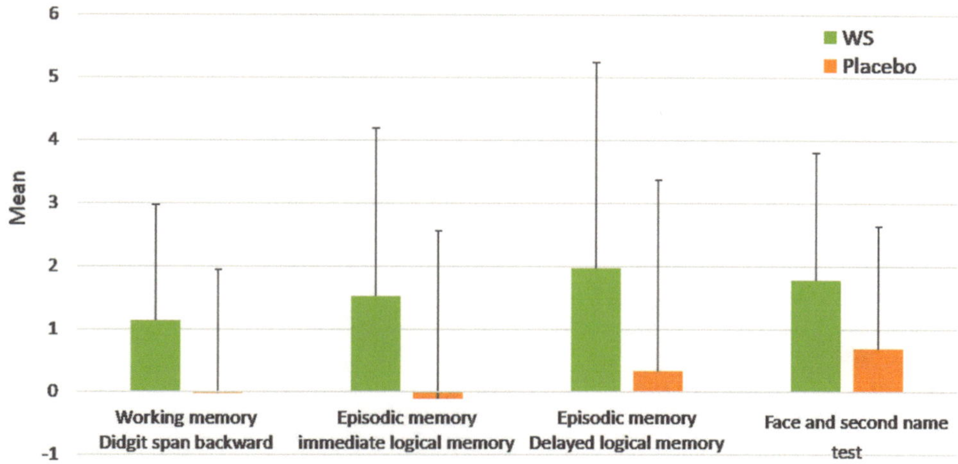


**Figure 7.**
*Change in scores of cognitive functions in both groups. * indicates significant differences versus placebo for multiple comparisons using the Bonferroni method (p < 0.05). This figure was constructed from Ref. ([55], Table 4). Source: Author.*

improvements in other cognitive functions, such as processing speed, Cd, short-term memory, attention, inhibition, rST, reasoning, and visuospatial performances.

There were no significant changes in the inhibition and processing speed performances after hexaraphane intake in older adults. This is inconsistent with a previous study (3.3.1) which showed a beneficial effect of hexaraphane on inhibition performance. However, they noted some differences in methods between this study and previous studies, such as the intervention period (12 weeks or 8 weeks), subjective cognitive complaints (healthy, subjective memory complaints, patients with chronic fatigue), psychological tests (trail making test, SS, or Cd), and participants' age (older adults or middle-aged adults). Therefore, further studies are required to assess the effect of hexaraphane on inhibition and processing performances.

They further noted that antioxidant and anti-inflammatory effects are involved in the improvement of cognitive function [56, 57]. The hippocampus is involved in working and episodic memory [58], and the antioxidant and anti-inflammatory effects of hexaraphane on the hippocampus may improve cognitive functions. Antioxidant and anti-inflammatory effects are believed to be important in maintaining brain function, but as hexaraphane is thought to have a variety of effects, further studies are required to elucidate the underlying mechanisms.

### 3.4 ME/CFS and brain fog improvement

The antioxidant and anti-inflammatory effects of hexaraphane may be beneficial in myalgic encephalomyelitis/chronic fatigue syndrome (ME/CFS).

ME/CFS is a disease with an unknown etiology and no known cure. However, chronic inflammation of the brain has been implicated in its causation [59, 60]. An estimated 17 million people worldwide suffer from the syndrome [61]. In addition to general malaise, low-grade fever, and muscle aches, patients also experience a loss of thinking ability, referred to as brain fog. Symptoms of brain fog include an inability to understand what others are saying, an inability to hold a conversation, an inability to read books and understand their meaning, and an inability to comprehend music.

| Performance status (PS) | Pre (n) | Post (n) | P-value |
|---|---|---|---|
| 0 | 0 | 0 | |
| 1 | 0 | 0 | |
| 2 | 0 | 2 | |
| 3 | 2 | 0 | |
| 4 | 0 | 1 | |
| 5 | 0 | 0 | |
| 6 | 1 | 1 | |
| 7 | 7 | 8 | |
| 8 | 4 | 2 | |
| 9 | 1 | 1 | |
| average | 6.8 ± 1.7 | 6.3 ± 2.1 | *0.001 |

**indicates significant differences between pre- and post-treatment PS using Fisher's exact test. A parametric paired Student's t-test and nonparametric paired Wilcoxon signed rank tests yielded similar significant differences ($p < 0.05$). This table was adapted from Ref. ([62], Table 2). Source: Author.*

**Table 1.**
*Change in PS after hexaraphane treatment.*

In a 12-week study by Oka et al., 15 patients with ME/CFS (3 males and 12 females, mean age 37.5 years [range 20–58], mean duration of illness 5.1 years) were administered 1.2 grams of WS (9.6 milligrams of hexaraphane per day). The results showed a significant improvement in the patients' performance scores (PS), with six patients improving their scores (two patients improved their scores by 2 or more grades) (**Table 1**) [62].

These findings demonstrate the potential of hexaraphane, considering that none of the patients in this study had responded to standard treatment for at least 3 months. Specifically, they showed improvement in headache frequency (4.1 to 3.0 times/week, p = 0.001), myalgia frequency (4.1 to 2.4 times/week, p = 0.019),

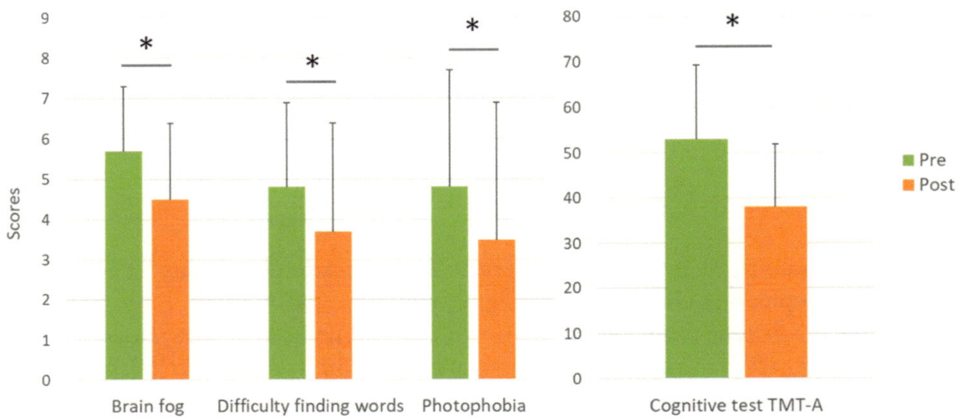


**Figure 8.**
*Changes in neurocognitive symptoms and TMT-A test scores after Hexaraphane treatment. * indicates significant differences between pre- and post-treatment scores using a parametric paired Student's t-test. ($p < 0.05$). This figure was constructed from Ref. ([62], Table 5). Source: Author.*

numerical rating scale brain fog scores (5.7 to 4.5, p = 0.011), difficulty in discovering the right words (4.8 to 3.7, p = 0.015), photophobia (from 4.8 to 3.5, p = 0.008), and mood state vigor scores (from 46.9 to 50.0, p = 0.045), right occipital pressure pain threshold (from 17.3 to 21.3, p = 0,01), and trail making test-A (TMT-A) (from 53.0 to 38.1, p = 0.007) (**Figure 8**). In particular, 14 out of 15 participants perceived improvement in brain fog and cognitive function, as well as significant improvement in general health perception, vitality, and vigor. While the mechanisms of improvement remain to be elucidated and need to be validated in a larger sample, 10 of the 15 patients continued to take WS at their own request after the completion of the study.

## 4. Conclusion

Wasabi is a pungent vegetable that contains several active ingredients, including hexaraphane. As wasabi is mainly grown and consumed in Japan, there is a paucity of research on its therapeutic effects in other parts of the world. Our basic research and clinical trials have demonstrated its efficacy in improving brain function, even in healthy individuals. The results of studies conducted in mice model of dementia also suggest its potential protective effect on the brain. It may not only help maintain brain function in healthy individuals but also improve symptoms in patients with MCI and dementia.

## Acknowledgements

We are very grateful to Dr. Nouchi and Dr. Oka for their energetic research on the functionality of wasabi, and to NOMON Corporation for their tremendous efforts not only in research but also in the marketing of wasabi. The research fees paid by Kinjirushi Co., Ltd., over the years have been the cornerstone of our current research. We are very grateful to Kinjirushi Co., Ltd.

## Conflict of interest

Isao Okunishi is an employee of Kinjirushi Co., Ltd., and received a research grant from the company.

**Author details**

Isao Okunishi
Kinjirushi Co., Ltd. Nagoya-City, Aichi, Japan

*Address all correspondence to: i-okunishi@kinjirushi.co.jp

## References

[1] Yamane K et al. Genetic differentiation, molecular phylogenetic analysis, and ethnobotanical study of *Eutrema japonicum* and *E. tenue* in Japan and *E. yunnanense* in China. Horticultural Journal. 2016;**85**:46. DOI: 10.2503/hortj.MI-065

[2] Chunichi Shinbun (Japanese newspaper). 2001; 26p.

[3] Hitomi H. Honchoushokukagami. Chiyoda, Tokyo, Japan: Heibonsya; 1976. 180 p

[4] Okada M. Newly Revised Illustrated Medicinal Plants of the World. Meguro, Tokyo, Japan: Hokuryukan; 2002. 154 p

[5] Izawa B, Aida T. Yakuso zukan. Shinjuku-ku, Tokyo, Japan: Ie-no-Hikari, Association; 1999. 277 p

[6] Dos Santos Szewczyk KDS et al. Chemical composition of extracts from leaves, stems and roots of wasabi (*Eutrema japonicum*) and their anti-cancer, anti-inflammatory and anti-microbial activities. Scientific Reports. 2023;**13**:9142. DOI: 10.1038/s41598-023-36402-y

[7] Hosoya T, Yun YS, Kunugi A. Five novel flavonoids from *Wasabia japonica*. Tetrahedron. 2005;**61**:7037. DOI: 10.1016/j.tet.2005.04.061

[8] Ina K. Volatile components of wasabi, horse radish and mustard on allyl isothiocyanate. Koryo. 1982;**136**:45

[9] Etoh H et al. ω -methylsulfinylalkyl isothiocyanates in Wasabi, *Wasabia japonica* matsum. Agricultural and Biological Chemistry. 1990;**54**:1587. DOI: 10.1080/00021369.1990.10870168

[10] Morimitsu Y et al. A sulforaphane analogue that potently activates the Nrf2-dependent detoxification pathway. The Journal of Biological Chemistry. 2002;**277**:3456. DOI: 10.1074/jbc.M110244200

[11] Fuke Y et al. Wasabi-derived 6-(methylsulfinyl)hexyl isothiocyanate induces apoptosis in human breast cancer by possible involvement of the NF-κB pathways. Nutrition and Cancer. 2014;**66**:879. DOI: 10.1080/01635581.2014.916322

[12] Yamada-Kato T et al. Inhibitory effects of wasabi isothiocyanates on chemical mediator release in RBL-2H3 rat basophilic leukemia cells. Journal of Nutritional Science and Vitaminology. 2012;**58**:303. DOI: 10.3177/jnsv.58.303

[13] Yoshida J et al. Glycogen synthase kinase-3β inhibition of 6-(methylsulfinyl) hexyl isothiocyanate derived from wasabi (Wasabia japonica matsum). Bioscience, Biotechnology, and Biochemistry. 2011;**75**:136. DOI: 10.1271/bbb.100507

[14] Nagai M, Okunishi I. The effect of wasabi rhizome extract on atopic dermatitis-like symptoms in HR-1 hairless mice. Journal of Nutritional Science and Vitaminology (Tokyo). 2009;**55**:195. DOI: 10.3177/jnsv.55.195

[15] Morroni F et al. Neuroprotection by 6-(methylsulfinyl) hexyl isothiocyanate in a 6-hydroxydopamine mouse model of parkinson's disease. Brain Research. 2014;**1589**:93. DOI: 10.1016/j.brainres.2014.09.033

[16] Yano T et al. The effect of 6-methylthiohexyl isothiocyanate isolated from *Wasabia japonica*

(wasabi) on 4-(methylnitrosamino)-1-(3-pyridyl)-1-buatnone-induced lung tumorigenesis in mice. Syokuhin-Kogyo. 2000;115(115):58. DOI: 10.1016/s0304-3835(00)00364-5

[17] Yano T et al. The effect of 6-methylthiohexyl isothiocyanate isolated from *Wasabia japonica* (wasabi) on 4-(methylnitrosamino)-1-(3-pyridyl)-1-buatnone-induced lung tumorigenesis in mice. Cancer Letters. 2000;**155**:115. DOI: 10.1016/s0304-3835(00)00364-5. S0304-3835(00)00364-5

[18] Yoshinori K et al. Modulation of Nrf2/Keap1 system by wasabi 6-methylthiohexyl isothiocyanate in ARE-mediated NQO1 expression. Molecular Nutrition & Food Research. 2013;**57**:854. DOI: 10.1002/mnfr.201200689

[19] Mochida K, Ogawa T. Anti-influenza virus activity of extract of Japanese wasabi leaves discarded in summer. Journal of the Science of Food and Agriculture. 2008;**88**:1704. DOI: 10.1002/jsfa.3268

[20] Yamada-Kato T et al. Anti-obesity effects of wasabi leaf extract on rats fed a high-fat diet are related to upregulation of mRNA expression of β3-adrenergic receptors in interscapular brown adipose tissue. Food Science and Technology Research. 2016;**22**:665. DOI: 10.3136/fstr.22.665

[21] Oowatari Y et al. Wasabi leaf extracts attenuate adipocyte hypertrophy through PPARγ and AMPK. Bioscience, Biotechnology, and Biochemistry. 2016;**80**:1594. DOI: 10.1080/09168451.2016.1179093

[22] Okamoto M et al. Preventive effects of wasabi leaf extract on carcinogenic initiation. Journal of Clinical Biochemistry and Nutrition. 2008;**43**:251

[23] Chiang HM et al. Wasabi leaf supplementation had antioxidant, anti-glycation, and improved skin melanin, spot and moisture. Journal of Functional Foods. 2023;**100**:105398. DOI: 10.1016/j.jff.2022.105398

[24] Isshiki K et al. Allyl isothiocyanate and wholesomeness of food. Japanese Journal of Food Microbiology. 1993;**10**:1. DOI: 10.14840/jsfm1984.10.1

[25] European Food Safety Authority (EFSA). Scientific opinion on the safety of allyl isothiocyanate for the proposed uses as a food additive1 EFSA panel on food additives and nutrient sources added to food (ANS). EFSA Journal. 2010;**8**:1943

[26] IARC. IARC monographs on the evaluation of carcinogenic risks to humans. In: Some Chemicals That Cause Tumours of the Kidney or Urinary Bladder in Rodents and some Other Substances. Vol. 73. LYON CEDEX, France: IARC; 1998

[27] Human Health Assessment Branch, Department of Pesticide Regulation California Environmental Protection Agency. Allyl Isothiocyanate Risk Characterization Document Occupational and Bystander Exposures. Human Health Assessment Branch, Department of Pesticide Regulation. Sacramento, CA: California Environmental Protection Agency; 2022

[28] Murata M et al. Content of 6-methylsulfinylhexyl isothiocyanate in wasabi and processed wasabi products. Nippon Shokuhin Kagaku Kogaku Kaishi. 2004;**51**:477. DOI: 10.3136/nskkk.51.477

[29] Kinjirushi Co., Ltd. Exploration of New Functionality of Wasabi Derived

Components and Development of Advanced Utilization Technology as Functional Food Ingredients, Development of Technologies to Improve the Functionality of Food Products. Chuo, Tokyo, Japan: New Food Creation Technology Research Association; 2004. pp. 69-100

[30] Tarozzi A et al. Neurodegeneration, neurogenesis, and oxidative stress. Oxidative Medicine and Cellular Longevity. 2013;**10**:415078. DOI: 10.1155/2013/415078

[31] Morimitsu Y et al. Antiplatelet and anticancer isothiocyanates in Japanese domestic horseradish, wasabi. BioFactors. 2000;**13**:271. DOI: 10.1002/biof.5520130141

[32] Okunishi I. Overview; functional studies on wasabi. In: The Annual Meeting of the Japanese Society for Food. Naka, Nagoya, Japan: Science and Technology Chubu branch; 2005

[33] Miyoshi N et al. Benzyl isothiocyanate inhibits excessive superoxide generation in inflammatory leukocytes: Implication for prevention against inflammation-related carcinogenesis. Carcinogenesis. 2004;**25**:567. DOI: 10.1093/carcin/bgh051

[34] Yamada-Kato T et al. Stimulatory effects of 6-methylsulfinylhexyl isothiocyanate on cultured human follicle dermal papilla cells. Food Science and Technology Research. 2018;**24**:567. DOI: 10.3136/fstr.24.567

[35] Yoshida S et al. Component analysis of wasabi leaves and an evaluation of their anti-inflammatory activity. Food Science and Technology Research. 2015;**21**:247-253. DOI: 10.3136/fstr.21.247

[36] Mashima K et al. Identification and characterization of apigenin 6-C-glucosyltransferase involved in biosynthesis of isosaponarin in wasabi (*Eutrema japonicum*). Plant & Cell Physiology. 2019;**60**:2733. DOI: 10.1093/pcp/pcz164

[37] Nagai M et al. The effect of isosaponarin isolated from wasabi leaf on collagen synthesis in human fibroblasts and its underlying mechanism. Journal of Natural Medicines. 2010;**64**:305. DOI: 10.1007/s11418-010-0412-y

[38] Okunishi I et al. Wasabi components 6-methylsulfinylhexylisothiocyanate and isosaponarin for hair growth. In: 136th Annual Meeting of the Pharmaceutical Society of Japan. Shibuya, Tokyo, Japan: The Pharmaceutical Society of Japan; 2016

[39] Satoh T et al. Activation of the Keap1/Nrf2 pathway for neuroprotection by electrophilic [correction of electrophillic] phase II inducers. Proceedings of the National Academy of Sciences of the United States of America. 2006;**103**(768):05057231. DOI: 10.1073/pnas.0505723102

[40] Yagishita Y et al. Nrf2 improves leptin and insulin resistance provoked by hypothalamic oxidative stress. Cell Reports. 2017;**18**:2030. DOI: 10.1016/j.celrep.2017.01.064

[41] Uruno A et al. Nrf2 suppresses oxidative stress and inflammation in app knock-in Alzheimer's disease model mice. Molecular and Cellular Biology. 2020;**40**:e00467-e00419. DOI: 10.1128/MCB.00467-19

[42] Sultana T et al. Comparison of flavour compounds in wasabi and horseradish. Journal of Food, Agriculture and Environment. 2003;**1**:117

[43] Nakajima R et al. Oral administration of 6-methylsulfinylhexyl isothiocyanate

extracted from is safe and improves the fatigue and sleep of healthy volunteers. BioPsychoSocial Medicine. 2023;**17**:30. DOI: 10.1186/s13030-023-00287-08

[44] Okunishi I et al. Safety evaluation of 6-(methylsulfinyl) hexyl isothiocyanate (hexaraphane) and wasabi sulfinyl, a hexaraphane-containing supplement. Food Science and Technology Research. 2020;**26**:813. DOI: 10.3136/fstr.26.813

[45] Solfrizzi V et al. Nutritional intervention as a preventive approach for cognitive-related outcomes in cognitively healthy older adults: A systematic review. Journal of Alzheimer's Disease. 2018;**64**:S229. DOI: 10.3233/JAD-179940

[46] Gardener SL, Rainey-Smith SR. The role of nutrition in cognitive function and brain ageing in the elderly. Current Nutrition Reports. 2018;**7**:139. DOI: 10.1007/s13668-018-0229-y

[47] Nouchi R et al. Effects of lutein and astaxanthin intake on the improvement of cognitive functions among healthy adults: A systematic review of randomized controlled trials. Nutrients. 2020;**12**:617. DOI: 10.3390/nu12030617

[48] Khazdair MR et al. Neuroprotective potency of some spice herbs, a literature review. Journal of Traditional and Complementary Medicine. 2019;**9**:98. DOI: 10.1016/j.jtcme.2018.01.002

[49] Jiang TA. Health benefits of culinary herbs and spices. Journal of AOAC International. 2019;**102**:395. DOI: 10.5740/jaoacint.18-0418

[50] Panickar KS. Beneficial effects of herbs, spices and medicinal plants on the metabolic syndrome, brain and cognitive function. Central Nervous System Agents in Medicinal Chemistry. 2013;**13**:13

[51] Mirmosayyeb O et al. Possible role of common spices as a preventive and therapeutic agent for Alzheimer's disease. International Journal of Preventive Medicine. 2017;**8**:5. DOI: 10.4103/2008-7802.199640

[52] Okunishi I et al. The effects of wasabi root-derived 6-(methylsulfinyl) hexyl isothiocyanate on neurocognitive functions in cognitively intact middle-aged and older adults—a randomized, double-blind, placebo-controlled trial. Japanese Pharmacology & Therapeutics. 2019;**47**:275

[53] Han H et al. Exercise improves cognitive dysfunction and neuroinflammation in mice through histone H3 lactylation in microglia. Immunity & Ageing. 2023;**20**:63. DOI: 10.1186/s12979-023-00390-4

[54] Valenzuela PL et al. Exercise benefits on Alzheimer's disease: State-of-the-science. Ageing Research Reviews. 2020;**62**:101108. DOI: 10.1016/j.arr.2020.101108

[55] Nouchi R et al. Benefits of wasabi supplements with 6-MSITC (6-methylsulfinyl hexyl isothiocyanate) on memory functioning in healthy adults aged 60 years and older: Evidence from a double-blinded randomized controlled trial. Nutrients. 2023;**15**:4608. DOI: 10.3390/nu15214608

[56] Hajjar I et al. Oxidative stress predicts cognitive decline with aging in healthy adults: An observational study. Journal of Neuroinflammation. 2018;**15**:17. DOI: 10.1186/s12974-017-1026-z

[57] Sartori AC et al. The impact of inflammation on cognitive function in older adults: Implications for healthcare practice and research. The Journal of Neuroscience Nursing. 2012;**44**:206

[58] Yonelinas AP. The hippocampus supports high-resolution binding in the

service of perception, working memory and long-term memory. Behavioural Brain Research. 2013;**254**:34

[59] Fukuda S et al. A potential biomarker for fatigue: Oxidative stress and anti-oxidative activity. Biological Psychology. 2016;**118**:88. DOI: 10.1016/j.biopsycho.2016.05.005

[60] Nakatomi Y et al. Neuroinflammation in patients with chronic fatigue syndrome/myalgic encephalomyelitis: An $^{11}C$-(R)-PK11195 PET study. Journal of Nuclear Medicine. 2014;**55**:945. DOI: 10.2967/jnumed.113.131045

[61] Lim EJ, Son CG. Review of case definitions for myalgic encephalomyelitis/chronic fatigue syndrome (ME/CFS). Journal of Translational Medicine. 2020;**18**:289. DOI: 10.1186/s12967-020-02455-0

[62] Oka T et al. Clinical effects of wasabi extract containing 6-MSITC on myalgic encephalomyelitis/chronic fatigue syndrome: An open-label trial. BioPsychoSocial Medicine. 2022;**16**:26. DOI: 10.1186/s13030-022-00255-0

Chapter 13

# Natural Anticarcinogens: The Efficacy of Herbs and Spices

*Biba Vikas, Vishnupriya K. Sweety, Sukumaran Anil and Senthilnathan Palaniyandi*

## Abstract

This chapter comprehensively reviews numerous herbs and spices' anticancer properties and mechanisms. Curcumin, derived from turmeric, exhibits cytotoxic, anti-proliferative, anti-invasive, and anti-metastatic effects against diverse cancer types by interacting with pathways controlling inflammation, survival, cell cycle, apoptosis, and angiogenesis. Ginger and its active compounds demonstrate antioxidant, anti-inflammatory, and anti-tumor activities in breast, skin, lung, and gastrointestinal cancer models, potentially through free radical scavenging, inhibiting inflammation, disrupting microtubules, and altering apoptotic gene expression. Garlic organosulfur compounds induce apoptosis, cell cycle arrest, and antioxidant activity in various malignancies by affecting signaling proteins like STAT3 and VEGF. Basil, caraway, cinnamon, clove, coriander, dill, cumin, rosemary, saffron, thyme, and oregano extracts restricted growth, prompted cell death and beneficially impacted proliferation, metastasis, and angiogenesis markers across numerous cancer cell lines. Black pepper, cayenne pepper, green tea, and milk thistle constituents displayed cytotoxic, immune-enhancing, and chemopreventive actions against diverse human cancer models through modulation of pathways connected to oxidative stress, chronic inflammation, detoxification, and malignant proliferation. The immunomodulatory herbs astragalus, ginseng, echinacea, St. John's wort, and cat's claw also exhibited direct anticancer effects in specific cellular and rodent experiments. Many dietary phytochemicals show promising anticancer potential, but more clinical trials are needed to substantiate therapeutic efficacy in humans.

**Keywords:** anticarcinogenic agents, phytochemistry, plant extracts, apoptosis, herbal medicine, curcumin, garlic, ginger, capsaicin

## 1. Introduction

The use of herbal medicine and natural products to treat disease has a long history in many cultures worldwide. There is a growing interest in the potential anticancer effects of compounds derived from herbs, spices, fruits, vegetables, and traditional medicinal plants. This introduction will provide background on the historical use of botanicals in cancer treatment and discuss recent research on natural anticarcinogens derived from dietary sources. References to the use of herbs for cancer treatment date back to ancient Egyptian papyri from 1900 BC, which described using garlic, opium poppy, and "breast tumors" (likely breast cancer) [1]. Traditional Chinese medicine, Ayurvedic medicine from India, and traditional healing practices in the Americas, Europe, and Africa have long used various plant extracts for managing different disease states [2]. Throughout history, herbal preparations were among the primary forms of cancer treatment before modern medicine shifted towards synthetic drugs in the nineteenth and twentieth centuries [3].

A resurgence of interest in traditional herbal medicine for cancer began in the latter half of the twentieth century once conventional cancer therapies' limitations and side effects became better recognized [4]. There is now considerable research interest in the potential anticancer effects of phytochemicals, biologically active compounds derived from plants. Dietary phytochemicals under scientific investigation include flavonoids from fruits and vegetables, curcumin from turmeric, resveratrol from red wine, genistein from soy, diallyl sulfide from garlic, and allicin from garlic [5]. These natural anticarcinogens are thought to suppress tumor growth by acting on various hallmarks of cancer, such as resisting cell death, enabling replicative immortality, inducing angiogenesis, activating invasion and metastasis, reprogramming energy metabolism, and evading immune destruction [6].

Numerous preclinical studies in cell culture and animal models have demonstrated the anticancer effects of herbal products and plant-derived compounds. For example, curcumin has anticancer properties in many types of cancer, including colorectal, pancreatic, breast, prostate, and lung cancers [7, 8]. Resveratrol exhibits anticancer effects in breast, stomach, and thyroid cancer models [9]. Many studies suggest that combinations of phytochemicals may have synergistic effects against cancer [10]. While the precise mechanisms of anticancer phytochemicals are still under investigation, they are thought to modulate molecular pathways and gene expression in cancer development and progression.

Despite promising preclinical data, there remains limited evidence from human clinical trials that definitively demonstrates the efficacy of many herbal medicines for cancer therapy [11]. Few clinical studies have shown the beneficial effects of some herbal preparations. For instance, combining herbal extracts decreased chemotherapy-induced toxicity in lung cancer patients [12]. Curcumin combined with chemotherapy improved the quality of life in advanced breast cancer [13]. However, more extensive rigorous trials are still needed. Challenges in advancing herbal extracts to mainstream cancer care include inconsistencies in herbal preparations, limited bioavailability, lack of profit incentive for drug development, and inadequacies in regulations governing herbal medicines [14].

Natural anticarcinogens from traditional medicinal plants are a promising area of cancer research. Phytochemicals have the advantages of low toxicity, low cost, and ease of availability. Preclinical studies demonstrate that numerous dietary agents have cancer-preventive and therapeutic effects mediated by various anticancer mechanisms. However, there remains a need for well-designed clinical studies to evaluate safety and efficacy in humans conclusively. Given the growing scientific basis supporting their anticancer properties, further research should continue to investigate botanical natural products' multifactorial mechanisms and clinical potential for managing cancer.

## 2. Herbs and spices with potential anticancer properties

### 2.1 Curcumin (*Curcuma longa L.*)

Curcumin is a polyphenol derived from the spice turmeric. For centuries, curcumin has been used in traditional Indian and Chinese medicine for its wide range of therapeutic effects [15]. In recent decades, extensive research has revealed that curcumin possesses potent anti-inflammatory and antioxidant properties that give it anti-cancer effects against numerous cancers [16]. Curcumin exhibits cytotoxic effects against multiple human cancer cell lines, including cancers of the skin, breast, lung, pancreas, prostate, bladder, kidney, cervix, liver, blood, and brain [17]. Treatment with curcumin hindered growth and proliferation while inducing apoptotic cell death through various signaling pathways depending on the cancer cell models studied [16]. Curcumin also beneficially impacted the invasion, metastasis, and angiogenesis behaviors of diverse malignancy types in preclinical analyses [18]. Oral curcumin administration suppressed tumor formation, restricted solid tumor growth, prevented metastasis development, and increased survival in chemically induced or xenograft models of skin, liver, lung, adrenal, bladder, intestinal, and hormone-related cancers [17, 19]. Despite poor bioavailability, a few human trials exhibit direct anticancer effects, as detailed in the next section. Curcumin displays broad pharmacological activity against myriad cancers; warranting continued translational research.

Curcumin is believed to modulate numerous molecular targets and signaling pathways pivotal in malignant progression. By interacting with transcription factors, enzymes, receptors, microRNAs, and additional upstream regulators, curcumin administration alters the expression or functioning of growth factors, inflammatory cytokines, adhesion molecules, angiogenic factors, and apoptosis-related proteins in diverse cancer cell lines and animal models [18, 20]. For example, curcumin disrupted nuclear factor kappa B (NF-kB), Wnt/β-catenin and STAT3 signaling to reduce proliferation, survival, invasion, metastasis, and inflammation markers in models of breast, pancreatic, prostate, and blood cancers [20, 21]. Curcumin halts the cell cycle of cancer cells, preventing them from proliferating uncontrollably [22]. It also activates apoptotic pathways like p53, caspase-3, and Bcl-2 that trigger programmed cell death in cancers [23]. Chronic inflammation perpetuates tumorigenesis, and curcumin can suppress pro-inflammatory cytokines like NF-kB and COX-2 that promote cancer growth [24]. In addition, curcumin prevents the development of new blood vessels that allow tumors to receive nutrients and oxygen for expansion [25]. It also reduces the expression of matrix metalloproteinases and cell adhesion molecules, allowing cancer to invade tissue and metastasize [18]. Furthermore, curcumin can re-sensitize resistant cancers to chemotherapeutics by inhibiting survival mechanisms [26]. Through these myriad mechanisms, curcumin can prevent cancer initiation and halt the progression of existing tumors.

Curcumin has been extensively studied as a cancer therapy for nearly every primary form of the disease. Research indicates curcumin has protective effects against gastrointestinal, genitourinary, gynecological, hematological, pulmonary, thyroid, breast, and skin cancers. Curcumin's efficacy is most potent against bowel, stomach, liver, and skin cancers [21]. Colorectal cancer is one of the cancer types most sensitive to curcumin [8]. Curcumin suppresses intestinal tumor formation in rodent models of colorectal cancer and inhibits the proliferation of human colorectal carcinoma cell lines [27]. In human clinical trials, doses between 0.45 g and 3.6 g daily reduced proliferation rates of colorectal cells and adenomatous polyp formation [28]. Analogously, curcumin demonstrates chemopreventive and chemotherapeutic capacities against stomach cancer through similar anti-proliferative and pro-apoptotic mechanisms [29]. Curcumin also exhibits protective effects against hepatocellular carcinoma (HCC), the most common form of liver cancer. Curcumin administration reduces HCC tumor volume and metastasis in murine models [30]. A dose-escalation trial found that doses up to 8 g per day were well-tolerated in HCC patients and produced downregulation of NF-kB and COX-2 levels [31]. Additionally, curcumin is promising for preventing and treating skin cancers like melanoma and squamous cell carcinoma. Topical curcumin formulations have been shown to reduce lesion size and tumor burden in skin cancer models via anti-proliferative and pro-apoptotic effects [32].

The preclinical evidence supporting curcumin's anti-cancer properties has prompted growing interest in clinical evaluation. Over 100 clinical trials are underway to investigate curcumin's therapeutic use in cancer patients [17]. Ongoing and upcoming trials assess curcumin across nearly all cancer types as a standalone and adjuvant therapy. Several Phase I trials have recently studied escalating doses to evaluate safety and tolerability [33]. Doses up to 12 g per day did not produce dose-limiting toxicity, suggesting the feasibility of further clinical development. Phase II trials now employ more specific patient populations to gather efficacy data. For example, a current Phase II study tests curcumin with gemcitabine in pancreatic cancer. Researchers are also recruiting for Phase II gastric, brain, and breast cancer trials [34]. In addition to primary treatment, curcumin is being trialed as an adjuvant for boosting chemotherapeutic efficacy and mitigating toxicity. Co-administration of curcumin with first-line therapy like FOLFOX or FOLFIRI is undergoing examination in colorectal cancer [35]. Using curcumin to reverse drug resistance is also being explored in refractory cancers. These clinical evaluations will help substantiate curcumin's role in cancer management.

Alongside treatment implications, routinely incorporating curcumin into diets or as a daily supplement may also boost cancer prevention. According to epidemiological data, higher antioxidant and anti-inflammatory compound consumption is associated with reduced cancer incidence [36]. Curcumin exhibits both antioxidant activity directly and enhances endogenous antioxidant defenses, which could mitigate DNA damage, enabling tumor promotion. It also beneficially regulates phase I and phase II detoxification enzyme expression to prevent the activation of procarcinogens [37] potentially. Curcumin administration prevented carcinogen-induced preneoplastic lesions and tumor development in rodent models' colon, liver, pancreas, stomach, and prostate, indicative of early-stage chemopreventative effects [37]. Low-dose curcumin intake favorably impacts multiple cancer-related mechanisms that may interrupt pathological processes before malignant transformations.

Here is the response in paragraph form:

While curcumin shows promising benefits for health and disease prevention, it is essential to note some potential risks and contraindications to consider when taking curcumin supplements. Curcumin may act as a blood thinner, so it could interact with medications like warfarin, which are also blood thinners. This may increase bleeding risk, so those on blood thinners should speak to their doctor before taking curcumin [38]. Curcumin may also decrease the body's absorption of certain oral drugs like contraceptives, diabetes medications, and chemotherapies. For this reason, anyone taking prescription medications should discuss curcumin use with their healthcare provider. Large doses of over 8 grams per day have caused side effects in some people, like nausea, diarrhea, and heartburn. Curcumin could also lower blood sugar, so diabetics may need monitoring [39]. Pregnant and breastfeeding women are advised to avoid curcumin supplements unless approved by a doctor due to the lack of research in these groups. In general, 400–600 mg per day is recommended for overall health, while 800–2000 mg daily divided into doses may benefit conditions like arthritis when taken under medical guidance. Doses over 2000 mg daily are not generally recommended without supervision from a health practitioner [40]. Most generally well-tolerated curcumin when taken as directed, but potential interactions and risks warrant caution, especially in those taking medications.

### 2.2 Ginger (*Zingiber officinale roscoe*)

Ginger (*Zingiber officinale*) is a plant used for culinary and medicinal purposes since ancient times. The underground rhizomes of ginger contain a complex array of nutrients, antioxidants, and bioactive compounds such as gingerols, shogaols, paradols, zingerone, and 6-gingerol, some of which have demonstrated anticancer activities [41]. Research over the past few decades indicates that ginger and its constituents can influence multiple mechanisms integral to cancer pathology, including proliferation, apoptosis, inflammation, angiogenesis, and oxidative stress. This suggests ginger may have therapeutic and preventative potential against malignancies [42]. A substantial body of *in vitro* research has demonstrated the potential anticancer properties of various crude extracts of ginger and specific compounds isolated from ginger. These studies have shown efficacy against a wide range of human cancer cell lines, including those from breast, ovarian, cervical, liver, skin, colorectal, pancreatic, bladder, prostate, and blood cancers. The mechanisms of action observed include direct cytotoxic effects and alterations in pathways related to apoptosis, cell cycle control, angiogenesis, invasion, metastasis, and inflammation [43]. The compound 6-shogaol has been noted to induce apoptosis via caspase-3 activation in breast tumor cells and ovarian cancer ascitic fluid samples [44]. Animal studies indicate that oral administration of gingerol or shogaol can inhibit the growth of tumors in the breast, skin, colon, pancreas, prostate, and lung in rodent xenograft models and lung cancer mouse models, without evident toxicity [45].

Laboratory models have revealed that ginger's anti-tumoral effects are likely due to multiple mechanisms. Both crude ginger extracts and specific compounds within ginger have demonstrated the ability to neutralize free radicals and enhance the body's antioxidant defense systems, potentially reducing DNA damage, a critical factor in the initial stages of cancer [46]. Moreover, ginger's compounds have anti-inflammatory properties, evidenced by their inhibition of enzymes like COX and reduction of specific cytokines that facilitate malignant growth. Other potential mechanisms include the disruption of tumor angiogenesis signaling, inhibition of metastasis-related enzymes, and alteration of cell cycle and apoptosis regulators, including proteins such as p53, NF-kB, and members of the Bcl-2 family, as well as markers of cancer stem cells [47]. Ginger's habitual consumption may offer protective effects against cancer development. Epidemiological evidence links higher dietary intake of antioxidant-rich plants to lower risks of various epithelial cancers. As a whole food, ginger contains vitamins, minerals, flavonoids, and phenolics that neutralize carcinogens and mitigate cellular damage [46]. Animal studies have shown that oral ginger can inhibit DNA damage and enhance detoxification enzymes in tissues exposed to carcinogens, suggesting a reduced risk of cancer-initiating mutations. Additionally, ginger has demonstrated the ability to suppress colon carcinogenesis in rat models, indicating potential chemopreventive properties [48].

People taking blood thinners or diabetes medications should use ginger cautiously due to potential additive effects. Some people may also experience heartburn, diarrhea, irritation, or nausea with high ginger intake [49]. The recommended dosage of ginger supplements depends on the intended therapeutic use, with most studies examining doses between 250 mg and 1 g per day of powdered ginger root. As a frame of reference, doses on the lower end of this range may provide antinausea effects. In contrast, anti-inflammatory effects are more commonly seen in studies using 1 g or more [50]. Ginger tea can be made by steeping 1–2 grams of fresh ginger in hot water and consumed 2–3 times daily.

### 2.3 Garlic (*Allium sativum L.*)

Garlic (*Allium sativum*), a member of the onion family, has been utilized in both culinary and medicinal contexts for millennia. Its essential bioactive compounds include organosulfur compounds (such as allicin and diallyl sulfides), flavonoids, saponins, and Maillard reaction products during cooking or processing [51]. Many *in vitro* studies have shown the anticancer potential of aged garlic extract and specific organosulfur compounds derived from garlic against various human cancer cell lines. These studies have demonstrated inhibited cell proliferation, disruption of cell cycle and proliferation signaling, induction of apoptotic cell death, and reduction in adhesion, invasion, angiogenesis, and metastasis in cancers such as lung, gastric, liver, pancreatic, breast, brain, bladder, cervical, bone, and blood cancers [52]. Specific compounds like diallyl trisulfide have been found to have dose-dependent antiproliferative effects in colorectal, lung, breast, and skin cancer cells, primarily by stimulating apoptosis [51]. Complementary in vivo studies using rodent models have confirmed these findings, with garlic administration leading to decreased tumor multiplicity, volume, weight, and progression and increased survival in gastric, colorectal, pulmonary, mammary, and hepatocellular carcinoma models [53].

The anticancer properties of garlic constituents are linked to their modulation of various cellular pathways critical to cancer proliferation and spread. Organosulfur compounds in garlic have shown potent abilities to neutralize free radicals and enhance cellular antioxidant defenses, thereby protecting against DNA damage that can lead to carcinogenesis. These compounds also affect tumor formation and angiogenesis through pathways like STAT 3 and receptor tyrosine kinase, potentially reducing tumor growth capacity [54]. Diallyl sulfides in garlic have been observed to influence cell cycle progression, stimulate immune cells, modulate arachidonic acid metabolism, and regulate histone activity, all contributing to inhibiting cancer-promoting processes [55].

Garlic can also have some risks and contraindications. Garlic can increase bleeding risk, so people with bleeding disorders or taking anticoagulant medications should use caution with garlic supplements. Some people may also experience heartburn, nausea, vomiting, diarrhea, or other gastrointestinal discomfort. The recommended daily dosage of garlic supplements for adults is 400–1200 mg of standardized allicin potential, approximately equal to the consumption of 1–2 fresh garlic cloves [56]. Consumption through dietary sources is generally recognized as safe, but garlic supplements have a higher allicin yield and, therefore, possibly a higher risk of side effects [57].

### 2.4 Basil (*Ocimum basilicum L.*)

Basil (*Ocimum basilicum*) has been an aromatic herb for centuries for culinary and medicinal purposes. Over 150 compounds have been identified in basil, including flavonoids, phenolic acids, terpenoids, and essential oils. In laboratory studies, Basil contains several bioactive compounds showing anti-inflammatory, antioxidant, and anti-cancer effects [58]. Several laboratory studies have indicated the potential anticancer effects of basil extracts and isolated compounds from basil. Extracts from basil leaves have been found to inhibit the growth and spread of specific cancer cells, including oral, skin, breast, cervical, prostate, pancreatic, and colon cancers. Specific compounds identified in basil, such as eugenol, linalool, ursolic acid, β-sitosterol, and β-element, have exhibited anti-proliferative effects on cancer cells and prevented tumor formation and metastasis in animal models [59].

Different extracts and compounds from basil have been shown to influence multiple mechanisms involved in cancer progression, including cell cycle arrest, apoptosis induction, inhibition of cell proliferation, migration, and metastasis, restricting blood vessel growth, and reducing inflammation [60]. Alcohol extracts from basil leaves inhibited cell growth and induced cell death in the mouth, cervical, prostate, and skin cancer cell lines [59]. Ursolic acid from basil inhibited cell proliferation in skin cancer, slowed tumor growth in melanoma, and prevented lung metastasis of skin cancer in mice. These effects appear to result from basil compounds modulating gene expression and signaling pathways associated with cancer [61]. Certain compounds in basil, such as rosmarinic acid, linalool, and eugenol, have demonstrated antioxidant activities that may help protect cells against DNA damage that can promote tumor growth. Inflammation plays a crucial role in cancer development, and basil extracts and isolates have reduced inflammatory signaling molecules like NF-kB in cancer cell lines [62].

Basil can also have some risks and contraindications. Due to its effects on blood clotting, people with bleeding disorders or taking anticoagulant medications should exercise caution when using basil supplements [63]. Basil essential oils or supplements could interact with other medicines like sedatives or hypoglycemics. Some people may experience side effects like low blood pressure or liver damage [64]. There is no established recommended daily dosage for basil. Basil is generally recognized as safe when used in average culinary amounts. However, concentrated supplements or essential oils extracted from basil provide higher levels of active compounds and have a higher risk of side effects and interactions [65]. Using fresh or dried basil leaves from dietary sources is typically not a safety concern for healthy adults, but the risks increase when taking concentrated basil supplements.

### 2.5 Caraway (*Carum carvi L.*)

Caraway (*Carum carvi*) is a biennial plant from the Apiaceae family. The dried "seeds" of caraway are commonly used as a culinary spice and also have a long history of use in traditional medicine. Over 100 compounds have been isolated from caraway seeds, including essential oils like carvone, limonene, carveol, flavonoids, fatty acids, polysaccharides, proteins, and more [66]. Research indicates that caraway has cytotoxic effects on stomach, breast, colon, cervical, liver, and lung cancer cells. Specific compounds in caraway oil, including carveol, carvone, and limonene, have shown growth inhibitory effects in human colon cancer cell lines, and colon cancer growth was reduced in mice given oral carvone from caraway [67]. Caraway was shown to induce cell cycle arrest and apoptotic cell death in human colon cancer cells, indicating modulation of pathways that regulate tumor cell proliferation and survival. Flavonoids and carvone reduced inflammation by affecting inflammatory signaling molecules COX2, TNF-α, and NF-κB in colon cancer cells and models [68].

The use of caraway may pose risks, mainly when consumed in excessive amounts or by individuals with certain health conditions. Adverse effects of caraway are generally rare, but it can interact with certain medications and health conditions. For instance, caraway oil's emmenagogue properties may pose risks during pregnancy. Additionally, individuals with a history of gallstones should exercise caution, as certain compounds in caraway can stimulate gallbladder contractions [67]. Regarding dosage, the recommended daily intake of caraway varies based on the form and purpose of use. In culinary uses, caraway seeds are generally considered safe when used in typical food amounts. Standardized extracts are often used for medicinal purposes, and dosages may vary. For example, in the form of an essential oil, a typical dose might range from 0.1 to 0.3 mL [69]. However, due to different product variations in strength and concentration, it's crucial to follow specific manufacturer recommendations or consult with a healthcare professional for appropriate dosing tailored to individual needs and health conditions.

## 2.6 Cinnamon (*Cinnamomum cassia L.*)

Cinnamon is a commonly used spice derived from the inner bark of trees from the Cinnamomum genus. The most widely used species are *Cinnamomum verum* J. Presl, Ceylon cinnamon or true cinnamon, and *Cinnamomum cassia* (L.) J. Presl is often referred to as Cassia or Chinese cinnamon. Over 100 bioactive phytochemicals have been found in cinnamon, including cinnamaldehyde, eugenol, linalool, camphor, and caffeic acid (1). Research over the past decade indicates that cinnamon and its compounds exhibit antioxidant, anti-inflammatory, antiproliferative, and immunomodulatory influences, suggesting the potential for cancer applications [70]. *In vitro* studies report the anticancer effects of cinnamon extracts and isolates such as cinnamaldehyde against various human cancer cell lines. This includes inhibitory impact on the growth, proliferation, migration, invasion, and metastasis signaling pathways of cervical, ovarian, prostate, liver, stomach, colorectal, breast, lung, and brain cancer cells [71].

Cinnamaldehyde altered microtubule functioning to block cell division and induce apoptotic cell death in human leukemia and lymphoma cells. It also exhibited immunomodulatory effects by enhancing natural killer cell activity against myeloma cells [72]. Animal studies further support cinnamon's antitumor influences. Orally administered cinnamon extract reduced tumor cell proliferation markers, increased cancer cell death through disrupted angiogenesis and blood flow, and decreased growth of melanoma and lymphoma tumors in mice [73]. Lab experiments indicate that cinnamon alters signaling proteins that regulate cell proliferation, survival, angiogenesis, adhesion, migration, and apoptosis in different cancer cell types [74]. Particular compounds like cinnamaldehyde modulate tumor necrosis factor, vascular endothelial growth factor, matrix metalloproteinases, and interleukin secretion from cancer cells. The observed antioxidant capacities of cinnamon extracts may also contribute by protecting against oxidative damage and suppressing chronic inflammation involved in malignant transformations [75].

Cinnamon, especially Cassia cinnamon, also contains coumarins, which can have risks and contraindications at high doses. High coumarin intake may negatively affect liver health, and caution people taking anticoagulant medications due to coumarin's blood thinning effects [76]. Generally, Ceylon cinnamon has a lower coumarin content than Cassia cinnamon. The recommended daily dosage of cinnamon supplements for adults is approximately 120–360 mg of a cinnamon extract standardized to 5–6% flavonoids. This equals approximately 1–3 g of cinnamon powder [77]. Intake above this level increases the risk of liver toxicity and negative interactions. High levels of coumarin intake have been associated with liver toxicity and damage in sensitive individuals [78]. Therefore, caution is advised, especially for those with pre-existing liver conditions or using medications that affect liver function.

### 2.7 Clove (*Syzygium aromaticum L.*)

Cloves are the aromatic flower buds of the *Eugenia caryophyllata* tree. Clove (*Syzygium aromaticum* (L.) Merr. & L.M. Perry) is an aromatic flower bud used as a spice and has a long history of traditional medicinal uses. They have been used as a culinary spice and traditional medicine for centuries. At least 30 compounds have been isolated from cloves, including flavonoids, terpenoids, and the phenolic compound eugenol, which makes up 70–85% of clove essential oil [75]. Several preclinical studies demonstrate the anticancer effects of clove extracts or eugenol against various human cancer cell lines. For instance, in culture, clove extracts exhibited cytotoxic actions and triggered cell death pathways in skin, breast, cervical, prostate, colon, blood, and bone cancer cells [79]. Eugenol also inhibited growth, proliferation, and spread while promoting apoptosis in models of melanoma, leukemia, gastric, breast, and mast cell cancer cells. Beneficial effects have also been shown in rodent models. Oral eugenol administration suppressed colon cancer in rats and reduced solid tumor size while extending lifespan in mice studies [80, 81].

One study provided 200 mg of an ethanolic clove extract containing 40% eugenol to 41 patients with oral submucous fibrosis, a premalignant lesion [82]. After up to 3 months of supplementation, patients taking clove extract had reduced burning sensations and increased mouth opening, and 71% demonstrated partial to complete clinical response, indicating anti-inflammatory benefits [83]. The abundance of eugenol in cloves has been found to influence gene expression of vascular endothelial growth factors, matrix metalloproteinases, and pro-inflammatory cytokines like NF-kB and COX2 in various cancer cell models [84]. Eugenol arrested cell division, prompted cancer cell death through p53 activation, caspase, and PARP cleavage, and disrupted mitochondrial functioning in human melanoma cells. The antioxidant capacity exhibited by eugenol and clove extracts may provide additional cancer-preventative influences by reducing DNA damage from oxidative stress [80]. Cloves contain polyphenols, flavonoids, and vitamin C that mitigate cellular damage from oxidation and resulting inflammation tied to malignant growths [85].

Compounds found in clove, such as eugenol, have some risks and contraindications, especially when taking clove essential oil. Clove may interfere with blood clotting and should be avoided by people with bleeding disorders or taking anticoagulant medications. Clove may also cause liver damage at high doses. Other side effects can include digestive upset [86]. The recommended daily dose of clove varies depending on the form of consumption and purpose of use. In a culinary context, clove is generally safe when used in average food amounts. As a supplement or medicinal remedy, the dosage needs careful consideration. A standard therapeutic dose for clove oil might range from 0.15 to 0.3 mL [87]. However, due to its potent nature, it is often diluted or used in a compound formulation. For whole or ground cloves, dosages are less well-defined but are considerably lower than those of the essential oil. Given the potency and potential risks of clove, it is advisable to start with minimal amounts and consult a healthcare professional for guidance on appropriate dosing, especially for therapeutic use. This is particularly important for individuals with pre-existing health conditions, those who are pregnant or breastfeeding, and young children.

## 2.8 Coriander (*Coriandrum sativum L.*)

Coriander (*Coriandrum sativum*) is an annual herb belonging to the Apiaceae family. The leaves and seeds of coriander have a long history of use as a food flavoring agent and in traditional medical systems. Over 100 phytochemicals have been identified in coriander, including essential oils like linalool, flavonoids, phenolic acids, and terpenoids [88]. Compounds like linalool and chlorogenic acid inhibited cell growth and spread while prompting cancer cell death through several signaling pathways in models of colon, neuroblastoma, and hepatocellular carcinoma [89]. Some in vivo animal experiments further support coriander's anti-tumor potential. Oral administration of coriander leaf extract before tumor induction decreased tumor incidence and burden in skin cancer and hepatocarcinogenesis rodent models [90].

Multiple mechanisms likely contribute to coriander's reported anticancer activities in laboratory models. Experiments indicate the modulation of essential signaling proteins connected to cancer cell proliferation, survival, inflammation, progression, and apoptosis after treatment with coriander extracts in various cancer cell lines [91]. Essential bioactive chemicals found abundantly in coriander seeds and leaves, including linalool, quercetin, and chlorogenic acid, have antioxidant and anti-inflammatory properties that may underlie the observed antiproliferative influences [92]. Coriander leaf extracts boosted the activity of carcinogen detoxification enzymes like glutathione-S-transferase and increased antioxidant status markers in liver cancer animal models, which could additionally block tumor promotion stages [93].

There are potential risks and contraindications associated with coriander use. Coriander can cause allergic reactions in some individuals, which may manifest as skin rashes, respiratory symptoms, or gastrointestinal upset. People with allergies to other members of the Apiaceae family, such as fennel or anise, may be more susceptible to a coriander allergy [94]. Additionally, coriander seeds' hypoglycemic effect might pose a risk for individuals with diabetes, especially when consumed in conjunction with diabetes medication, as it can potentially lead to hypoglycemia. The recommended daily dose of coriander varies depending on its form (fresh herb, dried seeds, or oil) and the intended use. In culinary applications, coriander is generally safe in standard food preparation amounts. For medicinal purposes, there is no universally established dosage, but typical recommendations might include:

- Fresh coriander leaves 1.5 to 5 grams per day.
- Dried coriander seeds: 0.5 to 1 gram per day.
- Coriander seed oil: Because of its concentration, this form should be used sparingly, and exact dosages should follow product-specific guidelines or healthcare professional advice.

Pregnant or breastfeeding women should exercise caution and seek medical advice before consuming coriander for medicinal purposes [95].

## 2.9 Dill (*Anethum graveolens L.*)

*Anethum graveolens*, commonly known as dill, is a perennial herb utilized historically for culinary and medicinal purposes. Its primary bioactive constituents include essential oils such as carvone, limonene, and apiol and flavonoids like kaempferol and vicenin [96]. *In vitro*, research reveals the anticancer capabilities of dill extracts and their specific bioactive components against various human cancer cell lines. Treatments with these extracts have shown cytotoxic effects, inducing apoptosis in breast, cervical, oral, liver, and colon cancer cells. Specifically, the apiol compound from dill impeded colon cancer cells' growth and metastatic potential [97]. Additionally, the flavonoid vicenin in dill demonstrated anti-proliferative effects by disrupting microtubule formation in lung cancer cells [98]. The anticancer effects of dill extracts and compounds may be attributed to multiple mechanisms. Research suggests that dill treatment modulates proteins in cancer cell cycle regulation, survival, proliferation, and apoptosis. Additionally, dill extracts and their component apiol exhibit antioxidant and anti-inflammatory activities, potentially preventing conditions conducive to tumor growth [99]. The antimutagenic properties of aqueous dill extracts may also offer preventive benefits by protecting cells from DNA alterations that can initiate cancer [100].

Due to potential effects on hormone and blood sugar levels, people with diabetes, endocrine disorders, or related medications should exercise caution with dill supplements. Dill may also cause sensitivity reactions in some people [101]. Additionally, due to its antispasmodic properties, dill might interact with medications that affect the central nervous system. Individuals with low blood sugar levels or those on diabetes medication should also exercise caution, as dill can have hypoglycemic effects [101]. The recommended daily dose of dill varies based on the form and purpose of use:

- For culinary purposes, using fresh or dried dill leaves in cooking is generally safe in typical amounts.

- In a medicinal context, there are no well-established dosage guidelines for dill. However, when used as an herbal tea, a standard preparation might include 1 to 2 teaspoons of dried dill per cup of hot water, steeped for 5 to 10 minutes.

- For dill oil, due to its concentrated nature, only a tiny amount (a few drops) is typically used, and it should be diluted if applied topically.

### 2.10 Cumin (*Cuminum cyminum L.*)

*Cuminum cyminum*, or cumin, is a flowering plant in the Apiaceae family, historically used in culinary and traditional medicinal contexts. Essential phytochemicals in cumin seeds include phenolic compounds, flavonoids, and oils such as cuminaldehyde, cymene, and terpenoids [102]. Studies have demonstrated that cumin and its constituent thymoquinone can inhibit proliferation and induce apoptosis in various cancer cell types, including cervical, breast, oral, stomach, colon, blood, pancreatic, and liver cancers [103]. Furthermore, thymoquinone has been shown to enhance the efficacy of natural killer cells against pancreatic and blood cancer cells. Preliminary rodent studies corroborate these findings, with oral administration of thymoquinone inhibiting tumor growth in colon and pancreatic cancer models without toxicity [104]. Preclinical studies suggest that cumin treatments inhibit cell proliferation and induce apoptosis by modulating cell division and survival signaling pathways [105]. These compounds exhibit anti-inflammatory properties by influencing molecules like NFκB and TNF-alpha, which are implicated in cancer progression. Their antioxidant potential also plays a role in preventing tumor formation by safeguarding cells from DNA damage leading to mutations.

Identifying the essential chemicals and processes responsible for these anticancer activities is crucial for developing cumin-based therapeutics [106].

Despite these positive attributes, cumin has potential risks and contraindications. Overconsumption of cumin, especially in its concentrated oil form, can lead to digestive upset, such as heartburn or stomach pain. It may also lower blood sugar levels, a concern for individuals with diabetes or those on medications that affect blood sugar [107]. No established recommended daily intake levels are specifically for culinary spices like cumin. Regular usage of cumin seeds or powder to season foods is generally considered safe by the U.S. Food and Drug Administration. As a spice, up to 1–3 grams of cumin powder per day is commonly used.

### 2.11 Rosemary (*Rosmarinus officinalis L.*)

Rosemary (*Rosmarinus officinalis*), a perennial herb belonging to the Lamiaceae family, has been widely recognized in traditional medicine and culinary practices. Rosemary is rich in bioactive compounds, including phenolic acids, flavonoids, and essential oils. Among these, carnosic acid, rosmarinic acid, and ursolic acid are particularly notable for their biological activities [108]. Carnosic acid has been identified as a potent antioxidant, while rosmarinic acid exhibits anti-inflammatory and antioxidant properties. Ursolic acid, a pentacyclic triterpenoid, has been shown to possess antitumor activities [109, 110]. Oxidative stress, resulting from an imbalance between free radicals and antioxidants in the body, is a known contributor to cancer development. Antioxidants can neutralize free radicals, thereby reducing oxidative stress and potentially preventing the initiation and progression of cancer. Rosemary's rich composition of antioxidants, particularly carnosic acid and rosmarinic acid, has been shown to protect against oxidative DNA damage, a precursor to cancer [68].

Rosemary's bioactive compounds, especially ursolic acid, have been extensively studied for their antitumor properties. These compounds exhibit various anticancer effects, including the induction of apoptosis (programmed cell death) in cancer cells, cell proliferation inhibition, and metastasis prevention. For instance, ursolic acid has been shown to induce apoptosis in various cancer cell lines, including breast, prostate, and colon cancers [111]. Cancer cells often exhibit altered metabolism, which supports their rapid growth and survival. Compounds in rosemary have been found to interfere with these metabolic pathways, thereby exerting antiproliferative effects on cancer cells [111]. Rosemary extracts have been studied for their potential to enhance the efficacy of conventional chemotherapy drugs. This synergistic effect is particularly significant as it may lead to more effective cancer treatments with lower doses of chemotherapeutic agents, thereby reducing their side effects [111]. A study by Gonzalez-Vallinas et al. [110] demonstrated that rosemary extracts enhanced the efficacy of 5-fluorouracil, a standard chemotherapy drug, in colon cancer cells.

Rosemary supplements may interact with anticoagulant or antihypertensive medications. High doses of some rosemary essential oil components like camphor can cause nausea, vomiting, and muscle spasms [112]. High doses of rosemary oil can cause gastrointestinal irritation and kidney damage and may lead to seizures if consumed in excessive amounts. Due to its potent effects, rosemary oil should be avoided by pregnant women and those with epilepsy or high blood pressure. Additionally, rosemary can affect iron absorption and should be used cautiously by individuals with iron deficiency. In culinary uses, rosemary is generally safe in standard seasoning amounts [113]. For medicinal purposes, a typical dose might include 4–6 grams of dried rosemary leaves.

### 2.12 Saffron crocus (*Crocus sativus L.*)

Saffron (*Crocus sativus*), a spice derived from the flower of the same name, has long been cherished for its culinary uses and medicinal properties. Saffron is known for its rich composition of bioactive compounds, including crocin, crocetin, safranal, and picrocrocin. These constituents are responsible for saffron's characteristic color, flavor, and medicinal properties. Crocin and crocetin, carotenoids that give saffron its distinct red color, have been extensively studied for their antioxidant properties. Safranal, which contributes to saffron's aroma, exhibits antioxidant and antitumor activities [114]. The antioxidants in saffron, particularly crocin, and crocetin, have demonstrated protective effects against oxidative DNA damage, a precursor to cancer [115]. Saffron's anti-inflammatory properties, primarily attributed to crocin and safranal, play a crucial role in its anticancer potential. By inhibiting vital inflammatory pathways and molecules like NF-kB and TNF-alpha, saffron can help prevent the inflammatory conditions that favor cancer development [116]. Saffron extracts can affect cancer cell metabolism, often altered to support rapid growth and survival of cancer cells. Compounds in saffron have been found to interfere with these metabolic pathways, exerting antiproliferative effects on cancer cells. Saffron extract could inhibit the critical enzyme activity in cancer cell metabolism, suppressing cancer cell growth [117]. Studies have shown that saffron extract can enhance the effectiveness of chemotherapeutic agents like cisplatin and 5-fluorouracil in various cancer models [118].

Saffron supplements may interact with anticoagulant, antihypertensive, or antidepressant medications. High doses can cause nausea, dizziness, and vomiting [119]. Pregnant women should avoid saffron supplementation due to uterine stimulation effects. In culinary uses, saffron is typically safe when used in small quantities for flavoring and coloring, generally a few strands or less than 0.5 grams. For medicinal purposes, studies often use dosages ranging from 15 to 30 milligrams per day. However, these dosages are specific to certain conditions and should not be generalized. Due to its potential influence on mood and the nervous system, individuals with mood disorders or those taking antidepressants should use saffron cautiously. Saffron may also lower blood pressure and should be used with care by individuals with hypotension or those on blood pressure medications.

### 2.13 Thyme (*Thymus vulgaris L.*)

Thyme (*Thymus vulgaris*) is an aromatic herb used for centuries as a culinary seasoning and traditional medicine. Over 300 compounds have been identified in thyme, including flavonoids, phenolic acids, tannins, terpenoids, and essential oils such as thymol, carvacrol, linalool, and geraniol (1). Emerging research reveals that thyme extracts and compounds demonstrate antioxidant, anti-inflammatory, antimutagenic, and anticancer effects, highlighting potential applications against malignancies [120]. Treatment with thyme preparations has displayed cytotoxic, apoptotic, anti-invasive, and anti-metastatic actions in models of oral, liver, breast, lung, skin, bladder, brain, bone, and blood cancers [121]. The phenolic monoterpene thymol inhibited cell proliferation and promoted cell death pathways in colorectal, liver, breast, cervical, and leukemia cancer cells [122]. Thyme honey mouthwash decreased the severity of mucositis and associated pain relative to saline solution, though direct antitumor impacts were not assessed [123].

Multiple mechanisms likely contribute to the observed anticancer effects of thyme extracts. Thyme polyphenols and essential oils have been found to interact with cell signaling, controlling inflammation, proliferation, angiogenesis, invasion, and survival in diverse human cancer cell culture models. For instance, the compound thymol targets molecules like NF-kB, PARP, and Bcl-2 proteins to prompt apoptosis and restrict growth signaling in various cancer types [124]. Thyme extracts also activated cellular antioxidant defense mechanisms in experiments, which could prevent initial carcinogenic damage.

Thyme may also have risks and uncertainties at higher supplementary doses. Thyme compounds can increase bleeding risks, so those with bleeding disorders or taking anti-coagulant medications should avoid concentrated thyme supplements [125]. Other side effects may include digestive upset, headache, and dizziness. The recommended daily dose of thyme varies depending on its form and intended use. In culinary applications, thyme is generally safe when used in typical cooking amounts. For medicinal purposes, a standard dosage is 1–2 teaspoons of dried or 2–3 teaspoons of fresh thyme, which can be steeped in hot water to make tea. This can be consumed 1–3 times a day [120].

### 2.14 Oregano (*Origanum vulgare L.*)

Oregano is an aromatic herb used for culinary and medicinal purposes. The most abundant phytochemicals identified in oregano include polyphenolic compounds like rosmarinic acid, flavonoids, and volatile oils such as thymol, carvacrol, limonene, linalyl acetate, linalool, and geranyl acetate [126]. Experiments indicate inhibited proliferation, cell cycle, and proliferation signaling disruption, apoptotic cell death induction, reduced adhesion, invasion, angiogenesis, and inflammation in colon, lung, breast, liver, bladder, cervical, blood, and thyroid cancer models. Carvacrol and rosmarinic acid compounds exhibited antiproliferative, cytotoxic, and anti-metastatic actions across ovarian, oral, prostate, and skin cancer cells [127]. The well-reported anticancer effects of oregano extracts in preclinical models have been attributed to the modulation of numerous cellular pathways integral to malignant proliferation. Oregano polyphenols and essential oils demonstrated potent free radical scavenging and cyclooxygenase enzyme inhibition, protecting against oxidative damage and inflammation and enabling carcinogenesis [128]. Additionally, oregano compounds affected cell cycle arrest, apoptosis induction, tumor suppressor genes, and cell proliferation/survival signaling in various cancer cell culture experiments [129].

Oregano due to its influence on blood clotting, those with bleeding disorders or taking anticoagulant medications should avoid concentrated oregano supplements. Other side effects can include gastrointestinal upset and irritation when applied to the skin. Oregano, particularly in its essential oil form, can irritate the skin and mucous membranes when used in high concentrations [56]. It may also interact with certain medications, such as anticoagulants, due to its potential to inhibit blood clotting. Individuals with bleeding disorders or those preparing for surgery should exercise caution. Additionally, the high phenol content in oregano may cause digestive discomfort if consumed in large amounts [130]. The recommended daily dose of oregano depends on its form and intended use: In culinary applications, oregano is generally safe when used in typical seasoning amounts. There is no standard dosage for medicinal purposes, but a general guideline is 1–2 teaspoons of dried oregano, which can be used to make tea. This can be consumed up to three times daily.

### 2.15 Sage (*Salvia officinalis L.*)

Sage (*Salvia officinalis*) is an aromatic herb used culinary and medicinally for centuries. The main bioactive components identified in sage include phenolic acids like rosmarinic acid, phenolic diterpenes such as carnosol and carnosic acid, and volatile oils such as 1,8-cineole, camphor, and α/β-thujone [131]. Recent *in vitro* studies demonstrate promising anticancer effects of various crude sage extracts and specific isolates against diverse human cancer cell lines. Experiments indicate inhibited proliferation, cell cycle, and proliferation signaling disruption, apoptotic cell death induction, reduced adhesion, invasion, angiogenesis, and inflammation in models of colon, melanoma, lymphoma, leukemia, liver, lung, prostate, cervical, and breast cancers. Both sage extracts, carnosol, and carnosic acid exhibited cytotoxic and antiproliferative actions across skin, head/neck, colon, leukemia, and brain cancer cells [132]. Oral carnosol administration restricted tumor weight and volume in prostate cancer xenografts and prevented DNA damage from carcinogens [133].

The anticancer effects of sage extracts in preclinical models have been attributed to the modulation of numerous cellular pathways integral to malignant proliferation. Sage polyphenols demonstrated potent free radical scavenging and anti-inflammatory activities, which protect against oxidative damage, enabling carcinogenesis [134]. Additionally, sage compounds exhibited regulatory effects on signaling proteins influencing cell cycle progression, apoptosis, nitrogen signaling species generation, and tumor invasion markers in various cancer cell culture experiments [135].

Sage may also have risks and contraindications, especially concerning components like α-thujone and β-thujone. The thujones may interact with medications due to influences on GABA and serotonin. High doses of thujones can potentially cause seizures, vomiting, and kidney damage [136]. Due to its potential estrogenic effects, sage is also advised to be used cautiously in individuals with hormone-sensitive conditions, such as certain types of breast cancer. Pregnant and breastfeeding women are typically advised to avoid high doses of sage due to the lack of safety data. Using sage in regular cooking is generally considered safe for culinary purposes. In a medicinal context, the appropriate dose can vary. As a general guideline, 4–6 grams of dried sage leaf or 1–2 teaspoons of sage leaf infusion (tea) taken once or twice daily can be used [131].

### 2.16 Fennel (*Foeniculum vulgare mill*)

Fennel (*Foeniculum vulgare*) is a flowering plant used for culinary and medicinal applications since ancient times. The primary phytochemicals identified in fennel include the volatile oil constituents trans-anethole, estragole, fenchone, flavonoids, phenolic acids, and fatty acids [137]. Specific fennel compounds like trans-anethole and limonene also exhibited dose-dependent antiproliferative actions across breast, liver, stomach, and skin cancer cells [138]. Studies indicate that fennel contains compounds that modulate signaling proteins connected to regulating cell cycle progression, proliferation, angiogenesis, and apoptosis in certain cancer cell types. Extracts and trans-anethole constituents also display antioxidant and anti-inflammatory properties, which can help prevent conditions that enable tumor promotion and growth [139]. The antimutagenic and free radical scavenging capacity exhibited by fennel components could provide additional protective influences by preventing DNA changes from triggering aberrant cell changes.

Fennel may also have some uncertainties and contraindications related to its effects and safety, especially in supplemental forms. One of the primary concerns with fennel, especially in its essential oil form, is the presence of estragole and anethole, which, in high doses, could be toxic and have been linked to potential carcinogenic effects. Due to its phytoestrogen content, fennel may interact with hormone-sensitive conditions, such as certain types of breast cancer or endometriosis [140].

Pregnant women are often advised to avoid excessive use of fennel due to a lack of conclusive safety data. Additionally, individuals with allergies to carrots or celery, which are in the same family as fennel, might also be allergic to fennel. Fennel may also interact with medications changed by the liver. Side effects can include sun sensitivity, gastrointestinal issues, or allergic reactions [141]. The recommended daily dose of fennel varies depending on the form and intended use. In culinary uses, fennel seeds and bulbs are generally safe when consumed in typical food amounts. Medicinally, the typical dosage for fennel seeds is 5 to 7 grams per day, which can be taken as an infusion (tea), tincture, or in capsule form.

### 2.17 Black pepper *(Piper nigrum L.)*

Black pepper (*Piper nigrum*) is a flowering vine that produces peppercorns, which have been used as a pungent spice and herbal medicine for centuries. The main bioactive components identified in black pepper include volatile oils like pinene, sabinene, limonene, and piperine alkaloids [142]. Recent *in vitro* studies reveal promising anticancer effects of various crude black pepper extracts and particular isolates against different human cancer cell lines. For instance, treatment displayed cytotoxicity, induced apoptotic cell death, caused cell cycle arrest, impeded invasion, and reduced metastasis signaling proteins in models of colorectal, breast, prostate, liver, blood, brain, bone, and skin cancers [143]. Key compounds piperine, β-caryophyllene, and linalool isolated from peppercorns also dose-dependently inhibited proliferation across oral, lung, cervical, and ovarian cancer cells. Oral piperine administration significantly suppressed mouse lymphoma tumor growth [144]. Piperine has demonstrated the capacity to influence numerous pathways integral to malignant proliferation, including apoptosis, angiogenesis, metastasis, cell cycle regulation, detoxification, and inflammation [145].

Excessive consumption of black pepper can lead to gastrointestinal irritation, including dyspepsia and gastroesophageal reflux. Individuals with ulcers or digestive tract inflammation should use black pepper cautiously. Piperine may inhibit medication breakdown, leading to potential overdoses of certain drugs, leading to adverse reactions. Monitoring for adverse reactions is crucial, and medical advice should be sought in case of any adverse symptoms or concerns [146]. There are no established standardized recommended daily intake levels specifically for black pepper. Using black peppercorn or powder to season food in small culinary amounts is generally safe.

### 2.18 Cayenne pepper (*Capsicum annuum var. annuum L.*)

Cayenne pepper is a hot chili from the Capsicum family that has been used as a spice and herbal medicine for centuries. The main bioactive components identified in cayenne include capsaicin, capsanthin, luteolin, and piperine [147]. In particular, the capsaicin, capsanthin, and luteolin compounds prompted apoptosis and restricted growth signaling proteins across leukemia, melanoma, gastric, bladder, and glioma cancer cells. Though limited in number, some rodent experiments support the idea that oral capsaicin administration suppressed lung tumor incidence and growth in mice without apparent toxicity [148]. Experiments indicate compounds like capsaicin can alter nuclear factor-kB (NF-kB) and other inflammatory signaling pathways, enabling runaway malignant cell proliferation. Additional proposed pathways include selective cancer cell toxicity through increased oxidative stress, cell cycle, and apoptosis regulator modulation, disruption of microtubule assembly dynamics, and inhibition of angiogenesis, invasion, and metastasis enzymes [149]. Oral administration of capsaicin and pepper extracts also prevented carcinogen-induced pulmonary and colon tumors in rodent models, indicative of protective influences during pathogenesis [150].

Ingesting cayenne can result in stomach pain, cramps, or diarrhea in sensitive individuals. When handling the peppers, skin irritation, and burns are also risks. It can also cause skin irritation when applied topically. Due to its potent nature, individuals with sensitive digestive systems, irritable bowel syndrome, or hemorrhoids should use cayenne pepper cautiously. Additionally, capsaicin can interact with certain medications, such as blood thinners and stomach acid reducers [151]. Currently, no standardized recommended daily intake levels are explicitly established for cayenne pepper. Using small culinary amounts of cayenne powder or flakes to add heat and flavor to dishes is generally considered safe for most healthy adults. For medicinal purposes, dosage recommendations can vary. A typical dosage for cayenne pepper supplements is between 30 and 120 mg, taken up to three times daily. Topical applications of capsaicin-containing creams usually contain 0.025–0.1% capsaicin and are applied several times daily [152].

### 2.19 Green tea (*Camellia sinensis L.*)

Green tea is produced from the leaves of the *Camellia sinensis* plant and has been consumed for centuries for its health benefits. The major bioactive components in green tea include epigallocatechin-3-gallate (EGCG), epicatechin, epicatechin-3-gallate, and epigallocatechin [153]. *In vitro* studies demonstrate promising anticancer effects of green tea extracts and catechins against diverse human cancer cell lines. Treatment has displayed inhibited proliferation, prompted apoptotic cell death, and reduced angiogenesis, invasion, and metastasis behaviors in models of skin, lung, colon, pancreas, breast, bladder, prostate, and blood cancers. The green tea polyphenol EGCG also reversed chemoresistance and impacted signaling pathways controlling tumor survival, proliferation, and apoptosis in leukemia, multiple myeloma, and glioblastoma cells. In vivo, rodent experiments lend further support as oral administration of green tea catechins restricted tumor growth and progression while increasing survival across the intestinal, prostate, breast, and colorectal cancer models [154]. Green tea catechins and extracts have been found to interact with multiple intracellular proteins and enzymes involved in the regulation of cancer cell proliferation, differentiation, apoptosis, angiogenesis, and metastasis. The abundant catechin EGCG potently deactivated proteins like VEGF, MMP, and telomerase are implicated in tumor survival and growth signaling across gastric, breast, and colorectal cancer models [155]. Additionally, green tea compounds enhanced the functioning of crucial tumor suppressors like p53 and preventative Phase II detoxification enzyme pathways.

High intake of green tea, especially in the form of supplements or concentrated extracts, can lead to liver toxicity and other adverse effects due to its high catechin content. Individuals with liver disorders should exercise caution or avoid high concentrations of green tea. Green tea also contains caffeine, which can cause insomnia, nervousness, irritability, and increased heart rate and blood pressure in sensitive individuals or when consumed in large amounts [156]. Additionally, the tannins in green tea can inhibit iron absorption, which concerns individuals at risk of iron deficiency. Green tea may also interact with certain medications, including blood thinners, stimulants, and certain chemotherapy drugs, due to its caffeine content and ability to affect drug metabolism. Pregnant and breastfeeding women are advised to limit their caffeine intake, including that from green tea [157]. Green tea beverages are safe for most when consumed reasonably, but extracts have more potential interactions and toxicity risks with inappropriate doses or sensitive individuals.

## 2.20 Milk thistle (*Silybum marianum L.*)

Milk thistle (*Silybum marianum*) is a flowering plant used for centuries as a herbal medicine. The major bioactive compounds identified in milk thistle seeds are the flavonolignans, including silymarin, silibinin, isosilibinin, silydianin, and silychristin [158]. Emerging research over the past few decades indicates that milk thistle extracts and isolates exhibit antioxidant, anti-inflammatory, hepatoprotective, and anticancer effects in preliminary cellular and animal studies [159]. Experiments indicate inhibited proliferation, disrupted cell signaling pathways, and reduced viability and apoptotic cell death induction in prostate, cervical, bladder, and breast cancer models [160]. Silibin also suppressed tumor growth, prevented metastasis, and increased chemotherapeutic efficacy in rodent models of lung, skin, colorectal, and bladder cancer [161]. A case-control pilot study associated higher milk thistle exposure with reduced risk of invasive bladder cancer [162]. Several phase I toxicity trials using intravenous silibinin found doses up to 20 g per week well-tolerated and associated with stabilized disease for 3–12 months in advanced hepatocellular, prostate, and lung cancer patients [163].

Milk thistle oil provides high concentrations of active silymarin compounds through seed extraction and concentration. Milk thistle oil may be used in supplement form and is often incorporated into dietary regimens for its potential benefits in supporting liver health, reducing inflammation, and providing antioxidant support [164]. This oil can be taken orally in capsules, similar to other extracts. However, research on these various milk thistle preparations' efficacy, safety, and pharmacokinetics requires further robust investigation in humans.

The most common side effects include gastrointestinal upset, diarrhea, and bloating. Less common effects can include rash, headaches, and muscle pain. High doses may have a blood sugar-lowering effect, so those with diabetes or on anti-diabetic medications should use caution and monitor closely when taking milk thistle [165]. There is also a lack of safety research on milk thistle in pregnant or breastfeeding women, so supplementation should be avoided. There currently needs to be standardized dosage recommendations for milk thistle due to insufficient evidence regarding efficacy and long-term safety in humans. Most studied doses range between 140 and 800 milligrams of silymarin daily, but clinical outcomes vary widely in the existing literature. Standardized milk thistle extract, typically 70–80% silymarin, is commonly used in doses ranging from 140 to 210 mg taken 2 to 3 times daily [166]. For milk thistle tea, 1 to 2 teaspoons of crushed seeds steeped in hot water for 5 to 10 minutes is a standard preparation, which can be consumed up to three times daily.

## 2.21 Astragalus (*Astragalus membranaceus (fisch.) bunge*)

*Astragalus membranaceus* is a flowering plant within the legume family that has been used in traditional Chinese medicine for centuries. The major bioactive constituents found in the astragalus root include polysaccharides, saponins, and flavonoids. Emerging research over the past few decades indicates that astragalus extracts and isolates exhibit immunomodulating, anti-inflammatory, antioxidant, and possible anticancer effects based on preliminary cellular and animal studies [167]. Experiments indicate inhibited proliferation, prompted apoptotic cell death, cell cycle disruption, and reduced invasiveness in lung, breast, gastric, pancreatic, liver, and cervical cancer models. Astragalus saponins also mitigated multidrug resistance mechanisms in chemotherapy-resistant colon cancer lines [168]. Studies demonstrated that oral astragalus extracts enhanced the efficacy of platinum-based chemotherapy against xenograft models of hepatocellular and gastric cancer [169].

A few trials investigated adjuvant use of astragalus injections alongside conventional therapy, primarily in reducing chemotherapy toxicity [170]. Case series data also associates astragalus usage with symptom relief, improved immune responses, and possibly increased survival times as an adjunct in advanced cancer patients. However, placebo-controlled confirmation still needs to be improved [171]. Experiments indicate astragalus saponins and polysaccharide fractions modulate cytokine secretion, T cell activity, and other aspects of immune functioning vital to early cancer defense [172]. Extracts also exhibited direct inhibitory effects on growth, invasion, angiogenesis, proliferation, survival, and apoptosis evasion pathways across multiple cancer cell models. The antioxidant capacity and activation of phase II detox enzymes noted with specific isolates like astragalosides may additionally prevent carcinogenic mutations [173]. As an adaptogenic herb, astragalus contains various polysaccharides and flavonoids that help counter oxidative damage, DNA mutations, and aberrant methylation patterns tied to carcinogenesis. Studies also reveal that specific isolates prevented chemical induction of preneoplastic liver lesions and reduced lung tumor incidence and severity when provided before and during carcinogen exposure in rodents [174, 175].

Astragalus may also have some risks and uncertainties with supplementation. It may interact with immunosuppressant medications due to possible immune-boosting effects. Also, it can interact with blood pressure medications due to potential diuretic effects [167]. There are currently no standardized dosing recommendations for astragalus supplements or extracts. Most studied doses for specific conditions range widely from 0.5–60 grams of root powder per day. As a standardized extract, dosages often range from 250 to 500 mg, taken three to four times daily [176].

### 2.22 Ginseng (*Panax ginseng L.*)

Ginseng refers to the roots of plants in the Panax genus, which have been used in traditional Chinese medicine for centuries. The most common bioactive constituents in ginseng include ginsenosides, polysaccharides, peptides, polyacetylene alcohols, and fatty acids. Considerable research in recent decades indicates that ginseng extracts and isolates exhibit immunomodulating, anti-inflammatory, antioxidant, and possible anticancer effects based on preliminary cellular and animal studies [177]. *In vitro*, studies reveal promising anticancer effects of various ginseng root extracts and specific isolates against different human cancer cell lines. Experiments indicate dose-dependent inhibition of growth and proliferation, cell cycle arrest, apoptosis induction, reduced viability, and prevention of multidrug resistance mechanisms in colon, liver, ovarian, breast, and prostate cancer models [178, 179]. Though limited in number and scope thus far, some rodent studies demonstrate that oral ginseng extracts enhance efficacy and reduce the side effects of chemotherapeutics against implanted hepatocellular and ovarian cancer cells [180]. A few trials explored ginseng's ability to mitigate adverse events and hepatotoxicity associated with chemotherapy in gastrointestinal, nose, and lung cancer patients, though with mixed results [181, 182].

Experiments indicate that ginseng saponins exhibit direct antiproliferative, cytostatic, and cytotoxic effects in various cancer cell models. According to cell line data, extracts also beneficially modulated signaling pathways connected to apoptosis, cell cycle regulation, angiogenesis, invasion, inflammation, and multidrug resistance mechanisms [183]. The antioxidant effects demonstrated by some ginseng isolates like gintonin may provide additional preventative influences by reducing DNA damage that can enable carcinogenesis [184]. As an herbal adaptogen, ginseng's polysaccharides, oligopeptides, phytosterols, and flavonoids help counter oxidative stress, inflammation, and hormonal dysregulation, enabling pathogenesis.

Ginseng may also have some risks, such as causing insomnia, nervousness, and hypertension in some people due to its stimulatory effects. Ginseng can interact with several medications, including blood thinners, antidepressants, antidiabetic drugs, and stimulants, potentially altering their effects. Due to its weak phytoestrogenic activity, individuals with hormone-sensitive conditions, such as breast, uterine, or ovarian cancers, should use ginseng cautiously. Ginseng's immune-stimulating properties may interfere with immunosuppressive therapy; thus, it should be used cautiously by individuals undergoing such treatments [185]. There are currently no standardized dosing recommendations for ginseng supplements or extracts to guide consumers. Asian ginseng doses studied typically range from 1 to 10 grams daily of powdered root and 200–500 mg daily for ginseng extract standardized to 4–7% ginsenosides content.

### 2.23 Echinacea (*Echinacea purpurea L.*)

Echinacea is a genus of flowering plants used medicinally for centuries. The most studied species include *E. purpurea*, *E. angustifolia*, and *E. pallida*. Numerous bioactive constituents have been identified in echinacea, including alkamides, caffeic acid derivatives, polysaccharides, and glycoproteins [186]. Over the past few decades, emerging research indicates that echinacea extracts and isolates exhibit immunomodulating, anti-inflammatory, and antioxidant effects. Experiments indicate dose-dependent inhibition of growth, proliferation, and viability in leukemia, brain, colon, pancreatic, and ovarian cancer cell models [187, 188]. Echinacea phytocompounds also increased the efficacy of chemotherapeutic drugs like etoposide and cisplatin across lung, ovarian, and pancreatic cancer cells [189, 190].

Clinical data examining echinacea's potential to impact cancer outcomes or development remains scarce. A few small trials explored echinacea's ability to alleviate complications associated with conventional chemotherapy and radiation therapy in cancer patients. However, they provided limited evidence around efficacy for outcomes like leukopenia, nausea, fatigue, or other symptoms [191]. Experiments indicate that specific echinacea isolates inhibit activators of cancer-promoting inflammatory pathways involving NF-kB and cytokines. According to cancer cell line investigations, extracts also beneficially impacted proliferation, cell survival, and apoptosis evasion pathways [192]. Some echinacea constituents' antioxidant and possible immune modulation effects may provide additional preventative influences but require further elucidation. Characterizing primary bioactive and metabolic dynamics can strengthen translational development into practical pharmacological applications as adjuvants or standalone therapies (Manali. As an immunomodulatory herb, echinacea's alkamides, polysaccharides, and flavonoids exhibit systemic anti-inflammatory influences that could defend against chronic inflammation, enabling pathogenesis [192].

Echinacea may cause allergic reactions in sensitive individuals—rashes, hives, facial swelling. It may interact with immunosuppressant medications or liver medications. Most studies use doses ranging from around 300–6000 mg of total echinacea daily, either in three to four divided doses using liquid extract or two divided doses using tablets or capsules. Prolonged use of Echinacea (beyond 8 to 10 weeks) is generally not recommended, as it may lead to reduced effectiveness and potential immune suppression [193].

## 2.24 St. John's wort (*Hypericum perforatum L.*)

St. John's wort (*Hypericum perforatum*) is a flowering plant used for medicinal purposes since ancient times. The major bioactive constituents in the aerial parts of St. John's wort include hypericin, hyperforin, flavonoids, phenolic acids, and tannins [194]. Emerging research over the past few decades indicates St. John's wort extracts exhibit anti-inflammatory, antiviral, antidepressant, and possible anticancer effects based on preliminary cellular and animal studies. *In vitro* studies reveal promising anticancer effects from various St. John's wort extracts and specific isolates against different human cancer cell lines. Experiments indicate dose-dependent inhibition of growth, proliferation, viability, and invasion signaling in models of prostate, glioblastoma, melanoma, and leukemia cancer cells [195, 196]. Hypericin disrupted survival pathways and increased chemotherapy efficacy in models of nasopharyngeal carcinoma and cholangiocarcinoma [197]. Though limited in number thus far, animal studies demonstrate that oral hypericin inhibited the growth of implanted glioma cells and increased bladder cancer chemotherapy effectiveness [198, 199]. Only a few preliminary human pilot studies with limited sample sizes have explored hypericin's ability to inhibit the growth of glioblastoma cells or reduce gastrointestinal cancer markers when provided around surgery or before chemotherapy [200, 201].

Multiple mechanisms may contribute to the observed anticancer effects of St. John's wort extracts in preclinical analyses. Experiments indicate that certain constituents like hypericin exhibit photosensitizing and photodynamic activities that prompt targeted apoptosis and necrosis of cancer cells when exposed to specific wavelengths and doses of light [202]. St. John's wort extracts also impeded proliferation, angiogenesis, and cell signaling pathways related to growth and spread in cultured cancer cell investigations [203]. The monoamine oxidase and cytokine modulation effects of hyperforin and other components may provide additional immunomodulating influences, though the dynamics involved require further elucidation [204]. As an herb with demonstrated antidepressant effects in humans, St. John's wort holds monoamine-modulating compounds that may mitigate systemic inflammation, which can drive the pathogenesis of migraines, Alzheimer's, and malignancies [205]. Compounds also beneficially regulated detoxification enzymes and protected DNA from free radicals according to rodent and cell analyses, though human chemopreventive applications require significant further inquiry [206]. Nonetheless, the pleiotropic pharmacological profile supports its potential as a safe, affordable prophylactic supplement that future investigations could help substantiate. Routine usage could still plausibly aid primary and secondary prevention of certain inflammation-linked cancers over the years through multiple protective pathways that placebo-controlled trials can help substantiate.

St. John's wort does have associated risks and contraindications, one of which is photosensitivity. Photosensitivity refers to increased sensitivity to sunlight, leading to a greater risk of sunburns or skin rashes when exposed to the sun. This reaction is primarily due to hypericin, a constituent of St. John's wort [207]. St. John's wort is known to interact with a wide range of medications due to its influence on enzymes involved in drug metabolism. This can lead to decreased effectiveness of many drugs, including birth control pills, antidepressants, anticoagulants, and immunosuppressants, among others. It may interfere with the effectiveness of anesthetics and other medications used during and after surgery. Standardized dosing recommendations are difficult to firmly establish due to wide variability in St. John's wort preparations and conflicting evidence in the literature. The most common doses studied range between 500 and 1200 mg daily of St. John's wort extracts standardized to hypericin content.

## 2.25 Cat's claw *(Uncaria tomentosa (Willd. Ex Roem. & Schult.) DC)*

Cat's claw refers to a woody vine belonging to the Uncaria genus that has been used in traditional South American medicine for centuries. The most researched species is Uncaria tomentosa. Over 50 constituents have been identified in cat's claw, including pentacyclic oxindole alkaloids, quinovic acid glycosides, polyphenols, plant sterols, and carotenoids [208]. Emerging research indicates that based on preliminary cell and animal studies, cat's claw bark extracts exhibit immunostimulating, anti-inflammatory, antiviral, antioxidant, and possible anticancer effects [209]. Studies reveal promising anticancer effects from various cat claw extracts and isolated compounds against different human cancer cell types. Experiments indicate inhibited proliferation, cell cycle disruption, and cell death induction in breast, lung, and brain cancer cell models [210]. Quinovic acid glycosides enhance the apoptotic effects of chemotherapeutics like paclitaxel across ovarian and pancreatic cancer cells [211]. Though currently limited in number and scope, one animal study reported that oral Uncaria tomentosa doses prevented increased blood vessel growth and spread to the lungs in a murine melanoma model [212, 213]. More pharmacokinetic animal studies optimizing bioactive cat claw preparations and human trials are necessary to verify if it warrants inclusion as a complementary therapy against malignancies. Clinical research examining a cat's claw's potential to impact cancer outcomes or development risk beneficially remains exceptionally scarce. Only one small human pilot study has reported data so far where cat's claw supplementation reduced DNA damage from chemotherapy drugs in leukemia patients [214]. However, rigorously controlled trials with larger sample sizes are still needed to demonstrate therapeutic efficacy or adequate dosing guidelines of standardized cat's claw formulations against the progression, recurrence, or prevention of particular cancers in humans conclusively.

In preclinical analyses, multiple mechanisms may contribute to the observed anticancer effects of specific cat's claw extracts. Experiments indicate that certain alkaloid and quinovic acid fractions stimulate anti-tumor immune activity through increased lymphocyte proliferation, phagocytosis, and cytokine modulation, vital for early detection and defense [215]. Antiproliferative, pro-apoptotic, antiangiogenic, and antimetastatic activities have also been reported in various human cancer cell culture experiments [216, 217]. As an antioxidant-rich herb, cat's claw compounds protect against DNA damage from oxidation, nitrous compounds, and inflammation tied to pathogenesis [218]. Using a cat's claw as an herbal supplement may also aid the prevention of particular cancers over the long term. Some surveys link cat's claw usage to perceived improvements in immune markers, DNA repair, antioxidant capacity, and lower infection rates, which could mitigate associated inflammation and cancer risks. However, placebo-controlled studies have been limited in humans so far [214, 219]. As an immunostimulant herb, cat's claw holds alkaloids and carboxyl alkyl esters that activate white blood cell functioning vital to clearing aberrant cells before malignancy develops. Quinovic acid glycosides prevented oxidative damage and cell death from toxin exposure in several cell assays, though human chemopreventive applications require more substantiation. The cellular and limited animal studies report anticancer mechanisms of specific Uncaria tomentosa extracts related to immunomodulation, antiproliferation, and preliminary metastasis and angiogenesis inhibition.

While Cat's Claw is considered safe for most people when used in moderate amounts, there are some risks and contraindications associated with its use. Due to its immunostimulant effects, Cat's Claw should be used cautiously by individuals with autoimmune diseases like lupus, multiple sclerosis, or rheumatoid arthritis, as it might exacerbate these conditions. Cat's claw may slow blood clotting, posing a risk for individuals with bleeding disorders or those on anticoagulant or antiplatelet medications. Due to its effects on blood clotting, it is advisable to discontinue the use of Cat's claw at least 2 weeks before scheduled surgery to avoid excessive bleeding. There are no standardized dosing recommendations for cat claw supplements or extracts. The studied dosages vary widely from 20 to 3000 mg daily. If consumed as a tea, 1000 to 4000 mg of the bark is boiled in 250 mL of water. Dosages for other forms, such as tinctures or capsules, should follow the manufacturer's recommendations or a healthcare provider's guidance.

## 3. Challenges and future directions

While preclinical evidence shows promise for plant-based cancer therapeutics, translating these to clinical practice faces hurdles like inconsistent quality/standardization of extracts, limited knowledge of mechanisms, potential interactions with other drugs, negative perceptions among clinicians, and inadequate clinical trial data substantiating safety and efficacy. However, the future holds exciting possibilities for expanding plant medicines' role in cancer care through innovations enhancing bioavailability via novel delivery systems, scaling production/purification of active compounds, developing synergistic phytochemical formulations, emergent technologies like nano-encapsulations, and increased adoption of traditional ethnopharmacology knowledge into modern research. If challenges around conducting more rigorous human trials, identifying optimal therapeutic combinations/dosages, navigating regulatory systems, and changing clinical attitudes can be overcome, phytochemicals may find expanding niche applications as cheaper, safer adjuvants amplifying chemotherapy/radiation efficacy and mitigating side effects, while boosting prevention through routine nutritional intake among high-risk groups.

## 4. Conclusion

This comprehensive review highlights the promising anticancer potential of numerous dietary phytochemicals derived from familiar herbs and spices. Extensive preclinical evidence demonstrates that bioactive compounds from plants such as curcumin, ginger, garlic, basil, and black pepper can influence multiple intracellular signaling pathways integral to the hallmarks of cancer, including resisting cell death, sustaining proliferation, enabling replicative immortality, inducing angiogenesis, and activating invasion and metastasis. By modulating key enzymes, transcription factors, cytokines, growth factors, and gene expression, these natural products exhibit cytotoxic, pro-apoptotic, anti-inflammatory, antioxidant, anti-angiogenic, and anti-metastatic effects across diverse cancer cell lines and animal models. While clinical research is still limited, early human trials support the safety, tolerability, and preliminary efficacy of specific herbal preparations against progression markers or chemotherapy side effects for various malignancies. As conventional treatment options become increasingly costly and toxic, botanical medicines may offer safer, affordable complementary approaches, amplifying therapeutic outcomes through multimodal mechanisms. Further pharmacokinetic, formulation, delivery, and clinical research can pave the way for plant-derived anticarcinogens to realize their full preventative and therapeutic potential. Integrating traditional ethnopharmacology wisdom with modern scientific rigor can usher in more holistic, patient-centered cancer care paradigms emphasizing natural products alongside conventional modalities.

## Acknowledgements

I would like to express my gratitude to the various AI tools that assisted me in preparing and editing this book chapter. GitHub Copilot, an AI-powered code completion tool, provided valuable suggestions and code snippets that streamlined the writing process. Research Rabbit, an AI-based research assistant, helped me discover relevant literature and organize my references efficiently. Furthermore, Plagiarism Checker, an AI-powered plagiarism detection tool, ensured the originality of the content and helped me maintain academic integrity throughout the writing process. While these AI tools were incredibly helpful, I emphasize that the final content, ideas, and arguments presented in this chapter are my own. The AI tools served as assistants, but the responsibility lies with me for the work's accuracy, clarity, and coherence.

## Conflict of interest

The authors declare no conflict of interest.

## Author details

Biba Vikas[1*], Vishnupriya K. Sweety[2], Sukumaran Anil[3] and Senthilnathan Palaniyandi[4]

1 Biomedical Research Centre, Tuskegee University, Tuskegee, Alabama, USA

2 Pushpagiri College of Dental Sciences, Medicity, Perumthuruthy, Tiruvalla, Kerala, India

3 Department of Dentistry, Hamad Medical Corporation, Oral Health Institute, Doha, Qatar

4 University of Missouri-Columbia, Columbia, MO, USA

*Address all correspondence to: bibavikas@gmail.com

## References

[1] Taibi K, Abderrahim LA, Ferhat K, Betta S, Taibi F, Bouraada F, et al. Ethnopharmacological study of natural products used for traditional cancer therapy in Algeria Saudi. The Pharmaceutical Journal. 2020;**28**:1451-1465. DOI: 10.1016/j.jsps.2020.09.011. Epub 20200921

[2] Yuan H, Ma Q, Ye L, Piao G. The traditional medicine and modern medicine from natural products. Molecules. 2016;**21**:1-18. DOI: 10.3390/molecules21050559. Epub 20160429

[3] Pandey MM, Rastogi S, Rawat AK. Indian traditional ayurvedic system of medicine and nutritional supplementation. Evidence-based Complementary and Alternative Medicine. 2013;**2013**:376327. DOI: 10.1155/2013/376327. Epub 20130623

[4] Ali M, Wani SUD, Salahuddin M, Sn M, Mruthunjaya K, Dey T, et al. Recent advance of herbal medicines in cancer—A molecular approach. Heliyon. 2023;**9**:e13684. DOI: 10.1016/j.heliyon.2023.e13684. Epub 20230214

[5] Gonzalez-Vallinas M, Gonzalez-Castejon M, Rodriguez-Casado A, Ramirez de Molina A. Dietary phytochemicals in cancer prevention and therapy: A complementary approach with promising perspectives. Nutrition Reviews. 2013;**71**:585-599. DOI: 10.1111/nure.12051. Epub 20130801

[6] Gutschner T, Diederichs S. The hallmarks of cancer: A long non-coding RNA point of view. RNA Biology. 2012;**9**:703-719. DOI: 10.4161/rna.20481. Epub 20120601

[7] Tomeh MA, Hadianamrei R, Zhao X. A review of curcumin and its derivatives as anticancer agents. International Journal of Molecular Sciences. 2019;**20**:1-26. DOI: 10.3390/ijms20051033. Epub 20190227

[8] Ojo OA, Adeyemo TR, Rotimi D, Batiha GE, Mostafa-Hedeab G, Iyobhebhe ME, et al. Anticancer properties of curcumin against colorectal cancer: A review. Front. Oncologia. 2022;**12**:881641. DOI: 10.3389/fonc.2022.881641. Epub 20220422

[9] Xu G, Chen J, Wang G, Xiao J, Zhang N, Chen Y, et al. Resveratrol inhibits the tumorigenesis of follicular thyroid cancer via ST6GAL2-regulated activation of the hippo signaling pathway. Molecular Therapy—Oncolytics. 2020;**16**:124-133. DOI: 10.1016/j.omto.2019.12.010. Epub 20200110

[10] Gano CA, Fatima S, Failes TW, Arndt GM, Sajinovic M, Mahns D, et al. Anti-cancer potential of synergistic phytochemical combinations is influenced by the genetic profile of prostate cancer cell lines. Frontiers in Nutrition. 2023;**10**:1119274. DOI: 10.3389/fnut.2023.1119274. Epub 20230307

[11] Yin SY, Wei WC, Jian FY, Yang NS. Therapeutic applications of herbal medicines for cancer patients. Evidence-based Complementary and Alternative Medicine. 2013;**2013**:302426. DOI: 10.1155/2013/302426. Epub 20130711

[12] Huang J, Li JX, Ma LR, Xu DH, Wang P, Li LQ, et al. Traditional herbal medicine: A potential therapeutic approach for adjuvant treatment of non-small cell lung cancer in the future. Integrative Cancer Therapies.

2022;**21**:15347354221144312. DOI: 10.1177/15347354221144312

[13] Farghadani R, Naidu R. Curcumin as an enhancer of therapeutic efficiency of chemotherapy drugs in breast cancer. International Journal of Molecular Sciences. 2022;**23**:1-19. DOI: 10.3390/ijms23042144. Epub 20220215

[14] Parveen A, Parveen B, Parveen R, Ahmad S. Challenges and guidelines for clinical trial of herbal drugs. Journal of Pharmacy & Bioallied Sciences. 2015;7:329-333. DOI: 10.4103/0975-7406.168035

[15] Sharifi-Rad J, Rayess YE, Rizk AA, Sadaka C, Zgheib R, Zam W, et al. Turmeric and its major compound curcumin on health: Bioactive effects and safety profiles for food, pharmaceutical, biotechnological and medicinal applications. Frontiers in Pharmacology. 2020;**11**:01021. DOI: 10.3389/fphar.2020.01021. Epub 20200915

[16] Giordano A, Tommonaro G. Curcumin and cancer. Nutrients. 2019;**11**:1-20. DOI: 10.3390/nu11102376. Epub 20191005

[17] Mansouri K, Rasoulpoor S, Daneshkhah A, Abolfathi S, Salari N, Mohammadi M, et al. Clinical effects of curcumin in enhancing cancer therapy: A systematic review. BMC Cancer. 2020;**20**:791. DOI: 10.1186/s12885-020-07256-8. Epub 20200824

[18] Bachmeier BE, Killian PH, Melchart D. The role of curcumin in prevention and management of metastatic disease. International Journal of Molecular Sciences. 2018;**19**:1-18. DOI: 10.3390/ijms19061716. Epub 20180609

[19] Abd, Wahab NA, Lajis NH, Abas F, Othman I, Naidu R. Mechanism of anti-cancer activity of curcumin on androgen-dependent and androgen-independent prostate cancer. Nutrients. 2020;**12**:1-34. DOI: 10.3390/nu12030679. Epub 20200302

[20] Farghadani R, Naidu R. Curcumin: Modulator of key molecular signaling pathways in hormone-independent breast cancer. Cancers (Basel). 2021;**13**:1-29. DOI: 10.3390/cancers13143427. Epub 20210708

[21] Zoi V, Galani V, Lianos GD, Voulgaris S, Kyritsis AP, Alexiou GA. The role of curcumin in cancer. Treatment Biomedicines. 2021;**9**:1-19. DOI: 10.3390/biomedicines9091086. Epub 20210826

[22] Sa G, Das T. Anti cancer effects of curcumin: Cycle of life and death. Cell Division. 2008;**3**:14. DOI: 10.1186/1747-1028-3-14. Epub 20081003

[23] Wang JB, Qi LL, Zheng SD, Wu TX. Curcumin induces apoptosis through the mitochondria-mediated apoptotic pathway in HT-29 cells. Journal of Zhejiang University. Science. B. 2009;**10**:93-102. DOI: 10.1631/jzus.B0820238

[24] He Y, Yue Y, Zheng X, Zhang K, Chen S, Du Z. Curcumin, inflammation, and chronic diseases: How are they linked? Molecules. 2015;**20**:9183-9213. DOI: 10.3390/molecules20059183. Epub 20150520

[25] Wang TY, Chen JX. Effects of curcumin on vessel formation insight into the pro- and Antiangiogenesis of curcumin. Evidence-based Complementary and Alternative Medicine. 2019;**2019**:1390795. DOI: 10.1155/2019/1390795. Epub 20190619

[26] Shaikh S, Shaikh J, Naba YS, Doke K, Ahmed K, Yusufi M. Curcumin:

Reclaiming the lost ground against cancer resistance. Cancer Drug Resistance. 2021;**4**:298-320. DOI: 10.20517/cdr.2020.92. Epub 20210619

[27] Johnson SM, Gulhati P, Arrieta I, Wang X, Uchida T, Gao T, et al. Curcumin inhibits proliferation of colorectal carcinoma by modulating Akt/mTOR signaling. Anticancer Research. 2009;**29**:3185-3190

[28] Patel VB, Misra S, Patel BB, Majumdar AP. Colorectal cancer: Chemopreventive role of curcumin and resveratrol. Nutrition and Cancer. 2010;**62**:958-967. DOI: 10.1080/01635581.2010.510259

[29] Paulraj F, Abas F, Lajis NH, Othman I, Naidu R. Molecular pathways modulated by curcumin analogue, Diarylpentanoids in cancer. Biomolecules. 2019;**9**:1-14. DOI: 10.3390/biom9070270. Epub 20190710

[30] Pan Z, Zhuang J, Ji C, Cai Z, Liao W, Huang Z. Curcumin inhibits hepatocellular carcinoma growth by targeting VEGF expression. Oncology Letters. 2018;**15**:4821-4826. DOI: 10.3892/ol.2018.7988. Epub 20180207

[31] Tan BL, Norhaizan ME. Curcumin combination chemotherapy: The implication and efficacy in cancer. Molecules. 2019;**24**:1-21. DOI: 10.3390/molecules24142527. Epub 20190710

[32] Ng CY, Yen H, Hsiao HY, Su SC. Phytochemicals in skin cancer prevention and treatment: An updated review. International Journal of Molecular Sciences. 2018;**19**:1-24. DOI: 10.3390/ijms19040941. Epub 20180322

[33] HegdeM,GirisaS,BharathwajChettyB, Vishwa R, Kunnumakkara AB. Curcumin formulations for better bioavailability: What we learned from clinical trials thus far? ACS Omega. 2023;**8**:10713-10746. DOI: 10.1021/acsomega.2c07326. Epub 20230313

[34] Dhillon N, Aggarwal BB, Newman RA, Wolff RA, Kunnumakkara AB, Abbruzzese JL, et al. Phase II trial of curcumin in patients with advanced pancreatic cancer. Clinical Cancer Research. 2008;**14**:4491-4499. DOI: 10.1158/1078-0432.CCR-08-0024

[35] Howells LM, Iwuji COO, Irving GRB, Barber S, Walter H, Sidat Z, et al. Curcumin combined with FOLFOX chemotherapy is safe and tolerable in patients with metastatic colorectal cancer in a randomized phase IIa trial. The Journal of Nutrition. 2019;**149**:1133-1139. DOI: 10.1093/jn/nxz029

[36] Wilken R, Veena MS, Wang MB, Srivatsan ES. Curcumin: A review of anti-cancer properties and therapeutic activity in head and neck squamous cell carcinoma. Molecular Cancer. 2011;**10**:12. DOI: 10.1186/1476-4598-10-12. Epub 20110207

[37] Sharma RA, Ireson CR, Verschoyle RD, Hill KA, Williams ML, Leuratti C, et al. Effects of dietary curcumin on glutathione S-transferase and malondialdehyde-DNA adducts in rat liver and colon mucosa: Relationship with drug levels. Clinical Cancer Research. 2001;**7**:1452-1458

[38] Liu AC, Zhao LX, Lou HX. Curcumin alters the pharmacokinetics of warfarin and clopidogrel in Wistar rats but has no effect on anticoagulation or antiplatelet aggregation. Planta Medica. 2013;**79**:971-977. DOI: 10.1055/s-0032-1328652. Epub 20130627

[39] Hewlings SJ, Kalman DS. Curcumin: A review of its effects on human health. Food. 2017;**6**:1-11. DOI: 10.3390/foods6100092. Epub 20171022

[40] Gupta SC, Patchva S, Aggarwal BB. Therapeutic roles of curcumin: Lessons learned from clinical trials. The AAPS Journal. 2013;**15**:195-218. DOI: 10.1208/s12248-012-9432-8. Epub 20121110

[41] Semwal RB, Semwal DK, Combrinck S, Viljoen AM. Gingerols and shogaols: Important nutraceutical principles from ginger. Phytochemistry. 2015;**117**:554-568. DOI: 10.1016/j.phytochem.2015.07.012. Epub 20150727

[42] Mao QQ, Xu XY, Cao SY, Gan RY, Corke H, Beta T, et al. Bioactive compounds and bioactivities of ginger (Zingiber officinale roscoe). Food. 2019;**8**:1-21. DOI: 10.3390/foods8060185. Epub 20190530

[43] Chen GQ, Nan Y, Huang SC, Ning N, Du YH, Lu DD, et al. Research progress of ginger in the treatment of gastrointestinal tumors. World Journal of Gastrointestinal Oncology. 2023;**15**:1835-1851. DOI: 10.4251/wjgo.v15.i11.1835

[44] Rhode J, Fogoros S, Zick S, Wahl H, Griffith KA, Huang J, et al. Ginger inhibits cell growth and modulates angiogenic factors in ovarian cancer cells. BMC Complementary and Alternative Medicine. 2007;**7**:44. DOI: 10.1186/1472-6882-7-44. Epub 20071220

[45] Lechner JF, Stoner GD. Gingers and their purified components as cancer Chemopreventative agents. Molecules. 2019;**24**:1-27. DOI: 10.3390/molecules24162859. Epub 20190807

[46] Mashhadi NS, Ghiasvand R, Askari G, Hariri M, Darvishi L, Mofid MR. Anti-oxidative and anti-inflammatory effects of ginger in health and physical activity: Review of current evidence. International Journal of Preventive Medicine. 2013;**4**:S36-S42

[47] Zadorozhna M, Mangieri D. Mechanisms of Chemopreventive and therapeutic proprieties of ginger extracts in cancer. International Journal of Molecular Sciences. 2021;**22**:1-18. DOI: 10.3390/ijms22126599. Epub 20210620

[48] Prasad S, Tyagi AK. Ginger and its constituents: Role in prevention and treatment of gastrointestinal cancer. Gastroenterology Research and Practice. 2015;**2015**:142979. DOI: 10.1155/2015/142979. Epub 20150308

[49] Khandouzi N, Shidfar F, Rajab A, Rahideh T, Hosseini P, Mir TM. The effects of ginger on fasting blood sugar, hemoglobin a1c, apolipoprotein B, apolipoprotein a-I and malondialdehyde in type 2 diabetic patients. Iranian Journal of Pharmaceutical Research: IJPR. 2015;**14**:131-140

[50] Lete I, Allue J. The effectiveness of ginger in the prevention of nausea and vomiting during pregnancy and chemotherapy. Integrative Medicine Insights. 2016;**11**:11-17. DOI: 10.4137/IMI.S36273. Epub 20160331

[51] Shang A, Cao SY, Xu XY, Gan RY, Tang GY, Corke H, et al. Bioactive compounds and biological functions of garlic (Allium sativum L.). Food. 2019;**8**:1-31. DOI: 10.3390/foods8070246. Epub 20190705

[52] Mitra S, Das R, Emran TB, Labib RK, Noor ET, Islam F, et al. Diallyl disulfide: A bioactive garlic compound with anticancer potential. Frontiers in Pharmacology. 2022;**13**:943967. DOI: 10.3389/fphar.2022.943967. Epub 20220822

[53] Wang Y, Huang P, Wu Y, Liu D, Ji M, Li H, et al. Association and mechanism of garlic consumption with gastrointestinal

cancer risk: A systematic review and meta-analysis. Oncology Letters. 2022;**23**:125. DOI: 10.3892/ol.2022.13245. Epub 20220217

[54] Pandey P, Khan F, Alshammari N, Saeed A, Aqil F, Saeed M. Updates on the anticancer potential of garlic organosulfur compounds and their nanoformulations: Plant therapeutics in cancer management. Frontiers in Pharmacology. 2023;**14**:1154034. DOI: 10.3389/fphar.2023.1154034. Epub 20230320

[55] Malla R, Marni R, Chakraborty A, Kamal MA. Diallyl disulfide and diallyl trisulfide in garlic as novel therapeutic agents to overcome drug resistance in breast cancer. Journal of Pharmaceutical Analysis. 2022;**12**:221-231. DOI: 10.1016/j.jpha.2021.11.004. Epub 20211110

[56] Duda-Chodak A, Tarko T. Possible side effects of polyphenols and their interactions with medicines. Molecules. 2023;**28**:1-46. DOI: 10.3390/molecules28062536. Epub 20230310

[57] Lawson LD, Hunsaker SM. Allicin bioavailability and bioequivalence from garlic supplements and garlic foods. Nutrients. 2018;**10**:1-49. DOI: 10.3390/nu10070812. Epub 20180624

[58] Azizah NS, Irawan B, Kusmoro J, Safriansyah W, Farabi K, Oktavia D, et al. Sweet basil (Ocimum basilicum L.)—A review of its botany, phytochemistry, pharmacological activities, and biotechnological development. Plants (Basel). 2023;**12**:1-25. DOI: 10.3390/plants12244148. Epub 20231213

[59] Hasan MR, Alotaibi BS, Althafar ZM, Mujamammi AH, Jameela J. An update on the therapeutic anticancer potential of Ocimum sanctum L.: "elixir of life". Molecules. 2023;**28**:1-28. DOI: 10.3390/molecules28031193. Epub 20230125

[60] Perna S, Alawadhi H, Riva A, Allegrini P, Petrangolini G, Gasparri C, et al. In vitro and In vivo anticancer activity of basil (Ocimum spp.): Current insights and future prospects. Cancers (Basel). 2022;**14**:1-16. DOI: 10.3390/cancers14102375. Epub 20220511

[61] Liu P, Du R, Yu X. Ursolic acid exhibits potent anticancer effects in human metastatic melanoma cancer cells (SK-MEL-24) via apoptosis induction, inhibition of cell migration and invasion, cell cycle arrest, and inhibition of mitogen-activated protein kinase (MAPK)/ERK signaling pathway. Medical Science Monitor. 2019;**25**:1283-1290. DOI: 10.12659/MSM.913069. Epub 20190217

[62] Anusuya C, Manoharan S. Antitumor initiating potential of rosmarinic acid in 7,12-dimethylbenz(a)anthracene-induced hamster buccal pouch carcinogenesis. Journal of Environmental Pathology, Toxicology and Oncology. 2011;**30**:199-211. DOI: 10.1615/jenvironpatholtoxicoloncol.v30.i3.30

[63] Ge B, Zhang Z, Zuo Z. Updates on the clinical evidenced herb-warfarin interactions. Evidence-based Complementary and Alternative Medicine. 2014;**2014**:957362. DOI: 10.1155/2014/957362. Epub 20140318

[64] Rabbani M, Sajjadi SE, Vaezi A. Evaluation of anxiolytic and sedative effect of essential oil and hydroalcoholic extract of Ocimum basilicum L. and chemical composition of its essential oil. Research in Pharmaceutical Sciences. 2015;**10**:535-543

[65] Jamshidi N, Cohen MM. The clinical efficacy and safety of Tulsi in humans: A systematic review of the literature. Evidence-based Complementary and Alternative Medicine. 2017;**2017**:9217567.

DOI: 10.1155/2017/9217567. Epub 20170316

[66] Wang X-J, Luo Q, Li T, Meng P-H, Pu Y-T, Liu J-X, et al. Origin, evolution, breeding, and omics of Apiaceae: A family of vegetables and medicinal plants. Horticulture Research. 2022;**9**:1-24. DOI: 10.1093/hr/uhac076

[67] Mahboubi M. Caraway as important medicinal plants in management of diseases. Natural Products and Bioprospecting. 2019;**9**:1-11. DOI: 10.1007/s13659-018-0190-x. Epub 20181029

[68] Peng BY, Singh AK, Chan CH, Deng YH, Li PY, Su CW, et al. AGA induces sub-G1 cell cycle arrest and apoptosis in human colon cancer cells through p53-independent/p53-dependent pathway. BMC Cancer. 2023;**23**:1. DOI: 10.1186/s12885-022-10466-x. Epub 20230102

[69] Kazemipoor M, Radzi CW, Hajifaraji M, Haerian BS, Mosaddegh MH, Cordell GA. Antiobesity effect of caraway extract on overweight and obese women: A randomized, triple-blind, placebo-controlled clinical trial. Evidence-based Complementary and Alternative Medicine. 2013;**2013**:928582. DOI: 10.1155/2013/928582. Epub 20131110

[70] Rao PV, Gan SH. Cinnamon: A multifaceted medicinal plant. Evidence-based Complementary and Alternative Medicine. 2014;**2014**:642942. DOI: 10.1155/2014/642942. Epub 20140410

[71] Caserta S, Genovese C, Cicero N, Gangemi S, Allegra A. The anti-cancer effect of cinnamon aqueous extract: A focus on hematological malignancies. Life (Basel). 2023;**13**:1-15. DOI: 10.3390/life13051176. Epub 20230512

[72] Banerjee S, Banerjee S. Anticancer potential and molecular mechanisms of cinnamaldehyde and its congeners present in the cinnamon plant. Physiologia. 2023;**3**:173-207. DOI: 10.3390/physiologia3020013

[73] Kwon HK, Hwang JS, So JS, Lee CG, Sahoo A, Ryu JH, et al. Cinnamon extract induces tumor cell death through inhibition of NFkappaB and AP1. BMC Cancer. 2010;**10**:392. DOI: 10.1186/1471-2407-10-392. Epub 20100724

[74] Liu Y, An T, Wan D, Yu B, Fan Y, Pei X. Targets and mechanism used by Cinnamaldehyde, the Main active ingredient in cinnamon, in the treatment of breast cancer. Frontiers in Pharmacology. 2020;**11**:582719. DOI: 10.3389/fphar.2020.582719. Epub 20201209

[75] Khedkar S, Ahmad KM. Aqueous extract of cinnamon (Cinnamomum spp.): Role in cancer and inflammation. Evidence-based Complementary and Alternative Medicine. 2023;**2023**:5467342. DOI: 10.1155/2023/5467342. Epub 20230511

[76] Iwata N, Kainuma M, Kobayashi D, Kubota T, Sugawara N, Uchida A, et al. The relation between hepatotoxicity and the Total Coumarin intake from traditional Japanese medicines containing cinnamon bark. Frontiers in Pharmacology. 2016;7:174. DOI: 10.3389/fphar.2016.00174. Epub 20160620

[77] Kizilaslan N, Erdem NZ. The effect of different amounts of cinnamon consumption on blood glucose in healthy adult individuals. International Journal of Food Science. 2019;**2019**:4138534. DOI: 10.1155/2019/4138534. Epub 20190304

[78] Pitaro M, Croce N, Gallo V, Arienzo A, Salvatore G, Antonini G.

Coumarin-induced hepatotoxicity: A narrative review. Molecules. 2022;**27**:1-19. DOI: 10.3390/molecules27249063. Epub 20221219

[79] Liu H, Schmitz JC, Wei J, Cao S, Beumer JH, Strychor S, et al. Clove extract inhibits tumor growth and promotes cell cycle arrest and apoptosis. Oncology Research. 2014;**21**:247-259. DOI: 10.3727/0965040 14X13946388748910

[80] Zari AT, Zari TA, Hakeem KR. Anticancer properties of eugenol: A review. Molecules. 2021;**26**:1-17. DOI: 10.3390/molecules26237407. Epub 20211206

[81] Jaganathan SK, Supriyanto E. Antiproliferative and molecular mechanism of eugenol-induced apoptosis in cancer cells. Molecules. 2012;**17**:6290-6304. DOI: 10.3390/molecules17066290. Epub 20120525

[82] Afrendi E, Prastya ME, Astuti RI, Wahyuni WT, Batubara I. Bioactivity of the ethanol extract of clove (Syzygium aromaticum) as antitoxin. International Journal of Food Science. 2023;**2023**:3245210. DOI: 10.1155/2023/3245210. Epub 20230921

[83] Kamkar Asl M, Nazariborun A, Hosseini M. Analgesic effect of the aqueous and ethanolic extracts of clove. Avicenna Journal of Phytomedicine. 2013;**3**:186-192

[84] Jaganathan SK, Mazumdar A, Mondhe D, Mandal M. Apoptotic effect of eugenol in human colon cancer cell lines. Cell Biology International. 2011;**35**:607-615. DOI: 10.1042/CBI20100118

[85] Rudrapal M, Khairnar SJ, Khan J, Dukhyil AB, Ansari MA, Alomary MN, et al. Dietary polyphenols and their role in oxidative stress-induced human diseases: Insights into protective effects, antioxidant potentials and mechanism(s) of action. Frontiers in Pharmacology. 2022;**13**:806470. DOI: 10.3389/fphar.2022.806470. Epub 20220214

[86] Flores B, Trivedi HD, Robson SC, Bonder A. Hemostasis, bleeding and thrombosis in liver disease. Journal of Translational Science. 2017;**3**:1-16. DOI: 10.15761/JTS.1000182. Epub 20170304

[87] Kumar, Pandey V, Shams R, Singh R, Dar AH, Pandiselvam R, Rusu AV, et al. A comprehensive review on clove (Caryophyllus aromaticus L.) essential oil and its significance in the formulation of edible coatings for potential food applications. Frontiers in Nutrition. 2022;**9**:1-24. DOI: 10.3389/fnut.2022.987674

[88] Laribi B, Kouki K, M'Hamdi M, Bettaieb T. Coriander (Coriandrum sativum L.) and its bioactive constituents. Fitoterapia. 2015;**103**:9-26. DOI: 10.1016/j.fitote.2015.03.012. Epub 20150314

[89] Naeem A, Hu P, Yang M, Zhang J, Liu Y, Zhu W, et al. Natural products as anticancer agents: Current status and future perspectives. Molecules. 2022;**27**:1-64. DOI: 10.3390/molecules27238367. Epub 20221130

[90] Sobhani Z, Mohtashami L, Amiri MS, Ramezani M, Emami SA, Simal-Gandara J. Ethnobotanical and phytochemical aspects of the edible herb Coriandrum sativum L. Journal of Food Science. 2022;**87**:1386-1422. DOI: 10.1111/1750-3841.16085. Epub 20220312

[91] Mechchate H, Costa de Oliveira R, Es-Safi I, Vasconcelos

Mourao EM, Bouhrim M, Kyrylchuk A, et al. Antileukemic activity and molecular docking study of a polyphenolic extract from coriander seeds. Pharmaceuticals (Basel). 2021;**14**:1-16. DOI: 10.3390/ph14080770. Epub 20210805

[92] Scandar S, Zadra C, Marcotullio MC. Coriander (Coriandrum sativum) polyphenols and their nutraceutical value against obesity and metabolic syndrome. Molecules. 2023;**28**:1-12. DOI: 10.3390/molecules28104187. Epub 20230519

[93] Biswas P, Dey D, Biswas PK, Rahaman TI, Saha S, Parvez A, et al. A comprehensive analysis and anti-cancer activities of quercetin in ROS-mediated cancer and cancer stem cells. International Journal of Molecular Sciences. 2022;**23**:1-33. DOI: 10.3390/ijms231911746. Epub 20221004

[94] Jensen-Jarolim E, Leitner A, Hirschwehr R, Kraft D, Wuthrich B, Scheiner O, et al. Characterization of allergens in Apiaceae spices: Anise, fennel, coriander and cumin. Clinical and Experimental Allergy. 1997;**27**:1299-1306

[95] Bommakanti V, Puthenparambil Ajikumar A, Sivi CM, Prakash G, Mundanat AS, Ahmad F, et al. An overview of herbal nutraceuticals, their extraction, formulation, therapeutic effects and potential toxicity. Separations. 2023;**10**:177. DOI: 10.3390/separations10030177

[96] Jana S, Shekhawat GS. Anethum graveolens: An Indian traditional medicinal herb and spice. Pharmacognosy Reviews. 2010;**4**:179-184. DOI: 10.4103/0973-7847.70915

[97] Aswathy M, Banik K, Parama D, Sasikumar P, Harsha C, Joseph AG, et al. Exploring the cytotoxic effects of the extracts and bioactive triterpenoids from Dillenia indica against Oral squamous cell carcinoma: A scientific interpretation and validation of indigenous knowledge. ACS Pharmacology & Translational Science. 2021;**4**:834-847. DOI: 10.1021/acsptsci.1c00011. Epub 20210309

[98] Baruah TJ, Kma L. Vicenin-2 acts as a radiosensitizer of the non-small cell lung cancer by lowering Akt expression. BioFactors. 2019;**45**:200-210. DOI: 10.1002/biof.1472. Epub 20181129

[99] Mohammed FA, Elkady AI, Syed FQ, Mirza MB, Hakeem KR, Alkarim S. Anethum graveolens (dill)—A medicinal herb induces apoptosis and cell cycle arrest in HepG2 cell line. Journal of Ethnopharmacology. 2018;**219**:15-22. DOI: 10.1016/j.jep.2018.03.008. Epub 20180309

[100] Sloczynska K, Powroznik B, Pekala E, Waszkielewicz AM. Antimutagenic compounds and their possible mechanisms of action. Journal of Applied Genetics. 2014;**55**:273-285. DOI: 10.1007/s13353-014-0198-9. Epub 20140311

[101] Goodarzi MT, Khodadadi I, Tavilani H, Abbasi OE. The role of Anethum graveolens L. (dill) in the Management of Diabetes. Journal of Tropical Medicine. 2016;**2016**:1098916. DOI: 10.1155/2016/1098916. Epub 20161018

[102] Mnif S, Aifa S. Cumin (Cuminum cyminum L.) from traditional uses to potential biomedical applications. Chemistry & Biodiversity. 2015;**12**:733-742. DOI: 10.1002/cbdv.201400305

[103] Khan MA, Chen HC, Tania M, Zhang DZ. Anticancer activities of Nigella sativa (black cumin). African Journal of Traditional, Complementary, and Alternative Medicines. 2011;**8**:226-232. DOI: 10.4314/ajtcam.v8i5S.10. Epub 20110703

[104] Asaduzzaman Khan M, Tania M, Fu S, Fu J. Thymoquinone, as an anticancer molecule: From basic research to clinical investigation. Oncotarget. 2017;**8**:51907-51919. DOI: 10.18632/oncotarget.17206. Epub 20170418

[105] Ansary J, Giampieri F, Forbes-Hernandez TY, Regolo L, Quinzi D, Gracia Villar S, et al. Nutritional value and preventive role of Nigella sativa L. and its Main component Thymoquinone in cancer: An evidenced-based review of preclinical and clinical studies. Molecules. 2021;**26**:1-27. DOI: 10.3390/molecules26082108. Epub 20210407

[106] Ravindran J, Prasad S, Aggarwal BB. Curcumin and cancer cells: How many ways can curry kill tumor cells selectively? The AAPS Journal. 2009;**11**:495-510. DOI: 10.1208/s12248-009-9128-x. Epub 20090710

[107] Hannan MA, Rahman MA, Sohag AAM, Uddin MJ, Dash R, Sikder MH, et al. Black cumin (Nigella sativa L.): A comprehensive review on phytochemistry, health benefits, molecular pharmacology, and safety. Nutrients. 2021;**13**:1-60. DOI: 10.3390/nu13061784. Epub 20210524

[108] Sharma Y, Velamuri R, Fagan J, Schaefer J. Full-Spectrum analysis of bioactive compounds in rosemary (Rosmarinus officinalis L.) as influenced by different extraction methods. Molecules. 2020;**25**:1-21. DOI: 10.3390/molecules25204599. Epub 20201009

[109] Loussouarn M, Krieger-Liszkay A, Svilar L, Bily A, Birtic S, Havaux M. Carnosic acid and Carnosol, two major antioxidants of rosemary, act through different mechanisms. Plant Physiology. 2017;**175**:1381-1394. DOI: 10.1104/pp.17.01183. Epub 20170915

[110] Gonzalez-Vallinas M, Molina S, Vicente G, de la Cueva A, Vargas T, Santoyo S, et al. Antitumor effect of 5-fluorouracil is enhanced by rosemary extract in both drug sensitive and resistant colon cancer cells. Pharmacological Research. 2013;**72**:61-68. DOI: 10.1016/j.phrs.2013.03.010. Epub 20130401

[111] Allegra A, Tonacci A, Pioggia G, Musolino C, Gangemi S. Anticancer activity of Rosmarinus officinalis L.: Mechanisms of action and therapeutic potentials. Nutrients. 2020;**12**:1-24. DOI: 10.3390/nu12061739. Epub 20200610

[112] Ghasemzadeh Rahbardar M, Hosseinzadeh H. Therapeutic effects of rosemary (Rosmarinus officinalis L.) and its active constituents on nervous system disorders. Iranian Journal of Basic Medical Sciences. 2020;**23**:1100-1112. DOI: 10.22038/ijbms.2020.45269.10541

[113] Sayorwan W, Ruangrungsi N, Piriyapunyporn T, Hongratanaworakit T, Kotchabhakdi N, Siripornpanich V. Effects of inhaled rosemary oil on subjective feelings and activities of the nervous system. Scientia Pharmaceutica. 2013;**81**:531-542. DOI: 10.3797/scipharm.1209-05. Epub 20121223

[114] Butnariu M, Quispe C, Herrera-Bravo J, Sharifi-Rad J, Singh L, Aborehab NM, et al. The pharmacological activities of Crocus sativus L.: A review based on the mechanisms and therapeutic opportunities of its phytoconstituents. Oxidative Medicine and Cellular Longevity. 2022;**2022**:8214821. DOI: 10.1155/2022/8214821. Epub 20220214

[115] Bandegi AR, Rashidy-Pour A, Vafaei AA, Ghadrdoost B. Protective effects of crocus Sativus L. extract and Crocin against chronic-stress induced oxidative damage of brain, liver and

kidneys in rats. Adv. Pharmaceutical Bulletin. 2014;**4**:493-499. DOI: 10.5681/apb.2014.073. Epub 20141231

[116] Wang CZ, Ma Q, Kim S, Wang DH, Shoyama Y, Yuan CS. Effects of saffron and its active constituent crocin on cancer management: A narrative review. Longhua Chinese Medicine. 2022;**5**:1-14. DOI: 10.21037/lcm-21-72. Epub 20221230

[117] Al-Hrout A, Chaiboonchoe A, Khraiwesh B, Murali C, Baig B, El-Awady R, et al. Safranal induces DNA double-strand breakage and ER-stress-mediated cell death in hepatocellular carcinoma cells. Scientific Reports. 2018;**8**:16951. DOI: 10.1038/s41598-018-34855-0. Epub 20181116

[118] Li S, Shen XY, Ouyang T, Qu Y, Luo T, Wang HQ. Synergistic anticancer effect of combined crocetin and cisplatin on KYSE-150 cells via p53/p21 pathway. Cancer Cell International. 2017;**17**:98. DOI: 10.1186/s12935-017-0468-9. Epub 20171030

[119] Bostan HB, Mehri S, Hosseinzadeh H. Toxicology effects of saffron and its constituents: A review. Iranian Journal of Basic Medical Sciences. 2017;**20**:110-121. DOI: 10.22038/ijbms.2017.8230

[120] Hammoudi, Halat D, Krayem M, Khaled S, Younes S. A focused insight into thyme: Biological, chemical, and therapeutic properties of an indigenous mediterranean herb. Nutrients. 2022;**14**:1-22. DOI: 10.3390/nu14102104. Epub 20220518

[121] Wang H, Khor TO, Shu L, Su ZY, Fuentes F, Lee JH, et al. Plants vs. cancer: A review on natural phytochemicals in preventing and treating cancers and their druggability anticancer agents. Medicinal Chemistry. 2012;**12**:1281-1305. DOI: 10.2174/1871520128038330 26

[122] Nagoor Meeran MF, Javed H, Al Taee H, Azimullah S, Ojha SK. Pharmacological properties and molecular mechanisms of Thymol: Prospects for its therapeutic potential and pharmaceutical development. Frontiers in Pharmacology. 2017;**8**:380. DOI: 10.3389/fphar.2017.00380. Epub 20170626

[123] Charalambous M, Raftopoulos V, Paikousis L, Katodritis N, Lambrinou E, Vomvas D, et al. The effect of the use of thyme honey in minimizing radiation—Induced oral mucositis in head and neck cancer patients: A randomized controlled trial. European Journal of Oncology Nursing. 2018;**34**:89-97. DOI: 10.1016/j.ejon.2018.04.003. Epub 20180430

[124] Chimento A, De Luca A, D'Amico M, De Amicis F, Pezzi V. The involvement of natural polyphenols in molecular mechanisms inducing apoptosis in tumor cells: A promising adjuvant in cancer therapy. International Journal of Molecular Sciences. 2023;**24**:1-26. DOI: 10.3390/ijms24021680. Epub 20230114

[125] Eroglu, Ozkan E, Ersoy E, Yesil Canturk Y, Mataraci Kara E, Cinar E, et al. The therapeutic potential of Ethnomedicinally important Anatolian thyme species: A phytochemical and biological assessment. Frontiers in Pharmacology. 2022;**13**:1-13. DOI: 10.3389/fphar.2022.923063

[126] Gutierrez-Grijalva EP, Picos-Salas MA, Leyva-Lopez N, Criollo-Mendoza MS, Vazquez-Olivo G, Heredia JB. Flavonoids and phenolic acids from oregano: Occurrence, biological activity and health benefits. Plants (Basel). 2017;**7**:1-23. DOI: 10.3390/plants7010002. Epub 20171226

[127] Baloghova J, Michalkova R, Baranova Z, Mojzisova G,

Fedakova Z, Mojzis J. Spice-derived phenolic compounds: Potential for skin cancer prevention and therapy. Molecules. 2023;**28**:6251. DOI: 10.3390/molecules28176251. Epub 20230825

[128] Leyva-Lopez N, Gutierrez-Grijalva EP, Vazquez-Olivo G, Heredia JB. Essential oils of oregano: Biological activity beyond their antimicrobial properties. Molecules. 2017;**22**:1-24. DOI: 10.3390/molecules22060989. Epub 20170614

[129] Nanni V, Di Marco G, Sacchetti G, Canini A, Gismondi A. Oregano Phytocomplex induces programmed cell death in melanoma lines via mitochondria and DNA damage. Food. 2020;**9**:1-27. DOI: 10.3390/foods9101486. Epub 20201017

[130] Abebe W. Review of herbal medications with the potential to cause bleeding: Dental implications, and risk prediction and prevention avenues. The EPMA Journal. 2019;**10**:51-64. DOI: 10.1007/s13167-018-0158-2. Epub 20190108

[131] Hamidpour M, Hamidpour R, Hamidpour S, Shahlari M. Chemistry, pharmacology, and medicinal property of sage (salvia) to prevent and cure illnesses such as obesity, diabetes, depression, dementia, lupus, autism, heart disease, and cancer. Journal of Traditional and Complementary Medicine. 2014;**4**:82-88. DOI: 10.4103/2225-4110.130373

[132] Cattaneo L, Cicconi R, Mignogna G, Giorgi A, Mattei M, Graziani G, et al. Anti-proliferative effect of Rosmarinus officinalis L. extract on human melanoma A375 cells. PLoS One. 2015;**10**:e0132439. DOI: 10.1371/journal.pone.0132439. Epub 20150715

[133] O'Neill EJ, Hartogh DJD, Azizi K, Tsiani E. Anticancer properties of Carnosol: A summary of in vitro and In vivo evidence. Antioxidants (Basel). 2020;**9**:1-22. DOI: 10.3390/antiox9100961. Epub 20201008

[134] Lv Q-z, Long J-t, Gong Z-f, Nong K-y, Liang X-m, Qin T, et al. Current state of knowledge on the antioxidant effects and mechanisms of action of polyphenolic compounds. Natural Product Communications. 2021;**16**:1934578X211027745. DOI: 10.1177/1934578x211027745

[135] Kubczak M, Szustka A, Rogalinska M. Molecular targets of natural compounds with anti-cancer properties. International Journal of Molecular Sciences. 2021;**22**:1-27. DOI: 10.3390/ijms222413659. Epub 20211220

[136] Pelkonen O, Abass K, Wiesner J. Thujone and thujone-containing herbal medicinal and botanical products: Toxicological assessment. Regulatory Toxicology and Pharmacology. 2013;**65**:100-107. DOI: 10.1016/j.yrtph.2012.11.002

[137] Barakat H, Alkabeer IA, Aljutaily T, Almujaydil MS, Algheshairy RM, Alhomaid RM, et al. Phenolics and volatile compounds of fennel (Foeniculum vulgare) seeds and their sprouts prevent oxidative DNA damage and ameliorates CCl(4)-induced hepatotoxicity and oxidative stress in rats. Antioxidants (Basel). 2022;**11**:1-22. DOI: 10.3390/antiox11122318. Epub 20221123

[138] Sadegh Roudbari L, Eslami M, Movahedi M, Golab F. Evaluation of the anti-metastatic effect of Foeniculum Vulgare on the protein expression of HSP 70 & 90 in Balb/c mice with 4t1 model of breast cancer. Asian Pacific Journal of Cancer Prevention. 2023;**24**:833-840. DOI: 10.31557/APJCP.2023.24.3.833. Epub 20230301

[139] Sung B, Prasad S, Yadav VR, Aggarwal BB. Cancer cell signaling pathways targeted by spice-derived nutraceuticals. Nutrition and Cancer. 2012;**64**:173-197. DOI: 10.1080/01635581.2012.630551. Epub 20111209

[140] Gori L, Gallo E, Mascherini V, Mugelli A, Vannacci A, Firenzuoli F. Can estragole in fennel seed decoctions really be considered a danger for human health? A fennel safety update. Evidence-based Complementary and Alternative Medicine. 2012;**2012**:860542. DOI: 10.1155/2012/860542. Epub 20120729

[141] Das B, Rabalais J, Kozan P, Lu T, Durali N, Okamoto K, et al. The effect of a fennel seed extract on the STAT signaling and intestinal barrier function. PLoS One. 2022;**17**:e0271045. DOI: 10.1371/journal.pone.0271045. Epub 20220708

[142] Ashokkumar K, Murugan M, Dhanya MK, Pandian A, Warkentin TD. Phytochemistry and therapeutic potential of black pepper [Piper nigrum (L.)] essential oil and piperine: A review. Clinical Phytoscience. 2021;**7**:52. DOI: 10.1186/s40816-021-00292-2

[143] Prashant A, Rangaswamy C, Yadav AK, Reddy V, Sowmya MN, Madhunapantula S. In vitro anticancer activity of ethanolic extracts of Piper nigrum against colorectal carcinoma cell lines. International Journal of Applied & Basic Medical Research. 2017;**7**:67-72. DOI: 10.4103/2229-516X.198531

[144] Han EJ, Choi EY, Jeon SJ, Lee SW, Moon JM, Jung SH, et al. Piperine induces apoptosis and autophagy in HSC-3 human Oral cancer cells by regulating PI3K signaling pathway. International Journal of Molecular Sciences. 2023;**24**:1-16. DOI: 10.3390/ijms241813949. Epub 20230911

[145] Guo L, Yang Y, Sheng Y, Wang J, Ruan S, Han C. Mechanism of piperine in affecting apoptosis and proliferation of gastric cancer cells via ROS-mitochondria-associated signalling pathway. Journal of Cellular and Molecular Medicine. 2021;**25**:9513-9522. DOI: 10.1111/jcmm.16891. Epub 20210831

[146] Myers BM, Smith JL, Graham DY. Effect of red pepper and black pepper on the stomach. The American Journal of Gastroenterology. 1987;**82**:211-214

[147] Batiha GE, Alqahtani A, Ojo OA, Shaheen HM, Wasef L, Elzeiny M, et al. Biological properties, bioactive constituents, and pharmacokinetics of some capsicum spp. and capsaicinoids. International Journal of Molecular Sciences. 2020;**21**:1-35. DOI: 10.3390/ijms21155179. Epub 20200722

[148] Gil YG, Kang MK. Capsaicin induces apoptosis and terminal differentiation in human glioma A172 cells. Life Sciences. 2008;**82**:997-1003. DOI: 10.1016/j.lfs.2008.02.020. Epub 20080310

[149] Chen M, Xiao C, Jiang W, Yang W, Qin Q, Tan Q, et al. Capsaicin inhibits proliferation and induces apoptosis in breast cancer by Down-regulating FBI-1-mediated NF-kappaB pathway. Drug Design, Development and Therapy. 2021;**15**:125-140. DOI: 10.2147/DDDT.S269901. Epub 20210112

[150] Popescu GDA, Scheau C, Badarau IA, Dumitrache M-D, Caruntu A, Scheau A-E, et al. The effects of capsaicin on gastrointestinal cancers. Molecules. 2021;**26**:94. DOI: 10.3390/molecules26010094

[151] Aniwan S, Gonlachanvit S. Effects of chili treatment on gastrointestinal and

rectal sensation in diarrhea-predominant irritable bowel syndrome: A randomized, double-blinded, crossover study. Journal of Neurogastroenterology and Motility. 2014;**20**:400-406. DOI: 10.5056/jnm14022

[152] Anand P, Bley K. Topical capsaicin for pain management: Therapeutic potential and mechanisms of action of the new high-concentration capsaicin 8% patch. British Journal of Anaesthesia. 2011;**107**:490-502. DOI: 10.1093/bja/aer260. Epub 20110817

[153] Musial C, Kuban-Jankowska A, Gorska-Ponikowska M. Beneficial properties of green tea Catechins. International Journal of Molecular Sciences. 2020;**21**:1-11. DOI: 10.3390/ijms21051744. Epub 20200304

[154] Dou QP. Molecular mechanisms of green tea polyphenols. Nutrition and Cancer. 2009;**61**:827-835. DOI: 10.1080/01635580903285049

[155] Singh BN, Shankar S, Srivastava RK. Green tea catechin, epigallocatechin-3-gallate (EGCG): Mechanisms, perspectives and clinical applications. Biochemical Pharmacology. 2011;**82**:1807-1821. DOI: 10.1016/j.bcp.2011.07.093. Epub 20110730

[156] Xu R, Yang K, Li S, Dai M, Chen G. Effect of green tea consumption on blood lipids: A systematic review and meta-analysis of randomized controlled trials. Nutrition Journal. 2020;**19**:48. DOI: 10.1186/s12937-020-00557-5

[157] Fan FS. Iron deficiency anemia due to excessive green tea drinking. Clinical Case Reports. 2016;**4**:1053-1056. DOI: 10.1002/ccr3.707. Epub 20161005

[158] Bijak M. Silybin, a major bioactive component of Milk thistle (Silybum marianum L. Gaernt.)-chemistry, bioavailability, and metabolism. Molecules. 2017;**22**:1-11. DOI: 10.3390/molecules22111942. Epub 20171110

[159] Surai PF. Silymarin as a natural antioxidant: An overview of the current evidence and perspectives. Antioxidants (Basel). 2015;**4**:204-247. DOI: 10.3390/antiox4010204. Epub 20150320

[160] Varghese L, Agarwal C, Tyagi A, Singh RP, Agarwal R. Silibinin efficacy against human hepatocellular carcinoma. Clinical Cancer Research. 2005;**11**:8441-8448. DOI: 10.1158/1078-0432.CCR-05-1646

[161] Deep G, Agarwal R. Antimetastatic efficacy of silibinin: Molecular mechanisms and therapeutic potential against cancer. Cancer Metastasis Reviews. 2010;**29**:447-463. DOI: 10.1007/s10555-010-9237-0

[162] Bertrand J, Siegler N, Murez T, Poinas G, Segui B, Ayuso D, et al. Impact of preoperative immunonutrition on morbidity following cystectomy for bladder cancer: A case-control pilot study. World Journal of Urology. 2014;**32**:233-237. DOI: 10.1007/s00345-013-1229-6. Epub 20131221

[163] Siegel AB, Narayan R, Rodriguez R, Goyal A, Jacobson JS, Kelly K, et al. A phase I dose-finding study of silybin phosphatidylcholine (milk thistle) in patients with advanced hepatocellular carcinoma. Integrative Cancer Therapies. 2014;**13**:46-53. DOI: 10.1177/1534735413490798. Epub 20130611

[164] Karimi G, Vahabzadeh M, Lari P, Rashedinia M, Moshiri M. "Silymarin", a promising pharmacological agent for treatment of diseases. Iranian Journal of Basic Medical Sciences. 2011;**14**:308-317

[165] Kazazis CE, Evangelopoulos AA, Kollas A, Vallianou NG. The therapeutic

potential of milk thistle in diabetes. The Review of Diabetic Studies. 2014;**11**:167-174. DOI: 10.1900/RDS.2014.11.167. Epub 20140810

[166] Hawke RL, Schrieber SJ, Soule TA, Wen Z, Smith PC, Reddy KR, et al. Silymarin ascending multiple oral dosing phase I study in noncirrhotic patients with chronic hepatitis C. Journal of Clinical Pharmacology. 2010;**50**:434-449. DOI: 10.1177/0091270009347475. Epub 20091019

[167] Zheng Y, Ren W, Zhang L, Zhang Y, Liu D, Liu Y. A review of the pharmacological action of Astragalus polysaccharide. Frontiers in Pharmacology. 2020;**11**:349. DOI: 10.3389/fphar.2020.00349. Epub 20200324

[168] Xie Y, Wang C. Herb-drug interactions between Panax notoginseng or its biologically active compounds and therapeutic drugs: A comprehensive pharmacodynamic and pharmacokinetic review. Journal of Ethnopharmacology. 2023;**307**:116156. DOI: 10.1016/j.jep.2023.116156. Epub 20230206

[169] Cheng M, Hu J, Zhao Y, Jiang J, Qi R, Chen S, et al. Efficacy and safety of Astragalus-containing traditional Chinese medicine combined with platinum-based chemotherapy in advanced gastric cancer: A systematic review and meta-analysis. Frontiers in Oncology. 2021;**11**:632168. DOI: 10.3389/fonc.2021.632168. Epub 20210804

[170] McCulloch M, See C, Shu XJ, Broffman M, Kramer A, Fan WY, et al. Astragalus-based Chinese herbs and platinum-based chemotherapy for advanced non-small-cell lung cancer: Meta-analysis of randomized trials. Journal of Clinical Oncology. 2006;**24**:419-430. DOI: 10.1200/JCO.2005.03.6392

[171] Auyeung KK, Han QB, Ko JK. Astragalus membranaceus: A review of its protection against inflammation and gastrointestinal cancers. The American Journal of Chinese Medicine. 2016;**44**:1-22. DOI: 10.1142/S0192415X16500014

[172] Zhang X, Qu X, Zou Y. The effect of Astragalus on humoral and cellular immune response: A systematic review and meta-analysis of human studies. Complementary Medicine Research. 2023;**30**:535-543. DOI: 10.1159/000534826. Epub 20231110

[173] Chandrasekaran CV, Sundarajan K, Edwin JR, Gururaja GM, Mundkinajeddu D, Agarwal A. Immune-stimulatory and anti-inflammatory activities of Curcuma longa extract and its polysaccharide fraction. Pharmacognosy Research. 2013;**5**:71-79. DOI: 10.4103/0974-8490.110527

[174] Panda SK, Sahoo G, Swain SS, Luyten W. Anticancer activities of mushrooms: A neglected source for drug discovery. Pharmaceuticals (Basel). 2022;**15**:1-25. DOI: 10.3390/ph15020176. Epub 20220131

[175] Auyeung KK, Cho CH, Ko JK. A novel anticancer effect of Astragalus saponins: Transcriptional activation of NSAID-activated gene. International Journal of Cancer. 2009;**125**:1082-1091. DOI: 10.1002/ijc.24397

[176] Latour E, Arlet J, Latour EE, Juszkiewicz A, Luczkowska K, Marcinkiewicz A, et al. Standardized astragalus extract for attenuation of the immunosuppression induced by strenuous physical exercise: Randomized controlled trial. Journal of the International Society of Sports Nutrition. 2021;**18**:57. DOI: 10.1186/s12970-021-00425-5. Epub 20210716

[177] Hyun SH, Kim SW, Seo HW, Youn SH, Kyung JS, Lee YY, et al. Physiological and pharmacological features of the non-saponin components in Korean red ginseng. Journal of Ginseng Research. 2020;**44**:527-537. DOI: 10.1016/j.jgr.2020.01.005. Epub 20200206

[178] Chen S, Wang Z, Huang Y, O'Barr SA, Wong RA, Yeung S, et al. Ginseng and anticancer drug combination to improve cancer chemotherapy: A critical review. Evidence-based Complementary and Alternative Medicine. 2014;**2014**:168940. DOI: 10.1155/2014/168940. Epub 20140430

[179] Kim S, Kim N, Jeong J, Lee S, Kim W, Ko S-G, et al. Anti-cancer effect of Panax Ginseng and its metabolites: From traditional medicine to modern drug discovery. PRO. 2021;**9**:1344. DOI: 10.3390/pr9081344

[180] Nag SA, Qin JJ, Wang W, Wang MH, Wang H, Zhang R. Ginsenosides as anticancer agents: In vitro and in vivo activities, structure-activity relationships, and molecular mechanisms of action. Frontiers in Pharmacology. 2012;**3**:25. DOI: 10.3389/fphar.2012.00025. Epub 20120228

[181] Zhu H, Liu H, Zhu JH, Wang SY, Zhou SS, Kong M, et al. Efficacy of ginseng and its ingredients as adjuvants to chemotherapy in non-small cell lung cancer. Food & Function. 2021;**12**:2225-2241. DOI: 10.1039/d0fo03341c

[182] Xu W, Lyu W, Duan C, Ma F, Li X, Li D. Preparation and bioactivity of the rare ginsenosides Rg3 and Rh2: An updated review. Fitoterapia. 2023;**167**:105514. DOI: 10.1016/j.fitote.2023.105514. Epub 20230420

[183] Kim SJ, Kim AK. Anti-breast cancer activity of fine black ginseng (Panax ginseng Meyer) and ginsenoside Rg5. Journal of Ginseng Research. 2015;**39**:125-134. DOI: 10.1016/j.jgr.2014.09.003. Epub 20141018

[184] Choi JH, Oh J, Lee MJ, Ko SG, Nah SY, Cho IH. Gintonin mitigates experimental autoimmune encephalomyelitis by stabilization of Nrf2 signaling via stimulation of lysophosphatidic acid receptors. Brain, Behavior, and Immunity. 2021;**93**:384-398. DOI: 10.1016/j.bbi.2020.12.004. Epub 20201210

[185] Lim SW, Luo K, Quan Y, Cui S, Shin YJ, Ko EJ, et al. The safety, immunological benefits, and efficacy of ginseng in organ transplantation. Journal of Ginseng Research. 2020;**44**:399-404. DOI: 10.1016/j.jgr.2020.02.001. Epub 20200207

[186] Burlou-Nagy C, Banica F, Jurca T, Vicas LG, Marian E, Muresan ME, et al. Echinacea purpurea (L.) Moench: Biological and Pharmacological Properties. A Review. Plants (Basel). 2022;**11**:1244. DOI: 10.3390/plants11091244. Epub 20220505

[187] Tsai YL, Chiu CC, Yi-Fu Chen J, Chan KC, Lin SD. Cytotoxic effects of Echinacea purpurea flower extracts and cichoric acid on human colon cancer cells through induction of apoptosis. Journal of Ethnopharmacology. 2012;**143**:914-919. DOI: 10.1016/j.jep.2012.08.032. Epub 20120901

[188] Tomko AM, Whynot EG, Ellis LD, Dupre DJ. Anti-cancer potential of cannabinoids, terpenes, and flavonoids present in cannabis. Cancers (Basel). 2020;**12**:1985. DOI: 10.3390/cancers12071985. Epub 20200721

[189] Rininger JA, Kickner S, Chigurupati P, McLean A, Franck Z. Immunopharmacological activity of

Echinacea preparations following simulated digestion on murine macrophages and human peripheral blood mononuclear cells. Journal of Leukocyte Biology. 2000;**68**:503-510

[190] Senchina DS, Martin AE, Buss JE, Kohut ML. Effects of Echinacea extracts on macrophage antiviral activities. Phytotherapy Research. 2010;**24**:810-816. DOI: 10.1002/ptr.2991

[191] Miller SC. Echinacea: A miracle herb against aging and cancer? Evidence in vivo in mice. Evidence-based Complementary and Alternative Medicine. 2005;**2**:309-314. DOI: 10.1093/ecam/neh118

[192] Nagoor Meeran MF, Javed H, Sharma C, Goyal SN, Kumar S, Jha NK, et al. Can Echinacea be a potential candidate to target immunity, inflammation, and infection—The trinity of coronavirus disease 2019. Heliyon. 2021;**7**:e05990. DOI: 10.1016/j.heliyon.2021.e05990. Epub 20210208

[193] Kemp DE, Franco KN. Possible leukopenia associated with long-term use of echinacea. The Journal of the American Board of Family Practice. 2002;**15**:417-419

[194] Barnes J, Anderson LA, Phillipson JD. St John's wort (Hypericum perforatum L.): A review of its chemistry, pharmacology and clinical properties. The Journal of Pharmacy and Pharmacology. 2001;**53**:583-600. DOI: 10.1211/0022357011775910

[195] Martarelli D, Martarelli B, Pediconi D, Nabissi MI, Perfumi M, Pompei P. Hypericum perforatum methanolic extract inhibits growth of human prostatic carcinoma cell line orthotopically implanted in nude mice. Cancer Letters. 2004;**210**:27-33. DOI: 10.1016/j.canlet.2004.01.031

[196] Dona M, Dell'Aica I, Pezzato E, Sartor L, Calabrese F, Della Barbera M, et al. Hyperforin inhibits cancer invasion and metastasis. Cancer Research. 2004;**64**:6225-6232. DOI: 10.1158/0008-5472.CAN-04-0280

[197] Kamuhabwa AR, Agostinis P, D'Hallewin MA, Kasran A, de Witte PA. Photodynamic activity of hypericin in human urinary bladder carcinoma cells. Anticancer Research. 2000;**20**:2579-2584

[198] Couldwell WT, Gopalakrishna R, Hinton DR, He S, Weiss MH, Law RE, et al. Hypericin: A potential antiglioma therapy. Neurosurgery. 1994;**35**:705-709; discussion 9-10. DOI: 10.1227/00006123-199410000-00017

[199] Zhang W, Lawa RE, Hintona DR, Su Y, Couldwell WT. Growth inhibition and apoptosis in human neuroblastoma SK-N-SH cells induced by hypericin, a potent inhibitor of protein kinase C. Cancer Letters. 1995;**96**:31-35. DOI: 10.1016/0304-3835(95)03914-i

[200] Miskovsky P. Hypericin—A new antiviral and antitumor photosensitizer: Mechanism of action and interaction with biological macromolecules. Current Drug Targets. 2002;**3**:55-84. DOI: 10.2174/1389450023348091

[201] Birge RB, Boeltz S, Kumar S, Carlson J, Wanderley J, Calianese D, et al. Phosphatidylserine is a global immunosuppressive signal in efferocytosis, infectious disease, and cancer. Cell Death and Differentiation. 2016;**23**:962-978. DOI: 10.1038/cdd.2016.11. Epub 20160226

[202] Martinez-Poveda B, Quesada AR, Medina MA. Hypericin in the dark inhibits key steps of angiogenesis in vitro. European Journal of Pharmacology. 2005;**516**:97-103. DOI: 10.1016/j.ejphar.2005.03.047

[203] Menegazzi M, Masiello P, Novelli M. Anti-tumor activity of Hypericum perforatum L. and Hyperforin through modulation of inflammatory signaling, ROS generation and proton dynamics. Antioxidants (Basel). 2020;**10**:1-25. DOI: 10.3390/antiox10010018. Epub 20201228

[204] Herraiz T, Guillen H. Monoamine oxidase-a inhibition and associated antioxidant activity in plant extracts with potential antidepressant actions. BioMed Research International. 2018;**2018**:4810394. DOI: 10.1155/2018/4810394. Epub 20180115

[205] Kenda M, Kocevar Glavac N, Nagy M, Sollner DM. Medicinal plants used for anxiety, depression, or stress treatment: An update. Molecules. 2022;**27**:1-19. DOI: 10.3390/molecules27186021. Epub 20220915

[206] Miekus N, Marszalek K, Podlacha M, Iqbal A, Puchalski C, Swiergiel AH. Health benefits of plant-derived sulfur compounds, glucosinolates, and organosulfur compounds. Molecules. 2020;**25**:1-22. DOI: 10.3390/molecules25173804. Epub 20200821

[207] Henderson L, Yue QY, Bergquist C, Gerden B, Arlett P. St John's wort (Hypericum perforatum): Drug interactions and clinical outcomes. British Journal of Clinical Pharmacology. 2002;**54**:349-356. DOI: 10.1046/j.1365-2125.2002.01683.x

[208] Sandoval M, Charbonnet RM, Okuhama NN, Roberts J, Krenova Z, Trentacosti AM, et al. Cat's claw inhibits TNFalpha production and scavenges free radicals: Role in cytoprotection. Free Radical Biology & Medicine. 2000;**29**:71-78. DOI: 10.1016/s0891-5849(00)00327-0

[209] Yepes-Perez AF, Herrera-Calderon O, Quintero-Saumeth J. Uncaria tomentosa (cat's claw): A promising herbal medicine against SARS-CoV-2/ACE-2 junction and SARS-CoV-2 spike protein based on molecular modeling. Journal of Biomolecular Structure & Dynamics. 2022;**40**:2227-2243. DOI: 10.1080/07391102.2020.1837676. Epub 20201029

[210] Garcia Gimenez D, Garcia Prado E, Saenz Rodriguez T, Fernandez Arche A, De la Puerta R. Cytotoxic effect of the pentacyclic oxindole alkaloid mitraphylline isolated from Uncaria tomentosa bark on human Ewing's sarcoma and breast cancer cell lines. Planta Medica. 2010;**76**:133-136. DOI: 10.1055/s-0029-1186048. Epub 20090901

[211] Pilarski R, Filip B, Wietrzyk J, Kuras M, Gulewicz K. Anticancer activity of the Uncaria tomentosa (Willd.) DC. Preparations with different oxindole alkaloid composition. Phytomedicine. 2010;**17**:1133-1139. DOI: 10.1016/j.phymed.2010.04.013. Epub 20100625

[212] Zari A, Alfarteesh H, Buckner C, Lafrenie R. Treatment with Uncaria tomentosa promotes apoptosis in B16-BL6 mouse melanoma cells and inhibits the growth of B16-BL6 tumours. Molecules. 2021;**26**:1-22. DOI: 10.3390/molecules26041066. Epub 20210218

[213] Dreifuss AA, Bastos-Pereira AL, Fabossi IA, Livero FA, Stolf AM, Alves de Souza CE, et al. Uncaria tomentosa exerts extensive anti-neoplastic effects against the Walker-256 tumour by modulating oxidative stress and not by alkaloid activity. PLoS One. 2013;**8**:e54618. DOI: 10.1371/journal.pone.0054618. Epub 20130207

[214] Sheng Y, Li L, Holmgren K, Pero RW. DNA repair enhancement of aqueous extracts of Uncaria tomentosa in a human volunteer

study. Phytomedicine. 2001;**8**:275-282. DOI: 10.1078/0944-7113-00045

[215] Mohamed AF, Matsumoto K, Tabata K, Takayama H, Kitajima M, Watanabe H. Effects of Uncaria tomentosa total alkaloid and its components on experimental amnesia in mice: Elucidation using the passive avoidance test. The Journal of Pharmacy and Pharmacology. 2000;**52**:1553-1561. DOI: 10.1211/0022357001777612

[216] Riva L, Coradini D, Di Fronzo G, De Feo V, De Tommasi N, De Simone F, et al. The antiproliferative effects of Uncaria tomentosa extracts and fractions on the growth of breast cancer cell line. Anticancer Research. 2001;**21**:2457-2461

[217] Pilarski R, Zielinski H, Ciesiolka D, Gulewicz K. Antioxidant activity of ethanolic and aqueous extracts of Uncaria tomentosa (Willd.) DC. Journal of Ethnopharmacology. 2006;**104**:18-23. DOI: 10.1016/j.jep.2005.08.046. Epub 20050930

[218] Blanck JJ, Huebner TM, Rolls AM, Cornell JS, Hwang CS. Comprehensive review of the components in cat's claw (Uncaria tomentosa) and their antibacterial activity. AppliedChem. 2022;**2**:1-29. DOI: 10.3390/appliedchem2010001

[219] Sheng Y, Bryngelsson C, Pero RW. Enhanced DNA repair, immune function and reduced toxicity of C-MED-100, a novel aqueous extract from Uncaria tomentosa. Journal of Ethnopharmacology. 2000;**69**:115-126. DOI: 10.1016/s0378-8741(99)00070-7